edra 1

Proceedings of the 1st annual Environmental Design Research Association Conference. Edited by Henry Sanoff and Sidney Cohn.

Copyright 1970 by Henry Sanoff and Sidney Cohn
Library of Congress Catalog Number 72-123774
Printed in the United States of America

All rights reserved. This book may not be reproduced, in whole or in part, in any form (except by reviewers for the public press) without written permission from the publisher.

Cover designed by Fred Eichenberger, North Carolina State University, Raleigh

ACKNOWLEDGEMENT

The papers in this volume are edited versions of those presented at the first annual conference of the Environmental Design Research Association (EDRA) which was held in conjunction with the Design Methods Group (DMG) and Man-Environment Systems (MES). This conference was co-sponsored by the School of Design at North Carolina State University in Raleigh and the Department of City and Regional Planning at the University of North Carolina in Chapel Hill and held in June 1969 at Chapel Hill. The substantive papers were read in the order in which they appear in the text.

The editors wish to acknowledge the invaluable assistance of Mrs. Julia Coates in the preparation of this volume.

PREFACE

In the last decade, along with the growth of knowledge, there appeared an increasing specialization in the sciences where disciplines were breaking up into isolated sub-cultures with only tenuous lines of communication between them. With increasing specialization and sub-group formation, not only is there less communication possible among disciplines, but there is a greater chance that the total growth of knowledge will be slowed down by the loss of relevant communication. Thus, the need emerges to enable a specialist to catch relevant communications from others. The sciences, in attempting to cope with this dilemma evolved the concept of a general systems theory whereby a specialist who, whether psychologist, economist or cristallographer, works with a growth concept, for example, may be sensitive to the contributions of other fields if he is aware of the many similarities of the growth process in widely different empirical fields.

The indicators of such interdisciplinary movements are the development of hybrid disciplines such as social psychology in the early part of the century and more recently cybernetics which grew out of electrical engineering, neurophysiology, physics, biology and economics, information theory, management science.

A similar development is presently taking place in the field of environmental design. Faced with increasingly complex environmental problems which defied satisfactory solution, a few designers came to realize that both their traditional problem-solving methodology and their knowledge of the man-environment system was highly inadequate. They realized that their training as designers left them un-equipped to understand the problem or to develop effective solutions. Seeking to improve this state, they sought the assistance of scientists, for example, in psychology, sociology, anthropology, medicine—all concerned with the study of man. They found for the most part the work of the sciences was unrelated to environmental design research and problem solving. Fortunately, however, they found a few scientists interested in studying this problem area.

Until a half-dozen years ago, only a mere handful of designers and even fewer scientists pursued what can be regarded as scientific research in this area. Since that time interest has increased significantly. Scientists from a variety of disciplines have joined the small but increasing nucleus of designers in this study of man-environment, and design and a new hybrid of environmental design researcher is emerging.

The newly developing field of environmental design research recognizes no arbitrary boundaries between "inside" and "outside" or large or small domains. Environmental design has come to connote a commitment to the evolution of the environment as an integral part of human systems. The environmental designer's task is to understand the environment, described by Walter Buckley as a complex adaptive or internally as well as externally open system, and to create effective new environmental solutions to meet man's needs. Analyzing and solving environmental problems requires multi-disciplinary collaboration, since a discipline by its very nature deals only with a segment of the environment.

In order to facilitate this multi-disciplinary collaboration, a forum for discussion was needed as a vehicle to bring individuals from the sciences together with environmental designers. To this end the First Annual Environmental Design Research Association Conference was held to strengthen as well as expand communications in this area.

The basic types of explorations which form the core of environmental design are: the externalization of the design process and the construction of models of these processes; the application of systematic and systems approaches to the design process; the stimulation of individual and group creative problem solving; and the development of social and behavioral knowledge as to the man-environment relationship. Today these basic streams of development are giving rise to new approaches generated by combinations of the basic types of investigations.

The major focus for this conference was the discussion of developing models and methods towards a framework of coherence and definable structure of environmental design. The papers in this book represent the proceedings of the conference and are concerned mainly with the contributions of scientific disciplines toward towards the creation of improved methods problem solving as well as understanding the nature of human responses to the environment. The more fundamental objective is towards achieving an optimum environment for man.

As with research in any new area of scientific research and particularly multi-disciplinary scientific investigation, the researchers are plagued with a variety of theoretical, methodological, and even philosophical problems, and errors will be made. All of the participants are vested with what can be justifiably be called an ideology, i.e. a belief that these relationships can be identified, but lack a comprehensive body of theory to explain the phenomenon, a common classification framework, and even a commonly accepted set of methods and tools. Each discipline tends to view the problem from its own particular and limited theoretical framework and each applies its own set of methods and techniques to studying it. Each tends to employ the dependent variables popular in the parent discipline and all fail to codify the environment into a systematic set of independent operational variables. While concern is manifested for the many variables which can modify this relationship, each discipline tends to focus almost exclusively upon its own area of concern, thus resulting in an incomplete and inconsistent investigation of the complex relationship. Too, faced with the complexity of the real world, the problem of environmental relevance which provided the impetus to this new effort, is frequently subverted in favor of scientific rigor.

Designers cum scientists, initially the stimulus for the effort, continue to provide a major source of manpower in the expanding field of investigation. Having recently come to the area of scientific investigation, however, they have contributed to these problems. Although more catholic and holistic in approach, they lack a strong background in the methods and theory of the sciences. There is a tendency to be strong on relevancy and weak on scientific rigor. There is a tendency to be unaware of an existing body of knowledge relevant to their problems and the need to rediscover and re-argue epistemological issues long since resolved or declared unresolvable in the social and behavioral sciences. As a result, they contribute little to resolving the major conceptual problems and simultaneously add to the total source of confusion and disarray which has pervaded this movement.

As a result, the products of these recent investigations into environmental design and behavior tends to be highly piecemeal in nature. The bits and pieces which indicate promise are not systematically or theoretically related and coordinated; each tends to exist as an entity in and of itself. Lacking a general framework of theory, it is difficult to relate the various bodies of ongoing research or to clearly identify the gaps in the body of knowledge. As Silverstone observed, in some cases the area of research is well formulated, but the research design is poor; in other cases, the research design is well conceived but the conceptualization of the phenomenon lacks relevance to solving the real world problems. Unfortunately, the "poor" research tends to further limit the heuristic value of the work.

Without question, the greatest significance of the conference and these proceedings lies in the recognition and identification of these problems by the participants. The need for well articulated theoretical models and systematic research design is frequently emphasized and questions are raised about the necessity of rediscovering knowledge and reinventing methodology. The relevance of specific investigations is challenged and the rigor of others is criticized. Questions are raised about the use of judgmental behavior as well as validity concealed in the guise of relevance or the reliability of that which is measured. Similarly, questions are raised with respect to the oversimplification of the phenomenon under study in the search scientific rigor. Given the state of knowledge in the field, the heavy reliance on accepted behavioral research tools is faulted and a plea issued for a more catholic approach. Lastly, the basic problem of imputing from research findings of from specific existing situations to the design of new solutions was considered including the role of the scientist versus the designer in the development of normative models for environmental programming.

While the papers which follow are subjected to severe criticism, they have great heuristic potential, particularly in the context of the excellent evaluation provided. The reader is urged to examine each paper in the context of the group and the comments by the chairman which reflect both his personal observations and those of the workshop participants. Given the complexity of the phenomenon under study and the state of existing knowledge they should be critically interpreted but not disregarded. Summarizing the conference, Carson noted the thoughts of Whitehead (*The Arms of Education,* MacMillan, 1929) who puts the rythm of education as follows: *the stage of romance,* in which we are motivated by wonder and necessity to search for answers; *the stage of precision,* in which we sharpen our tools and methods for gaining knowledge; and *the stage of generalization,* in which we apply our knowledge to the problem at hand, giving us the power over ourselves and our surround to live comfortably in that surround. We now, states Carson, seem to be moving into the atage of precision in this conference, a stage which is characterized by sharp and telling blows to the meretricious and enthralling. Can we hope that the next conference will move us closer to the stage of generalization?

Henry Sanoff
Sidney Cohn

Photographs by Stanley

TABLE OF CONTENTS

Preface.. V

Biographical Sketches.. XI

Introduction

Amos Rapoport, *An Approach to the Study of Environmental Quality*............ 1
Nicholas Negroponte, *Environmental Humanism Through Robots*................. 14
James Fitch, *Architectural Manipulation of Space, Time and Gravity*......... 18

Performance Systems

Peter Batchelor, *Residential Space Systems: Generic and Specific Concepts of Urban Structural Analysis*.. 33
Michael Brill and Richard Krauss, *Planning for Community Mental Health Centers: The Performance Approach*.. 45
Francis Duffy and James Friedman, *Patterns and Semiology*................... 60
Stuart Silverstone, *Chairman's Comments*.................................... 73

Environmental Attributes

Raymond Craun, *Visual Determinants of Preferences for Dwelling Environs*.... 75
Robert Hershberger, *A Study of Meaning and Architecture*.................... 86
George Peterson, Robert Bishop and Edward Neumann, *The Quality of Visual Residential Environments*.. 101
Dan Carson, *Chairman's Comments*... 115

Environmental Quality

William Mitchell, *Switching on the Seven Lamps*............................ 116
Fred Steel, *Problem Solving in the Spatial Environment*.................... 127
Richard Wilkinson, *Chairman's Comments*.................................... 137

Problem Solving Methods in Design

Charles Eastman, *Theoretical Premises of Problem Solving*.................. 138
Joseph Ballay, *Visual Information Processing in Problem Solving Situations*. 154
Isao Oishi, *Towards The Computerization of Architecture: A Theoretical Framework and a Partial Computer Program*................................. 165
John Grason, *Fundamental Description of a Floor Plan Design Program*....... 175
Charles Eastman, *Chairman's Comments*...................................... 181

Human Responses to the Environment

William Michelson, *Analytic Sampling for Design Information: A Survey of Housing Experiences*.. 183
Richard Worrall, George Peterson and M. J. Redding, *Toward a Theory of Accessibility Acceptance*..................................... 198
Ray Studer, *Chairman's Comments*... 211

Environmental Perception

Stuart Rose, *Towards a Stimulus-Response Theory of Environmental Design* 215
Gary Winkel, Roger Malek, and Philip Thiel, *A Study of Human Response to Selected Roadside Environments* 224
Dan Carson, *Chairman's Comments* 241

Attitudinal Responses to the Environment

Wolfgang Preiser, *Behavioral Design Criteria in Student Housing* 243
Man Sawhney, *Chairman's Comments* 260

Models in Planning

Wilferd Holton, Bernard Kramer, Peter New, Grazia Marzot, *Locational Decisions: Objectives and Consequences* 262
Edward Kaiser, *A Decision Agent Modeling an Approach to Planning for Urban Residential Growth* 280
George Hemmens, *Chairman's Comments* 292

Decision Models

Charles Burnette, *Toward a Theory of Technical Description for Architecture* 294
William Miller, *Matrix Method for Grouping an Interrelated Set of Elements* 304
Donald Watson, *Modelling the Activity System* 318
Murray Milne, *Chairman's Comments* 327

Participatory Planning

Ralph Brill, Eric Castro, A. J. Pennington, *The Community Development Workshop* 329
Henry Burgwyn, *Identifying Community Leaders* 334
David Godschalk, *Negotiate: An Experimental Planning Game* 345
Barry Jackson, *Chairman's Comments* 350

Design Education

John Peterson and Leonard Lansky, *An Experimental Approach to the Study of Critical Judgement in Design* 351
George Bireline, *Towards An Experimental Program in Design Education* 356
Marvin Sevely, *Chairman's Comments* 361

BIOGRAPHICAL SKETCHES

Peter Batchelor is an Associate Professor of Urban Design and Director of the Urban Design Program at the School of Design, North Carolina State University. Other teaching duties that he has performed include Sessional Lecturer in City Planning and Architectural History at the University of British Columbia and Instructor in Civic Design at the University of Pennsylvania. Mr. Batchelor has had eight years of professional practice in architecture and related fields and planning since 1951. He has also authored several articles which have been published in architectural journals. Mr. Batchelor has contributed to the following studies: The Potomac, Innovations in Building and Environmental Technology, and Performance Standards for Residency Design.

Joseph M. Ballay is Assistant Professor of Industrial Design in the College of Design, Architecture and Art at the University of Cincinnati, where he has been primarily responsible for the planning and teaching of design courses in the foundation years. In addition, he is a member of a research group which is designing portable shelter systems for the U. S. Air Force. During the academic year 1968-1969, Mr. Ballay has taken a leave of absence to engage in graduate studies at Carnegie-Mellon University. His study emphasized investigation of computer augmented design and design methodology.

George Bireline maintains an active involvement in painting and is an Associate Professor at the School of Design, North Carolina State University. He has been teaching in the basic design program for over ten years and is presently Director of Environmental Design, a 2-year core program. His interests lay in the area of cognitive and perceptual aspects of environmental design education.

Michael Brill is an Associate Professor and Chairman of the Department of Architectural and Urban Systems in the new School of Architecture and Environmental Design at the State University of New York at Buffalo. Until Spring 1969, Mr. Brill was the Senior systems Research Architect in the Building Research Division of the Institute for Applied Technology of the National Bureau of Standards, U. S. Department of Commerce. He was an adjunct Assistant Professor in the School of Architecture at Pratt Institute teaching design, the design and construction of buildings, and urban planning.

Ralph Brill is Vice President of Node 4 Associates, Inc. and is their Director of Research and Planning. He is currently developing a new technique for implementing community participation into the planning process with relation to housing systems that would be applicable to the flexible needs of a community. Mr. Brill has participated in multidisciplinary team efforts on town and community planning projects, both nationally and internationally. He is currently planning community facilities in Central Brooklyn. As a retained Advocate Planner, Mr. Brill has performed consulting services for many of the projects in Brownsville, New York. He is also often called upon by many city and federal government agencies to provide his opinion on inner city planning.

Henry King Burgwyn is a recent graduate of North Carolina State University, School of Design, Department of Architecture. As well, he holds a second degree in Political Science. His primary interests are in the area of social policy planning with emphasis on housing and community development.

Charles Hamilton Burnette is presently Director, Graduate Program in Community Design, Philadelphia College of Art; Head, Charles Burnette and Associates, a firm specializing in research, information management and architecturel systems design; Partner in *Polyplan,* a Philadelphia firm active in institutional planning; Hospital Architect on a research project at the Institute for Medical Research, Camden, New Jersey.

Previously, Research Associate and Principal Investigator at the Institute for Environmental Studies, University of Pennsylvania from 1963 until receiving the Ph.D. in architecture in 1969. The dissertation, "An Organization of Information for Computer-Aided Communication in Architecture" continues the development of ideas introduced in several publications, among them; *Residential Rehabilitation: The Pitfalls of Non-Profit Sponsorship,* Institute for Environmental Studies; "The ARC System: A functional Organization for Building Information", National Bureau of Standards; and the present issue of The Student Publication of the School of Design, North Carolina State University.

Daniel H. Carson is an Associate Professor of Environmental Science in the Division of Man-Environment Relations of the Pennsylvania State University. In addition to faculty positions at The Johns Hopkins University, University of California, Santa Barbara, University of Utah and University of Michigan, he has worked in various planning positions in San Mateo, California, Boston, Massachusetts and Bristol, Connecticut. His research papers have appeared in various journals, among which are: *Psychological Reports, Kybernetic, American Speech and Hearing Journal, and Consulting Engineer.* He has olso written several chapters to book on the environment. Hes present research concerns environmental stress and perception.

Eric Castro is presently President of Node 4 Associates, Inc., and is its Director of Construction. He was, for five years, a senior staff member of a leading consulting engineering firm in New York and worked with various leading construction firms in New York. Mr. Castro has assisted several sponsors in developing feasible proposals for non-profit and limited dividend housing sponsorship. Recently, acting as construction consultant, he was involved in structuring a community based construction organization in Bedford-Stuyvesa Brooklyn, New York. Mr. Castro's areas of professional interest are computer automation in urban problem and computers application to management and to business administration.

Sidney Cohn is currently an Associate Professor in the Department of City and Regional Planning at the University of North Carolina in Chapel Hill. He has also taught at the University of Washington and has served as a guest Architectural Critic at the Technical University Stuttgart in Germany. He has served as research assistant on several projects at the University of Washington and as Assistant Professor in Research at the same University. Also he has acted as research consultant to Robin M. Towne and Associates and has done independent research in Germany, Holland, Switzerland, Sweden, Great Britain and Denmark in a project by the Arnold W. Brunner Foundation. Dr. Cohn's current research interests are on the impact of visual dimensions of the urban environment on mental and physical well being, developing techniques for influencing urban form development in the planning process, and developing more systematic design methodology and techniques.

Raymond M. Craun, Jr. is a recent graduate of the Department of Architecture, North Carolina State University. One of his current interests is in the area of users' perception of the built environment and the ways which this could affect design decisions. Mr. Craun has worked with techniques for user evaluation of the environment, gaming simulation as a communication and educational tool for architectural decision-making. Mr. Craun presently works with Gary Stonebraker and the Advance Planning Research Group.

Francis C. Duffy, presently with the School of Architecture at Princeton University, has also acted as Assistant Architect at the National Building Agency, London and has worked on the NBA publication, *Generic Plans.* He also has served as editor of *Arena—the Architectural Association Journal.* Other publications include various articles and papers published in architectural journals in England and the United States.

Charles M. Eastman is currently Director of the Institute of Physical Planning, Carnegie-Mellon University, Pittsburgh. Previous to that, he was an Assistant Professor in the Departments of Computer Science and Architecture. He has also taught at the University of Wisconsin, Madison. Professor Eastman is interested in automated design and has published several papers on the subject. Two of these are: "Analysis of Intuitive Design Processes," to appear in *New Methods in Environmental Design and Planning,* G. Moore, (ed.), MIT Press, Cambridge, (in press), and "Cognitive Processes and ill-defined Problems: a Case Study from Design," *Proceedings of the Joint Conference on Artificial Intelligence,* Washington, D. C., 1969.

Jerry Finrow is presently a professor of Architecture and Director of the Center for Environmental Research at the Department of Architecture, University of Oregon. His primary focus of interest is in developing a comprehensive program in environmental research for both undergraduates and graduates. In the past, Mr. Finrow has been Associate Editor of the *Design Methods Group Newsletter* and is presently Regional Assistant editor of the same publication. Mr. Finrow has published articles in *DMG Newsletter* and the *Journal of Architectural Education.* In the near future Mr. Finrow hopes to become involved in research into urban patterns of human contact.

James M. Fitch is a professor in the School of Architecture at Columbia University in New York. In the past he has served with the Tennessee Planning Commission and as a low-cost housing analyst for FHS. Before going into education, he served for five years as associate editor of the *Architectural Record* and for four years as technical editor of *Architectural Forum.* Since 1935 he has authored five publications including: *Tennessee Population Trends, American Building: The Forces that Shape It,* (with F. F. Rockwell) *American Gardens, Architecture and Aesthetics of Plenty* and *American Building.*

John Grason is currently a full time instructor in electrical engineering at Carnegie-Mellon University where he is completing his Ph.D. in electrical engineering, majoring in systems and communications science. In September of 1969, Mr. Grason will assume duties of Assistant Professor in Electrical Engineering at Carnegie-Mellon.

David R. Godschalk is a lecturer in the Department of City and Regional Planning at the University of North Carolina, Chapel Hill, and Editor of the *Journal of the American Institute of Planners.* He has been a planning and design consultant, a city planning director, and a planning faculty member. His current research interests include participatory planning, gamin-simulation, and the new community design and development process.

George C. Hemmens is Professor and Director, Urban and Regional Systems Program, Department of City and Regional Planning, University of North Carolina at Chapel Hill. His interest is in the area of Urban Spatial Structure, planning/public policy process. Currently his activities include research on the analysis and simulation of daily activity patterns of urban area residents, research on residential choice carried out through a national interview study, and research on urban development programming.

Robert G. Hershberger Adjunct Assistant Professor of Architecture at Drexel Institue of Technology in Philadelphia. He is also serving as project architect and planner with Geddes, Brecher, Qualls, Cunningham: Architects in Philadelphia. Previously, he has been Assistant Professor of Architecture at Idaho State University and beginning in September, 1969 he will be Associate Professor of Architecture at Arizona State University.

Wilfred E. Holton is an instructor (Sociology) in the Department of Preventive Medicine, School of Medicine, Tufts University. He received his B.A. from Vanderbilt University, his M.A. from Boston University and is currently a Ph.D. candidate at Boston University. He was a National Science Foundation Research Fellow (1964) and a National Defense Education Act Fellow at Boston University (1964-1967). His proposed doctoral dissertation is: "Ecological Process in the Service Sector: An Exploratory Study."

Edward J. Kaiser is an associate Professor, Department of City and Regional Planning, at the University of North Carolina at Chapel Hill. He is an Asspciate Director at the Center for Urban and Regional Studies, Institute for Research in Social Science. Currently he is the principal investigator with Edgar Butler of the University of California at Riverside, of an NSF sponsored national survey of residential choice and moving behavior. In addition he is co-principal investigator on research projects in several other areas including "New Town Development: Convergence of Concept and Reality" (Public Health Service Grant) and "Multiple Reservoirs and Urban Development" (U. S. Department of Interior through the Water Resources Institute of the University of North Carolina).

Vahe Khachooni, presently Director of Computer Services with Daniel, Mann, Johnson and Mendenhall of Los Angeles, has also been engaged in the coordination and development of computer applications in the areas of architecture, engineering and planning. He has served as a computer consultant on an economic feasibility study for the Baltimore Regional Planning Council and for preparation of the Comprehensive General Plan for Alameda, California. He has also carried out research in the shielding of structures and slanting techniques for the U.S. Defense Department, and has taken part in urban planning and design of environmental systems in fallout shelters.

Bernard M. Kramer is an Associate Professor (Social Psychology) in the Department of Preventive Medicine School of Medicine, Tufts University. He received his B. A. from Brooklyn College and his M. A. and Ph.D. from Harvard University. He was formerly Research Social Psychologist, and Co-Principle Investigator, Rehabilitation Project, Massachusetts Mental Health Center, Boston. He was a Teaching Fellow and Research Assistant in the Department of Social Relations Harvard University. He has contributed a book and articles on medical sociology and intergroup relations.

Richard I. Krauss is Vice President of the Environmental Design Group in Cambridge. He is Consultant to the Institute for Applied Technology, National Bureau of Standards and is a faculty member in the Department of Architecture at MIT where he is involved in research on design process and computer techniques. Mr. Krauss has published numerous articles relating to computer applications in architecture and engineering.

Leonard M. Lansky is currently Professor of Psychology at the University of Cincinnati. Before assuming his duties there, he was lecturer and Senior Research Associate with the Graduate School of Education at Harvard University and was a Senior Research Associate with Fels Research Institute for the Study of Human Development at Antioch College.

Maria-Grazia Marzot is an Instructor (Sociology) in the Department of Preventive Medicine, School of Medicine, Tufts University. She received her B.A. from the Liceo Artistico Niccolo Barabino, her M.S.W. from the Scuola Superiore di Servizio Sociale and her M.A. from Indiana University. Her M.A. thesis was "Bollmington Welfare Organization: A Case-Study in the Intra-Community Systemic Linkage Concept."

William H. Michelson is presently an Associate Professor in the Department of Sociology, University of Toronto. He is the Director of a Study on "The Physical Environment as Attraction and Determinent: Social Effects in Housing, Center for Urban and Community Studies. He holds a Ph.D. from Harvard and has taught and conducted Research at Harvard, MIT, and Princeton in addition to professional consulting work. His research interests are in man and his urban environment and has articles published in *Journal of the AIP, Transaction, Social Forces and Sociological Inquiry* on this subject.

William R. Miller is presently systmes engineer with Arendt/Mosher/Grant, Architects in Santa Barbara, California. Mr. Miller is responsible for the design and development of in-house procedural systems as well as project director for various contracted research and development projects. His work has included systems studies for automated specification systems, relational problem solving and computerized map making. He has formerly served as structural engineer and consultant in various projects and has taught engineering graphics at Ventura College in Ventura, California.

Murray Milne is an associate professor in the School of Architecture and Urban Planning at UCLA. He previously taught at Yale in the Department of Architectuer where he had been involved in the Design Process and Computer Applications to Architectural Problems. Recently he has developed CLUSTR, an interactive computerized Design Partner which desplays the informational structure of the design problem. He recently edited a book, *Computer Graphics in Architecture and Design*.

William J. Mitchell, presently in the Master of Environmental Design Program, Yale University, has acted as Education Representative for the Australian Architecture Students' Association and as Student Representative on the Board of Architectural Education of the Victorian Chapter of the Royal Australian Institute of Architects.

Nicholas Negroponte is currently Assistant Professor of Architecture at MIT where he is teaching Architectural Communication. He previously taught courses in Computer-aided Urban Design. He is also involved in a research project, "Artificial Design Intelligence" under a Ford Foundation Grant. Other research activities at MIT and with IBM deal with the role of computers in urban design. Mr. Negroponte has published several articles dealing with machine technology (manipulation) toward the solution of architectural and urban problems.

Peter New is an Associate Professor (Sociology) in the Department of Preventive Medicine and the Department of Physical and Rehabilitation Medicine, School of Medicine, Tufts University. He received his B.A. from Dartmouth College and his M.A. and Ph.D. from the University of Missouri. He was formerly Project Director, Community Studies, Inc., Kansas City, Missouri, and Assistant Professor (Sociology), Graduate School of Public Health, University of Pittsburg. He was a Social Science Research Council Pre-doctoral Fellow (1955-1956) and an NIMH Post-doctoral Fellow at Harvard University (1960-1961). He has contributed articles on medical sociology and rehabilitation in a number of journals.

C. James Olsten is currently on leave of absence for California State Polytechnic College, School of Architecture where he has been teaching courses in architecture and is in charge of computer courses now offered and in those being developed. He is a member of the College ADP Advisory Committee, and the Benchmark Evaluation Team of Joint ADP Procurement Effort, California State Colleges. He was formerly assistant Civil Engineer and Programmer of the Construction Engineering Branch of the U.S. Army Cold Regions Research and Engineering Laboratory in New Hampshire. In addition, he was formerly an associate with Ethan Jennings, Jr., AIA and Associates, Inc. of Los Angeles.

A. James Pennington is currently Associate Professor of Electrical Engineering at Drexel Institute of Technology in Philadelphia. He has also taught electrical engineering at the University of Michigan. His professional interests include computers, control systems, electric power, urban technology and the philosophy of technology and society. Mr. Pennington has worked in industry for Dodco, Inc., Princeton, New Jersey, the Dupont Company, Wilmington, Delaware, and the Detroit Edison Company in Michigan.

John M. Peterson is currently Associate Professor of Architecture at the University of Cincinnati. He is presently director of Freshman Architectural Design and of Research Problems. He is involved in research in architectural education. Mr. Peterson has also exhibited and won awards for painting and sculpture.

Wolfgang F. E. Preiser is Research Assistant in the Division of Man-Environment Relations and Instructor in the Department of Architecture. Mr. Preiser received a research fellowship from Finland's Institute of Technology (1966), from Germany a Fulbright fellowship to the United States (1967) and an AISI fellowship (1968). He taught design at Virginia Polytechnic Institute and did consulting and architectural work with several firms in Germany, Austria and England. He is currently doing Ph.D. work at the Pennsylvania State University.

Amos Rapoport is currently a Lecturer in Architecture at University College, London and has recently been appointed the Senior Lecturer in Architecture at the University of Sydney, Australia. He has written and co-authored a number of articles and research reports and has recently published a book, *House Form and Culture*. He has taught at the University of Melbourne and the University of California at Berkeley. He has received a Fulbright Fellowship from Australia to the United States, and a French Government Study Fellowship as well as a number of other research awards.

Stuart Rose is currently Assistant Professor of Architecture in the School of Design, North Carolina State University. He has held previous teaching positions at Michigan State University, the University of Cincinnati, and the University of Nebraska. His main research interests are in the area of environmental perception particularly related to spatial sequence studies in terms of experience, imitation, simulation and behavioral response. Dr. Rose's publications include *A Notation/Simulation Process for Composers of Space,* publications dealing with the Effect on Behavior of the Qualitative Attributes of Space Establishing elements.

Henry Sanoff is chairman of the Steering Committee of EDRA and Associate Professor in the Department of Architecture, School of Design, North Carolina State University. He has taught at the Department of Architecture at the University of California, Berkeley, before coming to North Carolina. His research experience includes: Principal Investigator, University of California projects, Low Income Housing Demonstration, 1963-65 and Evaluation of three Case Study Dwellings, 1965-66, both under grants from the U.S. Department of Housing and Urban Development. Mr. Sanoff has served as Housing Consultant to the Office of Economic Opportunity and to VISTA. His varied research interests include the socio-psychological aspects of the physical environment and predictive and evaluative techniques for the measurement of performance. He has recently published a monograph "Techniques of Evaluation for Designers."

Man Mohan Sawhney is presently a faculty member in the Department of Sociology and Anthropology at North Carolina State University with a joint appointment in the Department of Architecture. His research experience includes work with the India Village Service and the Rockefeller Foundation in New Delhi where he was in charge of Field Survey and research. Dr. Sawhney has done extensive research and investigation into rural sociology.

Marvin Sevely, B. Arch. (Harvard) Ford Foundation Fellowship in Egypt 1953-55, United Nations Advisor to Turkish Government and Professor of Architecture Middle East Technical University 1956-59, Research Assistant in Department of Architecture at Princeton, 1956, Founder and Head, Graduate Program in Tropical Architecture, Pratt Institute, Senior Tutor, Tropical Department, Architectural Association School of Architecture, London 1967-68, Professor Architecture, College of Architecture Virginia Polytechnic Institute 1968-present.

Stuart M. Silverstone is currently teaching Architectural Communications in the Department of Architecture at MIT. Before assuming duties at MIT, Mr. Silverstone taught design courses at Boston Architectural Center and Texas A&M University. His research interests and activities include investigation of computer graphics for the Northeast Corridor Project at MIT and Communications in Urban Problem Solving at MIT. Mr. Silverstone has published several articles dealing with communication and information manipulation.

Fred I. Steele is presently Assistant Professor with the Organizational Behavior Group in the Department of Administrative Sciences at Yale University. He is also a partner in Environmental Dynamics, in Vermont. He was a behavioral science intern at the NTL Institute for Applied Behavioral Science and is presently an NTL Institute Associate.

Raymond Studer, educated as an architect, has a broad background of academic, professional and research activities. He is Head of the Man and Environment Division, the College of Human Development at Pennsylvania State University. Mr. Studer is also Consultant to the Research and Design Institute and to the New Cities project at Harvard. His research interests include the application of operant learning and interpersonal behavior systems theory to environmental design; the design of information and control systems for environmental planning; experimental analysis of behavior in the designed environment and computer simulator.

Donald Watson is presently practicing as an architect in Guilford, Connecticut and is Visiting Critic for the Department of Architecture at Yale University. From 1962-1964 he was an architect with the U.S. Peace Corps in Tunisia, North Africa and from 1964-1965 for the Government of Tunisia. From 1967-1969, Mr. Watson held an ACSA-AMAX Research Fellowship in Architecture.

Richard R. Wilkinson is currently head of the Department of Landscape Architecture at North Carolina State University. Prior to accepting that position, he was a professor in the Department of Landscape Architecture at the University of Michigan. While at the University of Michigan, he did research with the Nichols Arboretum. Mr. Wilkinson has served as Consultant to the Eastern Quebec Development Bureau, the City of Ann Arbor, Michigan; Urban Design Program Development. Mr. Wilkinson's field of professional interest is: regional development, physical growth, dynamics and physical development control policy.

Gary H. Winkel is a Research Associate in the Environmental Psychology Program at the City University of New York. He received his Ph.D. degree in Psychology from the Univesity of Washington. For therr he was associated with the College of Architecture and Urban Planning at the University of Washington and conducted research and teaching on human factors in environmental design. At the present time he is doing research on the perception of the large scale environment, the problem of locating urban colleges in ghetto neighborhoods and environmental quality. In addition, Dr. Winkel is editor of a new interdisciplinary journal concerned with the study, design and control of the physical environment and its interaction with human behavioral systems. The journal, *Environment and Behavior,* will appear in June 1969.

Richard D. Worrall is a manager in the office of Peat, Marwick, Livingston & Co. in Washington, D.C. Prior to assuming this position, he was on the faculty of Northwestern University and held positions as Senior Research Fellow at the University of Wales and Assistant Traffic Engineer for the City of Newcastle upon Tyne, England. At Northwestern he lectured in urban transportation planning, city planning, statistical and urban and regional systems analysis. Dr. Worrall's professional and consulting experience includes transportation planning and urban renewal projects in the United States and the United Kingdom. He has served as special consultant to a number of organizations including General Motors. Dr. Worrall has also published several papers on transportation planning and problems of urban travel.

Amos Rapoport
School of Architecture
University of Sydney

AN APPROACH TO THE STUDY OF ENVIRONMENTAL QUALITY

In the field of environmental design research and particularly in the study of environmental quality there has existed to dat˙ a rather restricted view of what can or should be considered pertinent sources of data. This paper deals wi͟h the possibility and desirability of employing materials and techniques not currently in use which are available and which can be of significant value to the study of environmental quality.

The term environmental quality has come to embrace many aspects of the environment. Here, its scope is confined to the symbolic, perceptual, and similar characteristics of the physical environment—built or natural—which a given group finds desirable. Specifically, the concept excludes the social environment and the like.

This use of the term implies that environmental quality will be defined and interpreted differently by different groups. Environmental choice is not a unitary phenomenon but a specific putting together of different proportions of a variety of traits selected from the whole possible array, the selection being largely based on the value system of the culture in question. These value systems can be seen as filters, screening out certain messages, reinforcing others, leading to different interpretations of what a "good" environment is seen to be.[1] Hence, there will probably exist cross-cultural differences in environmental choice due to different perceptions and interpretations of stimuli and the relative importance attached to them.

Although environmental quality cannot be defined *ab inito* but must be discovered, hypotheses about it can be made on the basis of previous experience and insight. One of the best ways of achieving such insight, particularly in a cross-cultural situation, is by studying the values, attitudes, and definitions of different groups in the context of a time and culture slice. If certain patterns are found to have a very wide distribution in space and time they will provide a reasonable starting point for hypotheses and lead to further research; they will help define desirable environmental characteristics which could be tested in various ways. The impact of this kind of approach can be seen in the development of human ecology from a rather simple mechanistic model based on the competition for space to the much more complex model involving socio-cultural variables.[2]

This type of study "should be grounded on intimate knowledge of the ways people think and feel about environment; this calls for substantial familiarity with social and intellectual history, with psychology and philosophy, with art and anthropology. All these field contribute to our knowledge of how we see the world we live in, how vision and value affect action and how action alters institutions."[3] This cannot be done without the analysis of many cultures and situations. This and the problem of analysis over long periods of time which it implies[4] raise methodological questions which constitute the main thrust of this paper, particularly those relating to the excluseive reliance on experimental techniques where momentary stimuli are often the main concern and where single variables are frequently studied.

The Problem of Methodology in Research Into Environmental Quality

The above discussion suggests that we need to examine the problems of environmental design research and determine whether techniques most commonly used are most suitable and whether additional methods might prove useful. In view of the often arbitrary nature of environmental design in the past, it is natural that there has been a reaction tending toward more rigour and objectivity and development of environmental design theory through research. Important and interesting as past introspection often was, designers' current research work has tended to swing to the other extreme and rely largely on experimental methods.[5]

This swing to experimental research is premature. Reliance on experimental work alone implies a more advanced stage of development than currently exists in the field of environmental quality. Such methodology implies the existance of a rather well-defined model of the phenomenon and the rigorous testing of ideas based upon this model. When these exist, the clearly defined methods for testing hypotheses come into their own. But such a model does not exist in this field and should not be implicitly assumed. We are in the empirical st of concept formation and hypothesis generating and there is still a need to define areas, formulate questions and hypotheses and to suggest models. While experimental verification could accompany this definition of the phenomenon, it should more correctly follow this phase.

In the process of generating hypotheses, we need to use the widest range of techniques and use the wide possible sample for classification, analysis and even mere inspection. Recently we have tended to neglect an a ray of techniques which could be most useful as an addition to experimental work. Material which is present neglected, for example, anecdotal material—may be extremely useful, particularly if it is not confined to desi ers anecdotes but a broad sample is obtained. Particularly if such material covers a large enough span of time and cultures, it can reveal shifts or constancies in environmental attitudes and preferences as well as unexpect regularities and connections. Even an inductive approach may sometimes be justified to help suggest likely h potheses which may then inform the further analysis of the material.

        1) Observational and indirect methods
        2) Analysis of indirect material such as written and pictorial materials as well as oral traditions such as songs, myths and the like.
        3) Analysis of physical environments themselves.

While the first of these has been neglected for a period, they have recently begun to be used rather more extensively. Indirect methods, while even more neglected have recently also begun to be considered. The second group is still almost completely neglected in environmental design research and form the main concer of this paper.

Observational Methods

These can be more useful than laboratory methods by giving real-life environmental situations and varia which may be missing in the laboratory situation. They include unforseen links,[6] combinations of smells, a sounds, textures, kinesthetics, associations and choices which give many more dimensions than the controlled of the laboratory.[7] These observational methods still present a number of difficulties in connection with the types of questions I have posed. Due to practical difficulties of time, resources and the like, it becomes diff to get a large enough sample, either cross culturally or over space. Another, and possibly the main, difficult is that *they cannot be used with material of the past* (which in themselves have a very wide space and time d bution). There is a basic assumption here that there is a need for cross-cultural studies as well as studies over Both experimental and observational studies are difficult for the former and impossible for the latter. They can neither reveal how people felt about different environments nor how these looked or were used.

The observation of overt behavior alone is not sufficienct for another reason. Daley, in her criticism of behaviouristic research in architectural psychology points out that one cannot separate "pure behaviour" fro attitudes, intentions or thought, nor can one confine discussion to physical or verbal actions alone.[8] The sa behaviour may then be due to different motivations based on different schemata which will affect how the e vironment is perceived and hence what meaning is given to it. A good example of this is given by Jahn in an context. He considers two groups of farmers—one in Italy, one Dahomey. Both are sowing crops and singing — their overt behaviour is identical, but different values and motivations are involved. For the Italians, sowing is the crucial activity and singing, at most, makes the work lighter. In Dahomey the seeds can do nothing until called upon to become active; singing is the indispensable activity needed to bring about growth, sowing merely an addition.[9]

Indirect Methods

These have recently received attention,[10] and a useful classification into indirect and direct methods has recently been proposed by Warr and Knapper.[11] Following this classification, most of the experimental work in our field which I have described, in which slides and photographs are involved would be regarded as indirect. Only work such as that by Lowenthal and Rivkin[12], both of whom rely on walks through real environments are direct according to the Warr and Knapper classification. The approach used by Siverts[13] and Lee who rely on people's memories of real environments as espressed through sketching, questionnaires and map drawing, is also direct in this sense.

The Use of Indirect Materials

The type of work just described is closely related to that using indirect information such as written, pictorial and verbal material which are only *more* indirect. The observer in both cases has already filtered the information provided by real environments; the use of indirect material is only more indirect than that of indirect method. The observers were exposed to a great range of sense modalities, and they observed, absorbed and commented not in connection with an experiment, not through questionnaires or being asked to sketch what they remembered, but very unselfconsciously—at least from the point of view of the study of the environment.[14] In Warr and Knapper's terms the use of such materials would fall into the category of *indirect, no intervention, the past*.[15] In Craik's terms, the proposed method is *direct experience of environment with free description*.[16]

Direct experience is always agreed to be best, but is usually too expensive. In the method proposed, a large sample of reactions is available for our inspection. Much of it is also written by non-designers which enables us to introduce the different schemata of the public (or at least sensitive or ararticulate members of it) as opposed to those of professional designers. It thus brings into play an additional set of filters and allows, by comparison, to see how different groups see the same environments—potentially a source of very useful information. As I already suggested it also covers a long time-span. Free description implies minimum constraints and selfconsciousness[17] but does make it more difficult to obtain quantitative comparisons. My argument here is that such non-quantitative methods are very useful for the area and problem defining and hypothesis forming functions already discussed, and that in any case there is no *prima facie* evidence (other than cultural bias) that quantitative studies are the only valid ones. Actually I would argue that the type of material I am describing lends itself rather better to the traditional historical methods. This method involves scrutinizing the material, putting in into some order and drawing conclusions from the patterns and regularities which emerge. Any conclusions generally enable us to ask the right questions rather than give immediate prescriptions for action—although they often suggest courses of action. Only by first asking these questions can we proceed to the non-historical stage—suggestions as to how to build better environments for current situations.

Analysis of Physical Environments

We may usefully wish to employ the physical forms of the past as evidence. These artifacts can be seen as embodying beliefs, values and the like[18] and they can be seen as "congealed communication."[19] The use of physical objects as evidence introduces problems of wrong interpretation[20] which can possibly be overcome by considering:

(a) How an environment looked at the time it was created (rather than *now*)
(b) How thie environment was actually used at the time.
(c) How the physical forms related to the value systems of the people, their world view, ethos, schemata, and ideal constructs.

For this purpose the more we can rely on literary and other written material, pictorial sources or verbal traditions such as songs and myths, the better. The more information of this kind is available, the more confidence can we have in the evidence of such physical forms as are available.

It therefore becomes clear that a fuller discussion of the materials which can be used for these purposes and some discussion of the techniques to be employed in analyzing them is required, it would also be useful to discuss the success of this approach in other disciplines where it has been used more widely than in environme al design research.

Indirect Materials in Er.vironmental Quality Research

The material which I am discussing comprises both written and pictorial material.

A) *Written material*—The use of written sources in analyses of this kind presupposes that language, goin beyond what seems to be said directly, reflects how the world is seen. This implies an acceptance of what has been termed *Whorfian hypothesis*[21] and underlies the use of content analysis[22] semiotic analysis[23] and oth similar techniques, some of which will be discussed later. Theoretical writings of designers, utopians and theo ticians have always been used by historians, but written materials have not been used to any extent in recent e vironmental design research; when occasional instances of the use of such materials can be found, they are nev analyzed systematically. Written material comprises:

1. Newspapers, articles, columns, letters to the editor, public controversies (such as the London Motor way Box and the Tate Gallery extensions in the London Press in late 1968-early 1969.) The numbe references to environmental matters in any newspaper over a period of time, and the deeply felt reactions to t are quite remarkably high. They provide a most useful source of insight, particularly in giving different sub-c tural samples depending on the type of newspaper.[24]

2. *Mass Media magazines*—These are useful for the light they throw on environmental attitudes and pr ferences—and hence quality—for a specific segment of the population.[25] A comparison between at titudes to the environment and differing definitions of desirable environmental qualities in these magazir and the professional architectural and planning journals can be particularly revealing.[26] Analyzing both kind magazines over time can reveal and illustrate changes in attitudes which clearly show the impact of images and cultural millieus on environmental decisions.[27]

3. *Novels*—Potentially these provide a very useful source of material since descriptions of different environments are frequently found. These do not constitute the prime concern of the author but are rather perceptive insights related to the mood and activities in question. These again can provide a cross-cultu sample (using novels from different countries) a sample of sub-cultures (comparing "literary" novels with "po ular" ones) and a time-sample (using novels from different periods in one culture. Although there have beer comments on the potential utility of this source of material[28] novels have not been explored systematically. am currently collecting such material preparatory to analyzing it.

4. *Poetry*—Since poetry often uses environmental descriptions symbolically or in relation to mood, it is particularly useful in revealing the attitudes of an age and the images which underlie environmen al features. It very clearly reveals changes in these over time.

5. *Travel Books*—There is a vast travel literature which extends back a long time. There are travel accounts from Ancient Greece, Ancient China, early Arab travellers and many others down to our own day. These books provide information both on changing attitudes to environments, implicitly giving an indication on "good" and "bad" environments, as well as providing descriptions of how places looked and were used.[29] An indication of how travellers reacted to these places and uses in terms of their own culture can also be obtained.

6. *Science Fiction*—This has become a rather extensive branch of contemporary writing. While it is restricted to a relatively short time span, it is useful in providing material showing current utopian thinking, attitudes to present day trends in environmental change and lines of development. It may also reveal the relative sophistication of environmental insights and visions as opposed to social ones.[30]

7. *Religious Texts*—In the case of primitive and traditional cultures, and many cultures before our own day, religious attitudes are crucial for any understanding of the relation of man and the physical environment. The Bible, The Koran, Indian, Chinese and other texts, as well as verbal traditions of non-literate peoples, reveal these attitudes and clarify their impact on natural and urban landscape preferences and formation.[31]

8. *Written Versions of Oral Traditions*—such as songs, tales, myths, jokes, and the like may reveal underlying schemata and ideals of non-literate peoples as well as popular reactions to environmental variables in literate cultures. It may also prove possible to use the evidence of graffiti and so on.

9. *Plays*, revues, night club monologues and the like also provide a source of material regarding areas of the environment which are of concern to people and which are seen as positive and negative. Some of these sources go back in time while others are more purely contemporary.

B) *Pictorial material*—The analysis of pictorial materials is even less well developed than the analysis of written sources. Pictorial materials have largely been used illustratively, to show how things looked, and they have been largely confined to photographs and prints, paintings and drawings. The list could, however, usefully be extended to include *cartoons* which show attitudes, values and images rather more than how things look. Cartoons have, in fact, occasionally been used to illustrate a point, but not in the study of attitudes and schemata.

Films have been analyzed to determine what their use of architecture and urban settings reveals about the films or the *film-makers*, but again nor for the purpose of studying environmental attitudes or preferences and not in any systematic way to generate questions or hypotheses.[32] Although films are limited to a shorter time span, it may be possible to obtain ideas on environmental attitudes by seeing how environments are used symbolically in given contexts and what is regarded as good and bad—and what grounds. This could occur in the context of science fiction (complementing written material) in films such as *Metropolis* or *2001*. More commonly it would reveal attitudes to present day environments: for example, vast office spaces in *The Trial* or *The Apartment;* industrial landscapes in *The Red Desert;* urban landscapes in Antonioni's films or in Tati's *Playtime;* Contrast of old and new in *Mon Oncle* and the *Man From Rio;* the use of pastoral and ideal landscapes in most films; the placelessness of many films, for example *Muriel* where the stranger asks where the town centre is only to be told he is there. Many other examples will, I am sure, occur to everyone.

The different kinds of materials and sources discussed may prove more suitable for specific types of studies: e.g. spatial behaviour, values and attitudes, unwritten rules or physical descriptions. They could then be used together and complement each other, increasing the depth and breadth. For example, in studying spatial behaviour over the ages we may look at pictorial evidence of how people have used space and arranged themselves in space going back to Roman and Hellenistic times and compare it with spatial behaviour in more recent observational studies.[33] Written sources will give descriptions of such behaviour but also attitudes toward different spatial arrangements in relation to mood and context. Poetry may be able to indicate the inner reality of different spatial arrangements and all these could be related to the physical forms themselves.

How reliable are these various materials likely to be? They may not be very reliable in themselves. Writing is done by people with their own biases; letters to the editor are selected and edited; painters rearrange groupings for compositional purposes and, as Gombrich has shown, even the way places actually looked may not be reliably reported by paintings and drawings.[34] Firstly, however, all these distortions may be discovered through comparisons and then become very revealing in themselves. Secondly no other type of material is available for the type of studies which I have been discussing and their use is worth-while. The attempt, I believe, will pay off.

Similar materials have in fact, been widely used, and with considerable success in a number of related fie A brief survey os some of these uses may illustrate the potential utility of this approach and give suggestions for applications.

1. The work of J. B. Jackson, both his own writing and in the contents of the seventeen years of *Landscap* magazine, to my mind generally provides some of the most illuminating insights into the problems of t environment. In this work the evidence of physical forms is related to that provided by written sources of all kinds which are widely used.

2. Psychology, from Freud onward, has relied heavily on the insights of literature. Kurt Lewin has spoken the value of a "historical" approach[35] and has suggested that a description of behaviour by Dostoyevsk may be more valuable than a scientist's description. In a different context, David Riesman has made a similar point[36] and a recent book applies this approach to the study of abnormal personality.[37]

3. In the study of competing uses of resources, literary sources have proved of value, particularly in tracing evolving attitudes to various features of the environment and also in identifying current key attitudes. Luten uses many such sources—novels, newspapers and the like—in an analysis of this type.[38]

4. There have been attempts to apply content analysis techniques to songs, poems and picturebooks as a way of deriving the image of a city.[39]

5. In France there has developed a body of work on the phenomenology of the dwelling which relies on an analysis of literature.[40] This is related to the growth of Semiotic Analysis in France, which attempts to find out people's attitudes to space and environment by analyzing the significance of words.[41] In some cases this technique is applied to open-ended questionnaires, and the notion of using literary sources criticised.[42] it does, however, seem that these methods may validly be applied to literary material.

6. An example which is of great significance for an understanding of urban attitudes, and through them on the form of cities, is the notion of the *anti-urban tradition* which has been treated by a number of peop These analyses of writings of many kinds disclose deep-running attitudes which clarify planners' decisions and public reactions.[44] This has clear implications for the study of environmental quality and also shows this me od used in a closely related field.

7. The analysis of written and verbal material has been user rather extensively in cross-cultural studies, in conjunction with more experimental work.[45] The analysis of this concept often relies on language (accepting the Whorfian hypothesis), books and newspapers (as well as the experimental use of TAT and other t

Yamamoto, in his analysis of Japanese national character relies on the evidence provided by a large number of books.[46] Khing Maung Win analyzed Burmese attitudes to space and time, among other apparent aspects of Burmese character, by examining the Burmese language.[47] Murphey uses written records of travellers and local people comparing them to child rearing manuals in the United States.[48] An indication of the close link between the concept of national character and urban form has been given by Meyerson[49] and this relationship will prove to be very useful in future studies of environmental quality.

8. In the general field of human geography and its various branches, material of the kind which I have been discussing have been used very extensively. Cultural geographers particularly have frequently analyzed written material to obtain insights into landscape attitudes and hence landscape formation. The interesting results which they have obtained suggest the utility of this approach and also shows how much remains to be done in relation to the built environment—providing some of the rationale for my current project.

Glacken traces the relation of culture to nature in Western thought from antiquity to the eighteenth century by analyzing literary religious and philosophical texts.[50] He throws much light on the topic by showing the interplay of a number of recurring themes. Sopher, in studying Indian landscapes and cities, relies on early travellers' accounts as well as religious texts to show that there has been a prevailing attempt to seek harmony with nature. He also quotes novels and books by hunters on attitudes to the wild and to animals, tracing the impact of these attitudes on the Indian landscape. He also uses newspapers and religious texts to show how these attitudes have affected Indian cities.[51] Yi-Fu Tuan uses literary sources from Ancient Greece onwards, as well as Oriental literary sources, to study the variety of Man-Environment relations and the impact of ideas on landscape appreciation and hence landscape formation.[52]

Lowenthal and Prince rely heavily on literary sources, as well as the evidence of photographs and paintings, both for a description of the characteristics of English landscape and to trace the influence of landscape tastes.[53] These tastes, they suggest, are responsible for landscape formation, since people make over landscapes in terms of visual prejudices and idealized images.[54] Prince refers to novels as a source of insight which, while difficult to test by objective inquiry, is useful as a tool of geographic inquiry.[55] Lowenthal analyzes historical responses to the American physical environment. These, he suggests, reflect idealized images and visual stereotypes and affect the American scene. In addition to written responses, paintings and photographs are used to derive categories of American physical environments and to study reasons for their occurrence.[56]

Conclusion

In most of the studies quoted, paritcularly those in psychology and geography, straightforward insight has been used to analyze the material. These fields are the closest to our own since they deal with human motivation and organization os space respectively. There have, in fact, been attempts recently to link psycological and geographical research in the same way that this is happening in environmental design research. Other fields, such as cross-cultural studies have relied more on measurement, and have analyzed very large numbers of samples to provide clusters of characteristics[57] and a large literature has developed on the sampling of large numbers of cultures and cultural variables, and their analysis. In this field there have also been attempts to apply content analysis techniques and to use experimental methods in conjunction with the others.

There is a constantly growing number of techniques which could be applied to the analysis of indirect data. In this paper I do not wish to get involved with this topic too deeply—rather do I wish to generate discussion on the merits of using such data in the first place. However it may be useful to just mention two promising developments which may enable such studies to be done more easily, and possibly more rigorously. The first are the categories and establishing dimensions of environmental quality.[58] Secondly, the newly emerging field of numerical taxonomy, with its use of polythetic rather than monothetic classifications may prove useful in arriving at sets of characteristics which are significant in defining environmental quality in different contexts.[59]

In the final analysis, however, at this stage of development of our field, the need will be for each research worker to develop his own methods. These need to be explicitly stated and should offer a reliable system for categorizing various characteristics which other people can then use, and attempt to establish whether any consistency begins to emerge. Once such classifications, categories and groupings have been established, hypotheses can be generated, some of which could be tested experimentally. This would then lead to more refined hypotheses in the usual manner. The first step, however, and the critical one at this stage, is not to allow methodological constraints and implicit models to narrow the search. Rather, the search space and the use of more free-ranging techniques should be broadened, using perhaps the suggestions mentioned above.

The possible utility of this approach and of the proposed definition of environmental quality can be shown by an example. The Greater London Council recently did a study in which they assessed the environmental quality of Thames-side within greater London.[60] This study is most useful and marks a great advance in the field of planning for environmental quality, but it suffers from a number of faults. The survey is done by designers and planners by inspection with no attempt to find out what environmental quality means for the public or how they see it in relation to the Thames. There is, in fact, no attempt to define environmental quality even for the designers—its nature is taken as self-evident and not in need of definition. If this study had attempted to define environmental quality by finding out what the Thames means for people in London, which characteristics are important and what sub-group differences exist, it would have gained a great deal. For this purposes an historical analysis of writings, drawings and paintings of the Thames over the ages; an analysis of songs and stories about the river and areas around it; present day reactions to it as shown by newspapers, films and television as well as consultations with people, would have been most important. For research purposes a comparison of this with similar analyses in other cities and countries would begin to show whether there are any constant elements in the qualities of rivers in cities and relate these, and also any differences, to images, schemata and the like.

Another example in the area of cross-cultural comparisons of urban space preferences, is related to emerging work on complexity.[61] Suppose that an analysis were done of writings about cities appearing in travel books, newspapers, novels and the like. If it was found that over a long period of time, and in many cultures, favourable reactions were related to complex environments and unfavourable references were in relation to monotonous and chaotic ones, this would be of the greatest importance. It would provide information more constant and fundamental than any single experiment or even any explication of experiment done currently. This analysis would provide a most important source of confirmatory evidence or, if this constancy were not found, it would suggest that this need was not of the fundamental nature that I have been suggesting.

Furthermore such an analysis, combined with an analysis of paintings, drawings and actual environments, would begin to generate ideas about which common physical characteristics were present. An analysis such as this would also show any cultural differences in the specific environments preferred but, as related to values, motivations and ideals, these differences could be part of a common striving-in the same way as the preference for certain urban settings may reflect a common search for optimum complexity.

These, and other matters arising from the study which I am here foreshadowing, and of which this paper is a small beginning, I hope to be able to report on in the future.

References

1   J. J. Gibson, *The Senses Considered as Perceptual Systems.* London, Methuen, 1968, p. 240; 266ff.

2   Amos Rapoport, *House Form and Culture,* Englewood Cliffs, N. J. Prentice Hall, 1969; "Some Aspects of the Organization of Urban Space" *Student Publication of the School of Design,* (North Carolina State University) (In Press).

3   Rapoport, "Facts and Models" *op. cit.*

4   Rapoport, *House Form and Culture, op. cit.*

5   For a discussion of possible research techniques see Kenneth H. Craik, "The comprehension of the everyday physical environment" *AIP Journal,* January 1968, p. 29-37. For a different discussion techniques see Henry Sanoff *Techniques of Evaluation for Designers,* Design Research Laboratory, School of Design, North Carolina State University, May 1968.

6   A. Chapanis, "The Relevance of Laboratory Studies to Practical Situations" *Ergonomics,* 1967, Vol. 10, No. 5, p. 557-577.

7   See Karl E. Weick, "Systematic Observationan Methods" in G. Lindzey (ed.), *Handbook of Social Psychology,* (2nd edition 1969) p. 357-437; The work of Professor R. G. Hopkinson at the Building Research Station and the Bartlett which combines laboratory and real life situations. The work of Esser on Mental wards, Lipman on old peoples wards, both of which revealed patterns which would not have been possible in laboratory situations. A number of other studies at the Bartlett—both by Graduate students and by researchers have used such methods with success. (Joiner on office furniture arrangements, Stilitz on movement spaces.) Both revealed patterns which experiment or questionnaires would not have revelaed, because the variables involved had not been anticipated. Once they are known, of course, experiments may prove useful.

8   Janet Daley, "Psychological Research in Architecture" *Architects' Journal,* London, August 21, 1968, p. 339-341.

9   Jahnheintz Jahn "Value Conceptions in Sub-Saharan Africa" in F. S C. Northrop and H. H. Livingston (ed.) *Cross-cultural Understanding,* (Epistemology in Anthropology), New York Harper and Row, 1964, p. 60.

10 For example as part of the survey by Webb, Campbell, *et. al. Unobtrusive Measures,* Chicago, Rand, McNally 1966 which deals mostly with observational and erosion methods—i.e. non-reactive rather than indirect methods. However, their discussion of the use of archives or physical traces would in the widest sense, come into the category of indirect methods.

11 P. B. Warr and C. Knapper, *The Perception of People and Events,* London, Wiley, 1968, p. 30 and elsewhere.

12 David Lowenthal, *et. al., An Analysis of environmental Perception,* 2nd interim report to Resources for the Future Inc., November 2, 1967, (unpublished). M. D. Rivkin, *Boylston Street Interviews,* (Mimeograph, unpublished).

13 Thomas Sieverts "Perceptual Images of the City of Berlin", in University of Amsterdam, Sociographical Department, *Urban Core and Inner City,* Leiden E. J. Brill, 1967, p. 282-285; this is, of course, related to Kevin Lynch's pioneering work *The Image of the City,* Cambridge MIT Press, 1960; Terence Lee,

"Urban Neighbourhood as a Socio-Spatial Schema", *Human Relations,* Vol. 21, No. 3, August, 1968, p. 241-267.

14 See Amos Rapoport "Some Consumer Comments on a Designed Environment" *Arena,* London, January 1967, p. 176 ff.

15 Warr and Knapper, *op. cit.,* p. 30.

16 Craik, *op. cit.,* p. 31-32.

17 The descriptions in this case are not always free—there are editors constraints, publishing policy, etc, but with a large enough sample, these should become manageable.

18 As one example, see Amos Rapoport, *House Form and Culture, op. cit.*

19 This is implicit in archaeology generally. For a specific statement, see David L. Clarke, *Analytical Archaeology,* London, Methuen, 1968, p. 120 and elsewhere.

20 See Horace Miner, "Body Ritual Among the Nacirema" *American Anthropologist,* LVIII, (1956), p. 503-507.

21 Benjamin Lee Whorf, *Language Thought and Reality,* Cambridge, MIT Press, 1964.

22 Ithiel de Sola Pool (ed.), *Trends in Content Analysis,* Urbana, University of Illinois Press, 1959; P. J. Stone *et. al., The General Inquirer.* Cambridge, MIT Press, 1966.

23 The series published by the Centre de Recherche d'Urbanisme in Paris: H. Raymond, *et. al., l'Habitat Pavillonaire,* 1966; N. Haumont, *Les Pavillonaires.*

24 I have used some isolated examples of newspaper material in an unpublished Masters thesis ("An Approach to Urban Design" Rice University, June, 1957). and in *House Form and Culture, op. cit.,* Chapter 6.

25 See Amos Rapoport "Whose Meaning in Architecture" *Arena/Interbuild,* October, 1967, p. 44-46.

26 At a graduate Seminar in Berkeley, (Fall, 1966) some of my students compared popular magazines dealing with houses, with the architectural press and found that the universes of discourse were totally different. A student at the Bartlett has been doing a similar analysis not yet complete.

27 One of my students at the Seminar mentioned in fn. 26, Mr. Stephen L. Quick, analyzed a planning journal over a period of 15 years. He found rather dramatically changing interests, areas of concern and underlying attitudes. These of course would be totally different to those in the popular press.

28 For example, John Westgaard comments on the potential relevance of novels in the study of urban social studies, but then excludes them from his survey. "The scope of urban studies in Scandinavian countries" *Current Sociology,* UNESCO, (Paris) Vol, IV, No. 4, (1955) p. 77. Since finishing this paper I have come across two instances of changing attitudes: a source book which presents written material of this type, Anselm Strauss, (ed.) *The American City—A Sourcebook of Urban Imagery,* London, Allen Lane, the

Penguin Press, 1968; Notice of a course on the Victorian City to take place at the University of Leicester (England) April, 9-19. Half the sessions will deal with literary sources including novels, poetry and ballads, pictorial sources and the physical forms themselves. The approach in these cases is rather different from the one I am suggesting.

29 This source has been used widely by geographers. I have used it in a rather simple way in "The Architecture of Isphahan" *Landscape,* Vol. 14, No. 2, Winter 1964, p. 4-11.

30 See Robert B. Riley, "Architecture and the Sense of Wonder" *Landscape,* Vol. 15, No. 1, Autumn, 1965, p. 20-24; Mark Muchnik is currently doing an analysis of science fiction views of the city for a Masters degree at the Bartlett School of Architecture.

31 See the field of Geography of Religion generally (for a bibliography see Rapoport, *House Form and Culture, op. cit.,*); for some specific examples, see Erich Isaac, "God's Acre" *Landscape,* Vol. 15, No. 3, Spring, 1966, p. 30-36; Lawrence Moss, "Space and Direction in the Chinese Garden" *Landscape,* Vol. 14, No. 3, Spring, 1965, p. 29-33.

32 Raymond Durgenat, "Movie Eye" *Architectural Review,* March, 1965, p. 186-193; Stanley M. Sherman, "How the Movies See the City," *Landscape,* Vol. 16, No. 3, Spring, 1967, p. 25-26; D. K. McGuire, "Last Year at Marienbad" *Landscape,* Vol. 15. No. 1, Autumn, 1965, p. 31-32.

33 For example, Derk de Jonge, "Applied Hodology," *Landscape,* Vol, 17, No. 2, Winter 1967-68, p. 10-11.

34 E. H. Gombrich, *Art and Illusion,* New York, Pantheon Books, 1961 (2nd revised Edition), p. 69-72 and elsewhere.

35 Kurt Lewin, *Topological Psychology,* p. 13, p. 30.

36 David Riesman, Preface to *Return to Laughter,* by Elenore Smith Bowen, American Museum of Natural History paperback, 1964, p. XVI.

37 Alan A. Stone and S. S. Stone, *The Abnormal Personality Through Literature,* Englewood Cliffs, N. J. Prentice-Hall, 1966.

38 Daniel Luten, "Engines in the Wilderness" *Landscape,* Vo. 15, No. 3., Spring 1966, p. 25-27. He also refers to Leo Marx, *The Machine in the Garden,* which in itself uses literary sources to study American attitudes to the environment.

39 Reported by W. F. Heinemeyer, "The Urban Core as a Centre of Attraction" in University of Amsterdam, Sociographical Department, *Urban Core and Inner City, op. cit.,* p. 83.

40 Gaston Bachelard, *The Poetics of Space,* New York, Orion Press, 1964.

41 Georges Matore, *l'Espace Humain,* Paris, La Colombe, 1962; Francois Choay, "Semiologie et Urbanisme" *Architecture d'Ajourd'hui,* June-July, 1967, p. 8-10.

42 *l'Habitat Pavillonaire* and *Les Pavillonaires, op. cit.*

43 On the United States Morton and Lucia White, *The Intellectual Versus the City,* New York, Mentor paperback, 1964; on Great Britain, Ruth Glass, "Urban Sociology in Great Britain; a trend report" *Current Sociology,* UNESCO, Paris, Vol IV, 1955, p. 14-19.

44 Amos Rapoport, "Some Aspects of the Organization of Urban Space" *Student Publication of the School of Design,* Kenneth Moffett, Gary Coates, (eds.) In Press.

45 See Fred Eggan, "Some Reflections on Comparative Method in Anthropology" and M. E. Spiro, (ed.), *Context and Meaning in Cultural Anthropology,* New York, The Free Press, 1965, p. 357-372; B. Kaplan, (ed.) *Studying Personality Cross-Culturally,* New York, Harper and Row, 1961; A. Inkeles and D. J. Levinson, "National Character: The Study of Modal Personality and Sociocultural Systems" in G. Lindzey (ed.) *Handbook of Social Psychology,* Cambridge Mass., Addison-Wesley, 1954, p. 972-1020. This last paper suggest three methods of study, the second of these is related to the one I have been discussing—the analysis of collective adult phenomena such as folklore, mass media institutional practices, etc.

46 T. Yamamoto, "Recent Studies on the Japanese National Character," in F. S. C., Northrop and H. H. Livingston, (ed.) *op. cit.,* p. 93-104.

47 Khing Maung Win, "The Burmese Language: An Epistemological Analysis" *Ibid.,* p. 223-234.

48 Murrey G. Murphey, "An Approach to the Historical Study of National Character," in M. E. Spiro, (ed.), *op. cit.,* p. 144-163.

49 Martin Meyerson, "National Character and Urban Form," *Public Policy,* Harvard, XII, (1963), p. 78-96.

50 Clarence J. Glacken, *Traces on the Rhodian Shore,* Berkeley and Los Angeles, University of California Press, 1967.

51 David Sopher, "Landscape and Seasons (Man and Nature in India)" *Landscape,* Vol. 13, No. 3, Spring, 1964, p. 14-19.

52 Yi-Fu Tuan, " Mountains, Ruins and the Sentiment of Melancholy" *Landscape,* Vol 14, No. 1, Autumn, 1964, p. 27-30; "Man and Nature" *Landscape,* Vol. 15, No. 3. Spring 1966, p. 30-36.

53 David Lowenthal and Hugh C. Prince, "The English Landscape" *The Geographical Review,* Vol. LIV, No. 3 1964, p. 309-346; "English Landscape Tastes" *op. cit.* Vol. LV, No. 2, 1965, p. 186-222.

54 *Ibid.,* p. 186.

55 Hugh C. Prince, "The Geographical Imagination" *Landscape,* Vol. 11, No. 2, Winter 1961-62, p. 22-25, (esp. p. 24-25.)

56 David Lowenthal, "The American Scene" *The Geographical Review,* Vol LVIII, No. 1, 1968, p. 61-88.

57 G. P. Murdock (ed.) *Social Structure,* New York and London, 1949; J. H. Steward, *Theory of Culture Change,* Illinois, 1955.

58 This is currently being done at the Bartlett School of Architecture in studying acoustic environmental quality of the Royal Festival Hall (London). Paper by Ron Hawkes at the British Acoustical Society meeting, Bartlett School of Architecture, December, 1968.

59 R. R. Sokal, "Numerical Taxonomy" *Scientific American,* Vol, 215, No. 6. December 1966, p. 106-117; Sokal and Sneath, *Principles of Numerical Taxonomy,* San Francisco and London, 1963; for an application in the field of archaeology, see David L. Clarke, *op. cit.*

60 Greater London Council, 1968.

61 Rapoport and Kantor, *op. cit.;* Amos Rapoport and Rone Hawkes, "The Perception of Urban Complexity" (unpublished).

Nicholas Negroponte
Department of Architecture
Massachusetts Institute of Technology

ENVIRONMENTAL HUMANISM THROUGH ROBOTS

Computer-aided design should not occur without machine-intelligence. Without it, it would be dangerous.

In our era, however, most people have serious misgivings about the feasibility and more importantly, the desirability of dubbing the actions of a machine as intelligent behavior. These people generally distrust the concept of machines that approach (and thus why not pass) our own human intelligence. In our culture, an intelligent machine is immediately assumed to be a bad machine. As soon as intelligence is ascribed to the artificial, some people believe that the artifact will become evil and strip us of our humanistic values. Or, like the great gazelle and the water buffalo, we will be placed on reserves to be pampered by a ruling class of automata.

Why ask a machine to learn, to understand, to associate courses with goals, to be self-improving, to be ethical, in short, to be intelligent?

A design machine must have an artificial intelligence because any design procedure, set of rules, or truism is tenuous, if not subversive, when used out of context or in disregard of context. It follows that a mechanism must recognize (and understand) the context before carrying out an operation. Therefore, a machine must be able to discern changes in meaning brought about by changes in context, hence, be intelligent. And to do this, it must have a sophisticated set of sensors, effectors, and processors to view the real world (directly and indirectly).

Intelligence is a behavior. It implies the capacity to add to, delete from, and use stored information. What makes this behavior unique and particularly difficult to emulate in machines is its extreme dependence on context -- time, locality, culture, mood, etc. For example, the meaning of a literary metaphor is conveyed through context; assessment of such meaning is an intelligent act. (Note that a metaphor in a novel characterizes the time and culture in which it was written.) One test for machine intelligence (not necessarily machine maturity, wisdon or knowledge) is the machine's ability to appreciate a joke. The punch line of a joke is an about-face in context; as humans we exhibit an intelligence by tracing back through the previous metaphors and we derive pleasure from the new (and surprising) meanings brought on by the shift in context. (Note that people of different cultures have difficulty understanding each other's jokes.)

Some architects might propose that machines cannot design unless they can think, cannot think unless they want, and cannot want unless they have bodies? and, since they do not have bodies, they therefore cannot want? thus cannot think, thus cannot design -- quod erat demonstrandum. This argument however, is usually emotional rather than logical. Nonetheless, the reader must recognize (if he is an "artificial intelligence" enthusiast) that intelligent machines do not exist today and that theories of machine intelligence, at this time, can at best be substantiated with such examples as computers playing a superb game of checkers and a good game of chess. And furthermore, architecture unlike a game of checkers (with fixed rules and a fixed number of pieces) and much like a joke (determined by context), is the croquet game in Alice in Wonderland, where the Queen of Hearts (society, technology, economics) keeps changing the rules.

In the past, when only humans were involved in the design process, the absence of resolute rules was not critical. Being an adaptable specie, we have been able to treat each problem as a new situation, a new context. But machines, at this point in time, are not very adaptable and are prone to encourage repetition in process and repetition in product. The result is often embodied in a simple procedure that is computerized, used over and over, and then proves to be immaterial, irrelevant, undesirable. Unfortunately, it is easier (perhaps economically advantageous) to keep the extraneous and unapropos computerized procedures.

Ironically, though it is presently difficult for a machine to have adaptable methods, machines can be employed in a manner that treats information individually and in detail. Imagine a machine which could respond to local situations (a family that moves, a residence that is expanded, an income that decreases). It could report on and concern itself specifically with the unique and the exceptional. It would concentrate on the particulars, for particulars, as Aldous Huxely states, "make for virtue and happiness; generalities are intellectually necessary evils". Human designers cannot do this; they cannot accomodate the particular, instead they accomodate the general. Britton Harris proposes that "He (the architect) is forced to proceed in this way because the effectuation of planning requires rules of general applicability and because watching each sparrow is too troublesome for any but God".

Consider a beach formed of millions of pebbles; each has a specific color, shape and texture. A discrete pebble could have characteristics, for example, black, sharp, hard. At the same time, the beach might be generally described as beige, rolling, soft. Because humans learn particulars and remember generalities, study the specific and act on the general, and because in this case the general conflicts with the particular, the outcome is unsatisfactory.

Our concern is therefore twofold: first, architects cannot handle large scale problems (the beaches), for they are too complex; second, architects ignore small scale problems (the pebbles), for they are too particular and individual. As a result, according to John Eberhard, "less than 5 per cent of the housing built in the United States and less than 1 per cent of the urban environment is exposed to the skills of the design professions".

But architects do handle "building-size" problems, a grain of concern that too often competes with general goals while, at the same time, couches personal needs in anti-human structures. The result is an urban monumentalism that, through default, we have had foisted upon us by opulent, self-important institutions; our period is a period of neohancockism and post-prudentialism. The cause is the distinct maneuverability gap that exists between the scale of the mass and the scale of the individual, the scale of the city and the scale of the room.

Because of this, an environmental humanism might only be attainable in cooperation with robots that have been thought to be inhuman devices, but in fact are devices that can respond intelligently to the tiny, individual, constantly changing bits of information that reflect the identity of each urbanite as well as the coherence of the city. These devices need the adaptability of humans and the specificity of present day machines. They must recognize general shifts in context as well as particular changes in need and desire.

Let us take a "pebble-prejudice". (Most computer-oriented tasks today are the opposite: the efficient transportation system, the public open space, the flow of goods and money). Our bias towards localized information implies two directions for the proposed relationship between designer and machine. The first is a do-it-yourselfism, where, as in the McLuhan automation circuit, consumer becomes producer, dweller becomes designer. Machines, located in homes, could permit each resident to project and overlay his architectural needs upon the changing frameowrk of the city. The same machine might report the number of shopping days left until Christmas as well as alert the inhabitant of potential transformations of his habitat.

The second direction presupposes the architect to be the prime interpreter between physical form and human needs. The machine's role, in this case, is to exhibit alternatives, discern incompatibilities, make suggestions and oversee the urban rights of individuals. In the nature of a public service, the architect-machine partnership would perform, to the utmost of each actor's respective design intelligence, the perpetual iteration between form and criteria. The two directions are not exclusive, their joint enterprise is actually one.

In either case, it is interesting to ponder what a human designer must do or the behavior he must exhibit in order to be a good architect, a talented architect, an ethical architect (not perforce a successful architect). We know that he must somehow contribute and promote physical environments that both house and stimulate the good life. But we do not know much about the good life (it has no "utility function" and cannot be optimized). We know that he must have an understanding of and ease with physical form. But we do not know how our own cognitive processes visualize shape and geometry. We know that he must interpret human needs and desires. But we do not know how to acquire these needs and desires.

What probably distinguishes a talented, competent designer is his ability both to provide and to provide for missing information. Any environmental design task is characterized by an astounding amount of unavilable or undeterminable information. Part of the design process is, in effect, the procurement of this information. Some is gathered by doing research in the preliminary design stages. Some is obtained through experience, overlaying and applying a seasoned wisdom. Other chunks of information are gained through prediction, induction and guesswork. Finally, some information is handled randomly, playfully, whimsically, personally.

It is reasonable to assume that the presence of machines, of automation in general, will provide for some of the omitted and difficult-to-acquire information. However, it would appear foolish to suppose that, when machines know how to design, there will be no missing information or that a single designer can give the machine all that it needs. Consequently, we need robots that can work with missing information. To do this, they must understand our metaphors, must solicit information on their own, must acquire experiences, must talk to a wide variety of people, must improve over time, and must be intelligent. They must recognize context, particularly changes in goals and changes in meaning brought about by changes in context.

In contrast, consider for a moment a society of designers built upon machine aides that cannot evolve, self-improve, and most importantly, cannot discern shifts in context. These machines would do only the dull ignoble tasks and they would do these tasks employing only the procedures and the information designers explicitly give them. These devices, for example, could indescriminately optimize partial information and generate simplistic solutions that minimize conflicts among irrelevant (due to context) criteria. Furthermore, since no learning is permitted in our not-so-hypothetical situation, these machines would have the built-in prejudices and "default options" of their creators. Such would be unethical robots.

Unfortunately most researchers seem to be opting for the above condition. As a result, many computer-aided design studies are relevant inasmuch as they present more fashionable and faster (though rarely cheaper) ways of doing what designers already do. And, since what designers already do does not seem to work, we will get inbred methods of work that will make bad architecture, unresponsive architecture, even more prolific.

Therefore let us (architects and computer scientists) take advantage of the professional iconoclasms that exist in our day - a day of evolutionary revolution; let us build machines equipped with at least those devices that humans employ to design. Let us build machines that can learn, can grope and can fumble.

In the Department of Architecture at MIT, we are doing just that. Employing Interdata Processors and Interdata Memory, we are building the beginnings of an ethical robot, as we call it, an Architecture Machine. Our goal is to enroll this machine in the Undergraduate Program at MIT by 1980.

James Marston Fitch
Columbia University

THE ARCHITECTURAL MANIPULATION OF SPACE, TIME, AND GRAVITY

Architectural Manipulation of Space, Time, and Gravity[1]

Human experience, through its metabolic and sensory perceptual systems, occupies its own "habitat" in the external world. Indeed, for the purposes of fruitful architectural and urbanistic analysis, each of these component systems must be conceived of as having its own private habitat (Figure 1). Each of these can be des cribed topologically in terms of modalities of stimulus and response. But each of these habitats must also be described in spatial and temporal terms as well, since all of them have dimension (height, width, depth) and all of them are experienced across time.

Spatially, these habitats may be visualized as being of three scales, microcosmic, mesocrosmic and macrocosmic, nested one inside the other with interfaces between each (Figure 2). For the human body, the interface between micro- and meso- environments is delimited by the continuous, three-dimensional envelope of the epidermis. For architecture, the interface between meso and macro- environments is delimited by the walls of the room or the walls of the building. Along both interfaces we commonly install artificial membranes to modulate the flow of forces across them—clothing along the body line, insulated walls along the building line.

However, though these systems occupy the same space, the spatial characteristics of the habitat of the metabolic system are quite different from those of the perceptual systems. Only the actual thermo-atmospheric conditions along the body's surfaces play any role in the heat exchange across the epidermis. Only the air, water and food actually ingested through the mucuous membranes of the body's cavities afford the fuel required for the whole metabolic process. Thus we can say that, though the habitat of metabolism has extension, it has no significant depth or thickness at all. Boundary, interface and habitat are one and the same. Heat, oxygen, water and food may exist in plenty around us, in either the meso- or macro- environments. Our sensory scanning systems (sight, smell, hearing) may bring us information about their distribution in space. But they will have no significance for metabolic process until actual body contact with them is achieved. Moreover, such a relationship between metabolism and its habitat exists independently of any perception of it.[2] Thus, the heat exchange across the epidermis will always critically affect the body's metabolic posture, even if for some reason—sleep, coma, anaesthesia—the body's perceptual systems are unable at the time to perceive it.

This dual nature of animal existence is the source of most of the fundamental paradoxes of architecture. In order for the building to be experientially satisfactory, it must afford a good "fit" for the body—the metabolic system and its habitat, on the one hand; the sensory/perceptual systems and their special habitats, on the other. These latter will have altogether different spatial characteristics, varying greatly in cross-section: from microns (for the velvet we touch), to inches (for the rose we smell), to feet (for the painting we admire), to miles (for the church bells we hear or the distant view we admire), to interstellar space (for the stars we gaze at) Well-being, amenity, ultimately beauty itself, depend upon the architectural manipulation of all these environmental factors, with their varying scales and dimensions.

The boundaries of all architectural volumes are delimited by surfaces (floors, walls, ceilings) which constitute the second interface between man and the macrocosmic world of nature (Figure 3). These bounding surfaces play a decisive role in the way we respond to and behave in the spaces they enclose. Thus a wall may, at a given moment, be acting quite effectively to insulate us against the mechanical force and bitter cold of a winter gale. Together with the heating system, the building may thus afford optimal thermo-atmospheric conditions for our metabolic well-being. But this same wall, at the same time, may have an inner surface whose color or brightness is an outrage to the eye or whose acoustical response is an insult to the ear. Similarly, a given room (cafe, class room) may be equipped with comfortable chairs and tables, properly organized in spatial terms to support a successful luncheon meeting or seminar; but the same room may well have too low an illumination level, too high an ambient noise level or too poor a ventilation system (the smoke-filled room) to permit satisfactory interpersonal communication. In short, the problem of a good fit between the occupant and the room extends from his skin right out to the walls and beyond, and implies the satisfactory manipulation of every habitat requirement simultaneously.

The design of any architectural component thus becomes a problem of successful adjustment between the organism and its environment—the buttocks and the chair seat; the eye and the lighting system; the student and the classroom; the pedestrian and the city street. In each case, we deal with a set of spatial parameters which describe the contours of the special vessel or container they require. However, this vessel must be understood as not merely containing a given action but also as actually generating a specific mode of behavior (chair makes for sitting, light makes for reading, classroom makes for study, sidewalk makes for walking, etc.) As Studer has aptly expressed it

> "The designed environment can be analyzed as a prosthetic phenomenon. It functions prosthetically in two distinct, but interrelated modes: 1) it is physiologically prosthetic in that it supports behavioral goals through maintenance of required (behaviorally correlate) physiological states, and 2) it is behaviorally prosthetic (Lindsley, 1964) in that it intentionally configures specific behavioral topographies."[3]

The architectural fit, in order to support and/or elicit the desired behavior, must therefore be analyzed from several distinct but intimately related points of view:

1. ergonomics: the study of the expenditure of physical energy required to occupy space and to overcome the gravity, friction and inertia implicit in all physiological work;

2. anthropometrics: the study of the spatial patterns which the body describes in the performance of work—walking, sitting, reaching, lifting, pulling, resting, sleeping, etc.;

3. proxemics: the study of the behavioral consequences of spatial relationships for inter-personal relationships of all scales and types.

Man has often been described—especially in the Puritan tradition—as being fundamentally a "lazy" animal. The term is pejorative, involving an ethical judgement of experiential reality—man's compelling need to husband his limited supplies of energy. There may indeed be such a thing as a "lazy man"; but there is no doubt at all that all animals, including man, are compelled by experiential circumstances to expend large amounts of energy just to stay alive. Resisting the forces of gravity and inertia, of atmospheric pressure and environmental heat, is of itself stressful, as Selye has established. All animals develop characteristic modes of behavior in response to this circumstance. Thus cattle grazing on a hillside will follow the isoplethic contours so consistently that paths will be developed there. In just such fashion, college students will leave the formal patterns of paved walks to take diagonal shortcuts across the quadrangles. Cattle or college student, the trajectory of their movement represents a resolution of all the vectors of force acting upon them at that time. Some of these forces are exogenous (change of level, choice of sun or shade, mud or dry paving, etc.) and some are endogenous (late for an appointment, desire to be seen or not seen, etc.). Both sets of forces are equally "real," of course; and consciousness often compels men to take the less easy or more hazardous path (to jump into heavy seas to save a drowning child, to attack across an open field in battle, etc.).

Each trace or path represents the end result of a complicated process, in which we are continually weighing advantages against disadvantages, costs against possible returns, of this or that method of achieving our objectives. Exactly how this process is effectuated in human behavior is a complex and not fully understood question. All movement in space involves work, whether physiological or societal. Since movement is costly of energy, all animals have developed monitoring systems to aid in its conservation. In all higher animals, and above all in man, this is expressed in an extraordinary capability for orienting one's movements with reference to the outside world. It involves all the body's sensory perceptual systems but it clearly goes far beyond mere perception to include an ordered response. It gives the body the capacity to discriminate between movement and non-movement, or between its own movement and that of other bodies in the exterior world. Such capacity is critically important to survival—too important, as J. J. Gibson has put it, to be entrusted to any single set of sensory receptors.

"There are many kinds of movement to be registered. There is articular kinesthesis for the body framework; vestibular kinesthesis for the movement of the skin relative to what it touches; and visual kinesthesis for perspective transformations of the field of view. In all these perceptions, the sensory quality arising from the receptor type is difficult to detect but the information is perfectly clear. Kinesthesis is the registering of such information without being sensory; it is one of the best examples of detection without a special modality of sensation."[4]

The operation of these various forms of kinesthesis is quite apparent in the behavior of people in architectural or urbanistic space. Articular kinesthesis will be expressed in the way we sit in a chair or lie in a bed. If either is a bad fit, it will be expressed in the way we twist and squirm in an effort to find a more comfortable resolution. Vestibular kinesthesis determines our behavior on ramps and stairs. The spiral-ramped galleries of the Guggenheim Museum, for example, force a continuous disequilibrium on the visitor which stresses his vestibular apparatus. Cutaneous kinesthesis govern our response to surfaces—e. g., whether we follow the path or cut across the lawn will depend upon whether the grass is dry and resilient or wet and muddy. Visual kinesthesis enables us to anticipate, and hence to avoid, hazardous discontinuities in surface; it is at the base of our tendency to draw back from floor to ceiling glass walls in high-rise buildings. Of course, vision over-rides all other forms of perception with its capacity for transmuting other forms of sensory input into visual data. Thus it is thanks to visual kinesthesis that we can avoid mud without stepping in it; or steer clear of rough walls before we scrape an elbow on them; or avoid falls down unprotected changes in grade; or discriminate between soup and salad without having to taste either.

Each time the architect or urban designer erects a wall or paves a street, he intervenes in the behavioral modes of the population of that space. The consequences of his intervention may be major or minor, benign or malignant; they will always be real. Malfunction will be expressed in incidents or accidents among the population—in other words, by their not behaving the way they were supposed to. Conventionally, these are always blamed on the individual: he "wasn't looking where he was going"; he "slipped on the paving"; he "stumbled on the stairs"; "tripped over the chair"; "got dizzy and fell out the window"; etc., etc. But accident statistics are enough to suggest that this is a simplistic explanation of events, since different types of accidents are associated with specific sets of spatial configurations. It is true that the inner nature of accidents connected with buildings often seem inexplicable, because they involve movement. Careful research, especially time-lapse and slow-motion photography, is necessary to explicate even the simplest incident, since cause-and-effect are too entangled for the naked eye. But such research will usually reveal that despite the operation of articular, vestibular, cutaneous and visual kinesthesis, the user could not adapt his actions to the spatial set involved. In other words, the architectural element could neither support nor elicit the sequence of motions it was nominally designed to expedite.

Ergonomics: The Cost-Accounting of Work and Fatigue

What, specifically, does the animal body require of architectural space? Not at all that it provide a controlled environment of absolutely uniform or unvarying qualities, as is so often assumed. To the contrary: the requirements of the animal body vary constantly as it goes about its various tasks. For optimal well-being, it requires equilibrium, dynamic balance between internal needs and external means of meeting them.

Although this is true for all environmental components, it is most easily demonstrated for the thermal since, as Winslow and Herrington have put it, "the whole life process is a form of slow combustion." The body is only about 20% efficient in the conversion of fuel into work; it must therefore always dissipate a great deal of "waste" energy in the form of heat. And since it is very sensitive to its accumulation, this heat must be dissipated at the appropriate rate. The rate will vary with each level or type of activity—how widely is clear from this table of metabolic rates for men:[6]

type of activity	calories per hour
sleeping	65
sitting, at rest	100
typing rapidly	140
walking (2.6 mph)	200
walking (3.7 mph)	300
stone working	400
swimming	500
walking up stairs	1100

It is obvious that we deal here with ratios, not fixed quantities: hence the building should be regarded as a flexible instrument for maintaining these ratios at levels established by the body. How important these ratios become not merely to human well-being and efficiency but ultimately to life itself, may be seen from this table showing the correlation between temperature, work and physical condition:[7]

effective te temperature	total work in foot pounds	increase in rectal temp. degrees F p h	increase in pulse rate beats p.m.p.h.
70	225,000	0.1	7
80	209,000	0.3	11
90	153,000	1.2	31
100	67,000	4.0	103
110	37,000	8.5	237

These illustrations, drawn from the relationship between the body and the thermal component of its environment, have analogies in other areas of experience. Thus we can be sure that any task involving critical acuity in seeing (classroom study, fine machine work, microsurgery) will require a luminious environment precisely structured to that end; and we can be certain that both work and worker will suffer if it is not so structured. Similarly any activity demanding critical discrimination in hearing or communication (concert listening, airport control, radio broadcasting) will require its own special sonic environment and will suffer qualitatively from the lack of it. The proper task of good architecture, in short, is to organize space and environment in such a way that the eye, ear, tongue, or hand can accomplish its task or activity—seeing, hearing, speaking, tasting, touching and lifting—with the greatest ease and precision, subject to the least interference or friction from extraneous factors.

But other factors, equally "real" but much harder to isolate and define, are also involved. "Work itself seldom leads to the chronic condition we call fatigue," says one research. It is axiomatic that worry, not work, (is what) kills." Moreover, there is growing evidence that the very process of living, quite apart from environment and work, produces measurable stress which leads in turn to measurable fatigue. Though it is only under laboratory conditions that we can observe man at rest and even partially isolated from external stimuli, it is clear that even here the body is still under stress—the heart and muscles resisting gravity and atmospheric pressure, etc. The Canadian physiologist, Hans Selye, has gone so far as to try to isolate and measur such factors, giving them the status of a syndrome called "the stress of life."[8]
The task of architecture is not to "eliminate stress" or abolish fatigue. It is rather to eliminate or at least minimize those environmental pressures which are peripheral, tangential or only accidentally associated with the task at hand. Then stress will be the by-product of productive work and fatigue will come from the accom plishment of human tasks and from not mere animal exertion. Good architecture, by this standard, would be that which permits man to focus his energies on whatever he is doing, whether running a lathe or listening to a concert, writing an essay or recovering from surgery. The ultimate criterion would be not lack of stress but optimal well-being under concrete conditions.

It is obvious that we deal here with complex equations in which variables of time and energy are involved An important variant in this equation is the status of the individual himself—age, sex, health, emotional and intellectual attitudes. Just as the young man at rest has different thermal requirements from the young man at hard work, so a young man at hard work will have different needs from those of his father at the same task. In just such a fashion, women will have different reactions from men to the same task in the same set of environmental conditions. There are always idiosyncratic individuals whose requirements will vary from the norm. Their variations are real and cannot be ignored in the design of buildings involving them. But they should not obscure the fact that human response to heat, cold, glare, noise and stench are generally quite uniform. When plotted on a graph, all of us will lie on or quite near the general curve of humanity. The architect should accept them—as the physiologist and the psychologist do—as the basis, the point of departure, of his designs.

It goes without saying that the degree of precision of environmental control which will be required of a building will vary with its purpose. We demand a much lower order of performance from a bus shelter than from a bus terminal, from a circus tent than from an opera house, from a hospital waiting room than from the sugrical thearers of that same hospital. The "margin for error" in architectural design is in inverse ratio to the criticalness of the process to be housed.

Most of what we know about fatigue comes from the factory. Yet research in the field has been complicated, in a sense, by the one-sided investigations of its external or objective aspects. This was perhaps the natural result of research largely subsidized by the manufacturers themselves as a means of increasing productivity. However important this approach may be, it is far from complete. For one of the most puzzling aspects of fatigue is its duality: it is at once objective and subjective, physiological and psychological. Thus, if we observe and measure it from the standpoint of management alone, we are ignoring the fact that the relationship between the two halves is not merely complex; it is dialectic.

Of course, amny aspects of work and fatigue are being made obsolete by the rise of the new technology of mechanization, automation and computerization. Processes which once were studied by the experts either to save the energy of the worker or to increase his productivity have been leap-frogged. Now the process is automated and computerized and the worker is either reduced to a minor role in production or eliminated altogether.

Though this may be a perfectly valid line of development for many linear industrial processes, it has little direct application to many complex, non-linear activities which are little subject, or not at all subject, to such forms of rationalization. Theaters, schools, hospitals, libraries, court-rooms and residential types require a wide range of furniture, equipment and tools to facilitate their respective processes. All of this could stand vast improvement in terms of both design and manufacture. But the interface between the worker and his task requires the colsest and most sophisticated scrutiny if efficiency, amenity and safety are to be improved.

Anthropometry: The Mensuration of Movement

Just as, in everyday experience, we equate intellectual comprehension with the sense of sight, (I see what you mean), so emotional response is equated with the sense of touch (He has no feeling for the needs of others). This is no mere accident of language but describes rather a fundamental distinction. In many ways, the sense of touch is the most vivid capacity which the individual has for interaction with the world immediately outside his own body. It not only plays a central role in the satisfaction of sensual appetites (e.g., the sexual act). It is also the source of absolutely indidpensable factual information about the qualities of the spacial environment. For the shape, size, texture, density and temperature of surrounding surfaces is perceived first of all by that marvellously complex system of nerve-endings in the skin and muscles which furnish us with tactile, haptic and kinesthetic sensation.

Thus, our knowledge of the special properties of a material like marble derives largely from our tactile-haptic exploration of it. We learn its cold smoothness from stroking it, its hardness from sitting on it, its density and weight from trying to lift it. Only its color and pattern are perceived visually at this initial stage of exploration and discovery. After a quite limited experience with marble, we are able to transmute this tactile information into visual terms. We say "It looks like marble to me": we forget that the blind would, each time, have to stroke it, punch it, lift it before he could safely conclude it was a marbelline material.

The capacity of vision to metamorphosize other sorts of sensory data—data which are not primarily or even not at all visually-derived is vision's most splendid characteristic. It is a great time-and-energy saving mechanism, especially in the design fields. It enables us quickly to retrieve from memory that particular combination of sense-perceptible properties which goes to make up "marble." It enables us to say: "I don't need to see a sample—I remember quite clearly what it looks like."

Nevertheless, this same syncretic power of vision has inherent dangers for all designers, and most of all architects. For in real life, vision is only one channel of perception among many and in many experiential circumstances of no greater importance than others. Many architectural problems involve the manipulation of phenomena which are literally invisible (e.g., poisonous gases, high noise levels, high temperatures). Many environmental hazards are not perceptible by any sensory means (e.g., carbon monoxide, a structural member just before collapse). But neither working drawings of the proposed building nor photographs of the finished one give any visual information about the possible existence of such factors or their relative seriousness. Because such phenomena are hard to visualize they are often given low priority in relation to those which are easily visible. In short, the architect has a built-in bias in favor of the visually perceived; and it must be rigorously examined if his buildings are to be authentically successful. Nowhere is this more true than in the habitat of touch.

The sense of touch has a more complex capacity than the scanning senses of vision or hearing, depending as it does upon several types of neural sensors in the skin and muscles. These sensors, working in various combinations, bring us an astonishing range of information on the properties of surfaces around us:

haptic information	shape and volume dimension and size
tactile information	smooth-rough blunt-sharp wet-dry hard-soft elastic-plastic pressure-suction

thermal information   hot-cold ("pregnant center point")

As in the case of other mechanisms of sensory perception, that of touch can be seen as occupying its own habitat, bounded by thresholds of perception and pain and centered by some "golden" zone of satisfaction. The analogy is only approximate, however, as the properties of matter explored by this sense are not strictly comparable with the others. For the thermal sensors, the habitat would extend upward (to intolerably hot) and downward (to intolerably cold) from a neutral centerpoint—the temperature of the body itself. For the pressure sensors, the limits would lie between the gentlest perceptible air movement and the crushing blow, or the touch of velvet and the prick of a needle. Many other properties of matter would produce tactile sensation without stress: to handle a hard or a wet object is not, of itself, more or less stressful than handling a soft or dry one.

Thus, in general terms, we can deliniate the boundaries of the habitat of touch. It has upper and lower limits of stress or lack of stimulation; and some central zone of optimal stimulation. But here again, the limiting factor is that of movement and change in position along time. To discriminate between velvet and paper it is not enough merely to touch; one must also stroke. Any tactile-haptic experience, no matter how pleasureable initially, will cease to be satisfying if extended beyond its own built-in limits. Indeed, it will tend to cease to be perceptible at all. Here, as everywhere in the domain of the senses, change and variety is essential to perception itself.

An understanding of these tactile-haptic phenomena should be an integral part of the designer's expertise. For the boundaries of architectural and urbanistic spaces consist of surfaces—man-made or man-modified—which by their very presence, effect tactile-haptic sensation in important (and often in unanticipated or ambiguous) ways. Much of the reason for success or failure of any space lies in the impact of the bounding surfaces. We have already seen how our behavior is affected by the way in which a wall reflects light or heat or sound; analagous relationships between us and the wall in the field of tactile-haptic response. Surfaces which appear smooth polished or velvety to the eye will often lead us to stroke them, as if to confirm the tactile pleasure they afford. Very rough surfaces, on the other hand, cause us to shy away from them, irrespective of how handsome they may be in purely formal (e.g., visual) terms.

The tactile-haptic properties of floors and pavements have a similarly powerful influence upon the behavior of the people who traverse them. We have already discussed the effect of gross changes in level or direction: the effect of such surface characteristics as texture, resilience, cleanliness, wet/dryness is equally profound. Highly polished pavements (marble, terrazzo, faience tile may be handsome visually and pleasurable emotionally because of their connotation of wealth, ceremony, urbanity. But they look slippery and often are slippery; a duller matte finish would not only look safer, from this point of view, it actually is safer.

Resilience is a critical factor in paving materials, especially in museums and art galleries, where the perambulating picture-viewer is subject to unusual skeleto-muscle stress. In such situations, marble or terrazzo floor materials would be least desirable from an ergonomic point of view (even though, for formal reasons, they are very commonly used.) Cork or plastic would be better than lithic materials; carpeting on wood flooring installed on wood sleepers would perhaps be best of all. An inevitable experiential aspect of all these flooring materials is their acoustical behavior; since silence is a requirement in museums, galleries and libraries, the absorptive capacity of carpeting makes it a suitable material. The feeling of "hushed luxury" of carpeted spaces is thus as much an acoustical as a tactile-haptic phenomenon.

The exact way in which pedestrians behave on floors and pavements is subject to a number of environmental variables. Indoors, these are subject to very precise manipulation—light, heating, air movement, humidity, noise etc. But out-of-doors, a host of climatic variables affect the comfort, usability and safety—hence the patterns of use—of the pavement. While predictable, these variables are only predictable in gross terms—e.g., average monthly rainfall, sun and shadow patterns, wind directions and velocities, mean temperatures, etc. The maintenance of amenity in outdoor spaces is therefore very complex, involving dozens of meterological phenomena varying in intensity with time and space. The successful manipulation of these phenomena constitutes one of the most challenging and least explored aspects of landscape design. Perhaps because of its very complexity, the problem is evaded: instead of designing outdoor spaces to meet at least the gross impact of the seasons, they are conceived of as though for some ideal Platonic climate. It is always sunny and summer; there is no rain or snow, no dust or mud; no extremes of heat or cold, no wind, no cloudy days or stormy nights; no noise or fumes from the automobile traffic which is built into the design.

This tendency to regard outdoor spaces as purely aesthetic constructs, voids empty of micro-climatic reality, is nowhere more apparent than in the big urban plazas which are increasingly a feature of urban development. This conceptual error is obvious from their malfunction in real life. In experiential reality, there is no spot within the continental USA in which some or all of these meteorological factors, will not, for some part of the year, combine to render these handsome and costly plazas literally uninhabitable. They thus fail in their central pretension—that of eliminating gross differences between architectural and urbanistic spaces, of extending in time the areas in which urban life could freely flow back and forth between the two. For much of the country and for part of the year, rain, sleet, snow and/or ice will be the principal obstacles to easy, urbane movement across these plazas. under any combination of them, they become unpretty, unpleasant or unsafe.

Yet technically it is entirely possible to minimize or eliminate many of these microclimatic phenomena, scale models could be tested for aerodynamic behavior; cumulative shade patterns for cold weather months could be simulated; and the whole complex adjusted to minimize winds and shading, maximize exposure to the sun. Drainage patterns could be correlated to snow-plow paths, areas allocated for mounding surplus snow. Snow melters embedded in paved surfaces would obviously be optimal. In climates without snow or ice but with heavy rainfall (e. g., Florida, Gulf Coast, Puget Sound) the main problem becomes one of effective shelter along the lines of high pedestrian traffic.

The problem of year-round visibility in open urban spaces has its obverse side-protection from excessive insolation, with the resulting heat and glare, during hot weather months. All pavements tend to be much hotter than the natural surfaces they replace, because of their heat-absorbing and heat-holding characteristics. Unless adequately shaded and ventilated, they will create desert-like micro-climates, often with astonishing extremes of heat or glare. Thus, the urban plaza must be analyzed just as carefully for summertime amenity. Plant material of all sorts are more desirable shade makers than architectural constructions, since they maintain low surface temperatures, converting much of the solar energy which falls upon them into other forms of energy. Whatever the devices employed, however, hot weather viability implies minimizing ambient heat and light, maximizing shade and ventilation.

Proxemics: Human Deployment in Space

The senses of vision, touch and hearing bring us vital information about the objects which furnish, and the surfaces which enclose, terrestrial space. These sense-data are so important, in fact, that we are apt to think of them as being the exclusive source of information as to where, at any given moment, we "are." But the neural sensors embedded in our joints and muscles give us an altogethr different type of spatial information—i.e., our kinesthetic-proprioceptive orientation is space itself. These sensors afford us a marvelous control both of our over-all posture with reference to gravity and also the disposition of our various bodily members in relation to each other at any given moment. This kinesthetic-proprioceptive capacity is what makes bodily movement possible in higher organisms. But it does not explain, except in purely mechanical terms, why and how we move as social beings.

Any movement in space requires an outlay of energy and we have an easily understandable tendency to conserve it. However, this energy cost is only one factor in the complex equation of our physical movement in space. The actual pattern or trajectory which we trace in any given action must be viewed as the resolution of a whole system of vectors of force acting upon us, the algebraic sum of which determines that pattern or trajectory. Some of these forces are subjective, internally-generated, some are objective:

Psychological—the motivation of the individual, his incentive to accomplish the action in question.

Physiological—the physical condition of the individual with reference to the energy required for the action.

Socio-cultural—the type of behavior which the space is designed to elicit (playing field, work place, place of study or worship, etc.)

Micro-climatic—the actual environmental conditions obtaining on the site (rain or snow, sun or shade, temperature, wind, etc.)

Topographic—the contours, textures and shape of the surface on which the action transpires.

    The relative value of these various kinds of force, all of them simultaneously acting upon the individual, will vary with the circumstances actually obtaining. Thus a man in a desperate hurry will cut across a muddy field or run up a steep flight of stairs which a casual stroller would circumnavigate or reject. Social convention forbids the male to use the women's toilet, no matter how great the physiological pressure, just as it requires him to walk slowly and quietly in churches, libraries and museums. A bargain sale or an accident may pull pedestrians to the sunny side of a street which, otherwise, would be deserted because of heat and glare.

    The precise fashion in which architectural and urban space is to be organized, and the way in which furnishings and equipment are deployed within it, should be determined by a clear understanding of vectors of force described above. Unfortunately, architects and planners have had to rely upon a quite primitive theoretical approach to these complexities and a quite eclectic methodology for handling them.

    The behavior-eliciting function of architectural space is now being explored by many investigators, from many disciplines and from many points of view. For example, one group of experimental psychologists has studied the effects of spatial organization of certain housing projects upon interpersonal relationships which developed among the inhabitants.[9] The study established a dramatic distinction between what they call physical and functional distance. In the projects under study, they found that two apartment houses might be only thirty feet apart (physical distance); but if their entrances faced in opposite directions and gave out onto different streets, the chance of interaction (and hence of friendship) between the tenants of the two houses is as radically reduced as if thousands of feet intervened. The functional distance may be the equivalent of miles.

    Such investigations should have a sobering effect upon the architect, indicating as they do the socio-cultural consequences of all his spatial sets, large or small. He cannot erect a wall or enclose a room or pave a path without affecting some of those vectors of force which play a role in modifying the behavior of the occupants of his space. But there are subjective forces here too—internally-generated forces over which he has little or no control. These he can only try to understand and respect. The anthropologist Edward Hall has explored this aspect of human behavior.[10] Hall attempts to quantify the spatial dimensions of various levels of human interaction. He visualizes each individual as centered in a concentric series of "balloons" or "bubble" of private space, territorial extension of his own body in which he moves through space. These concentric spheres represent optimal distances for a hierarchy of inter-personal relationships: intimate, up to 1½ feet; personal, 1½ to 4 feet; social-consultive, 4 to 10 feet; public, 10 to 30 feet, and beyond. Any violation of these optima by other individuals will, according to Hall, be reflected by a stress upon that relationship.

    The scale of this heirarchy is partly a mere quantification of our powers of perception—how well we can hear the words of the actor, how clearly we can see the face of a friend, whether or not we can touch the person we love. But the governing factor is not only one of acuity of perception. We want contact with the actor, the friend, the lover: but we do not want to be as close to the actor as to the friend, nor as close to the friend as to the lover.

Even these spaces are not absolute, according to Hall: they will vary somewhat with culture, each society establishing its own norms. Thus male friends in the Middle East will want to be able to touch each other while in England such close contiguity would be considered distasteful. In just such a fashion, body odors might be considered attractive in one cultural milieu (e.g., Elizabethan England) while in another they would be a social liability (e.g., contemporary USA).

Technology, too, has made it possible to modulate some of these spatial balloons. The telephone permits verbal contact between lovers across miles of space. The footlight and spotlight extend the radius within which we can clearly see the actor's facial expression, just as the microphone has extended the distance across which we can hear the political speaker. And the air-conditioned, floodlighted arena has extended in time and in space the number of people who can follow the hockey game or volleyball game within acceptable limits of perceptual acuity.[11]

The express purpose of many types of architectural spaces (classrooms, offices, stores, workshops) is to expedite or facilitate interactions, contact or communication between certain groups of people engaged in certain types of activity. But privacy and isolation are desiderata in other types of spaces (bedrooms, toilets, libraries). In an interesting study of how university students actually use library reading rooms, significantly called *The Ecology of Privacy*, the behavioral psychologist Robert Sommer reports that:

"Of those students who entered the room alone, 64% sat alone, 26% sat diagonally across from another student, while only 10% sat directly opposite or beside another student.[12]

Chairs at the end of the table were invariably most popular because this automatically gave the student one protected flank. Moreover, elaborate strategems were employed to reinforce this privacy. At the same time, the students did not want absolute isolation while reading. Some felt more comfortable near people, although avoiding direct eye contact.

The Sommers study indicated a number of fairly simple measures aimed at ensuring the privacy which students require in library work, suggesting that attention to size, design and arrangement of library reading tables would go a long way toward meeting these requirements. Of course, the design and arrangement of furniture in a room can be used to bring people together as well as to separate them. In a now-famous study of the behavior of patients in mental institutions, Dr. Humphry Osmond found that certain seating arrangements in the day rooms tended to "bring people together" (sociopetal) and others tended to keep them apart (sociofugal). In spaces where interaction and communication was itself therapeutic, Osmond urged the use of sociopetal designs.[13]

One factor which is of great significance for architecture and urbanism emerges from these studies of spatial and territorial behavior. *For each type of social activity or process, there are upper, lower and optimal limits of size and density appropriate to that activity.* This is partly purely quantitative, though it assumes qualitative aspects, as Osmond says:

"Among ten people there are 45 possible two-person relationships; among fifty people there are 1225 possible relationships. The complexity of society has gone up by a factor of 27 at least.[14]

Four persons are required for a game of bridge, no more, no less: easy access to and recognition of the cards dictates a table no more than 42 inches across; tables should be at least 8 feet apart; but there is apparently no optimal number of tables. A regional shopping center also has a minimal (and almost certainly, a maximal) size and requires for its economic existence 50,000 people within a radius of 20 miles. In urbanistic terms there is clearly a "critical mass" below which that form of social invention and innovation which characterizes the city simply cannot occur. A village of 1,000 souls can neither produce nor support a symphony orchestra or a repertory theater. And yet, it is also clear that this state of critical mass creates only the preconditions, the socio-cultural climate for social creativity: it is, by itself, no guarantee of it. Neither Periclean

Athens nor Medicean Florence were as large as Chattanooga, Tennessee. Parameters of size appear to be relative, not absolute. Optima seem to vary with culture and especially with technology.

Optimal levels of density, on the other hand, appear to be relatively fixed, regardless of the size of the city. These density optima are functions of the physical limits of pedestrian range and personal contact. Such density factors have a decisive effect on the quality of life afforded by the city. The catastrophic drop in the effectiveness of American cities is due to, among other things, the fact that density has dropped in inverse proportion to rising size. One of the major factors here is technology—specifically, mechanized transportation by rail and, most recently, by private automobile. The sheer presence of large numbers of autos, in urban spaces which were calibrated to pedestrian speed and reach, has a negative impact on pedestrian behavior. The climate of social intercourse is at once diluted and polluted by the presence of these foreign bodies; the number, ease and frequency of face-to-face contacts is diminished; the sites of such contacts are more thickly dispersed in space; the minimal critical density necessary for self-propagating social intercourse is not achieved.

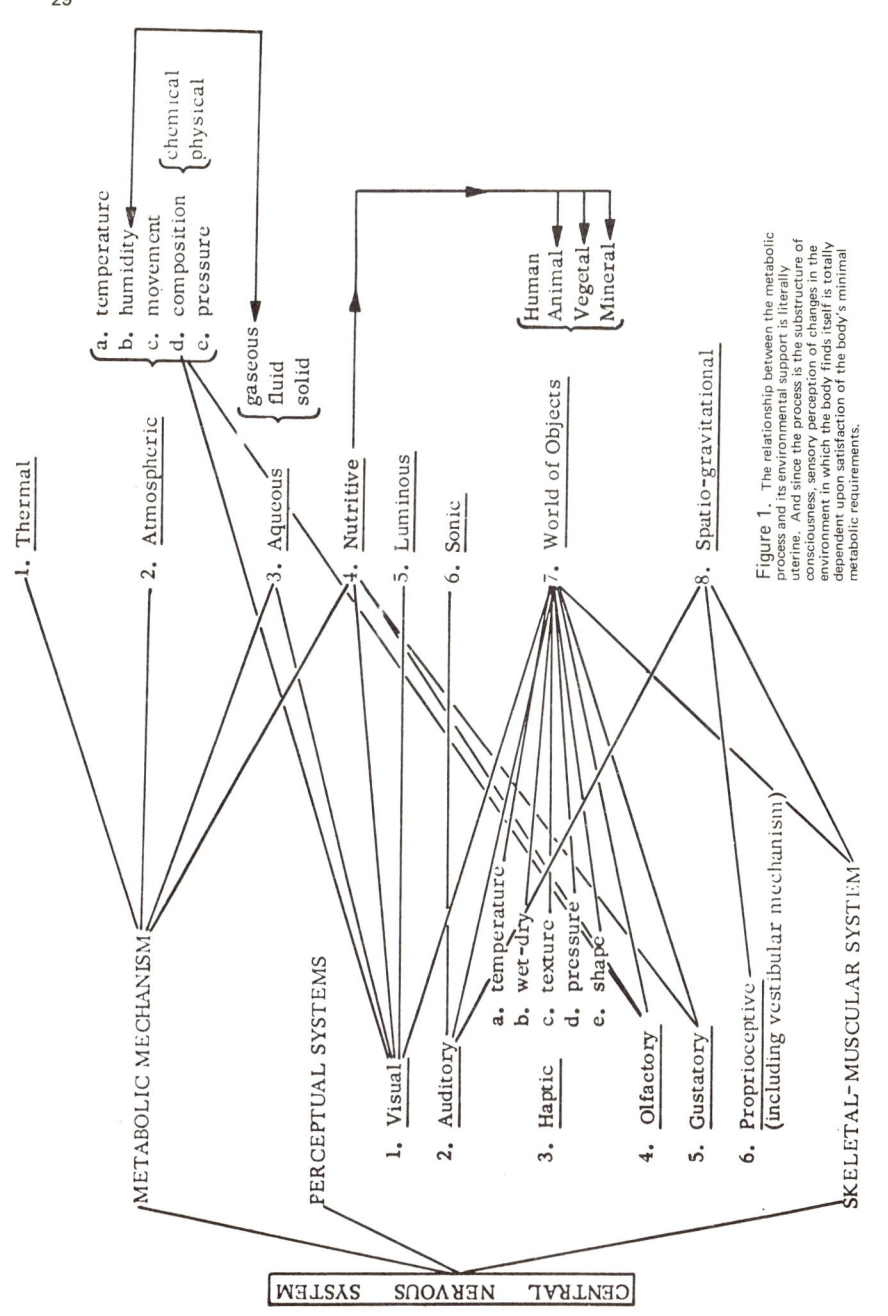

Figure 1. The relationship between the metabolic process and its environmental support is literally uterine. And since the process is the substructure of consciousness, sensory perception of changes in the environment in which the body finds itself is totally dependent upon satisfaction of the body's minimal metabolic requirements.

INTER FACE

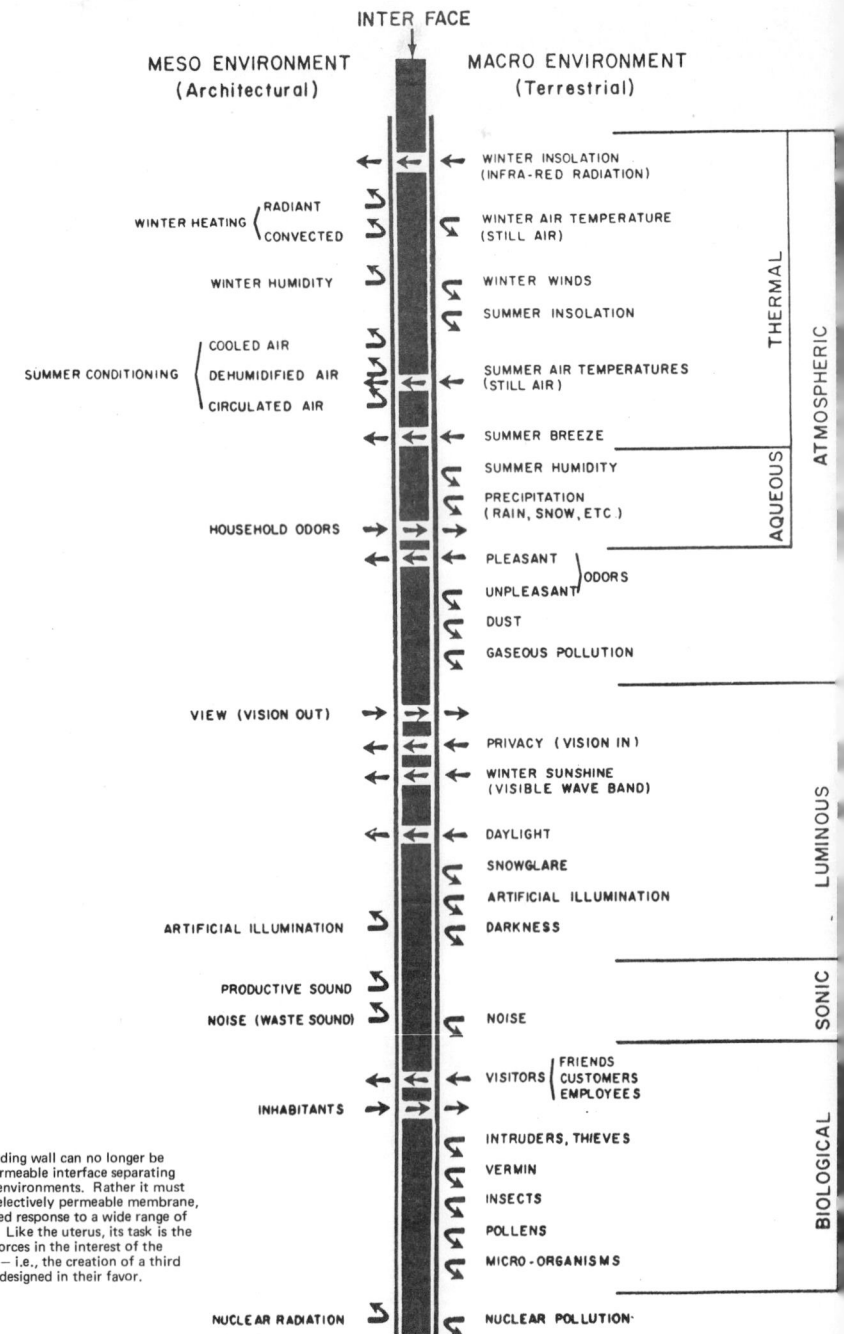

MESO ENVIRONMENT (Architectural) | **MACRO ENVIRONMENT** (Terrestrial)

MESO ENVIRONMENT		MACRO ENVIRONMENT	
	←	WINTER INSOLATION (INFRA-RED RADIATION)	THERMAL / ATMOSPHERIC
WINTER HEATING { RADIANT, CONVECTED }	→	WINTER AIR TEMPERATURE (STILL AIR)	
WINTER HUMIDITY		WINTER WINDS	
		SUMMER INSOLATION	
SUMMER CONDITIONING { COOLED AIR, DEHUMIDIFIED AIR, CIRCULATED AIR }	←	SUMMER AIR TEMPERATURES (STILL AIR)	
	←	SUMMER BREEZE	
		SUMMER HUMIDITY	AQUEOUS
		PRECIPITATION (RAIN, SNOW, ETC.)	
HOUSEHOLD ODORS	→		
	←	PLEASANT / UNPLEASANT ODORS	
		DUST	
		GASEOUS POLLUTION	
VIEW (VISION OUT)	→		LUMINOUS
	←	PRIVACY (VISION IN)	
	←	WINTER SUNSHINE (VISIBLE WAVE BAND)	
	←	DAYLIGHT	
		SNOWGLARE	
		ARTIFICIAL ILLUMINATION	
ARTIFICIAL ILLUMINATION	→	DARKNESS	
PRODUCTIVE SOUND	→		SONIC
NOISE (WASTE SOUND)	→	NOISE	
	←	VISITORS { FRIENDS, CUSTOMERS, EMPLOYEES }	BIOLOGICAL
INHABITANTS	→		
		INTRUDERS, THIEVES	
		VERMIN	
		INSECTS	
		POLLENS	
		MICRO-ORGANISMS	
NUCLEAR RADIATION	→	NUCLEAR POLLUTION	

Figure 2. The building wall can no longer be considered as an impermeable interface separating the meso- and macro-environments. Rather it must be conceived of as a selectively permeable membrane, capable of sophisticated response to a wide range of environmental forces. Like the uterus, its task is the modulation of these forces in the interest of the building's inhabitants — i.e., the creation of a third or meso-environment designed in their favor.

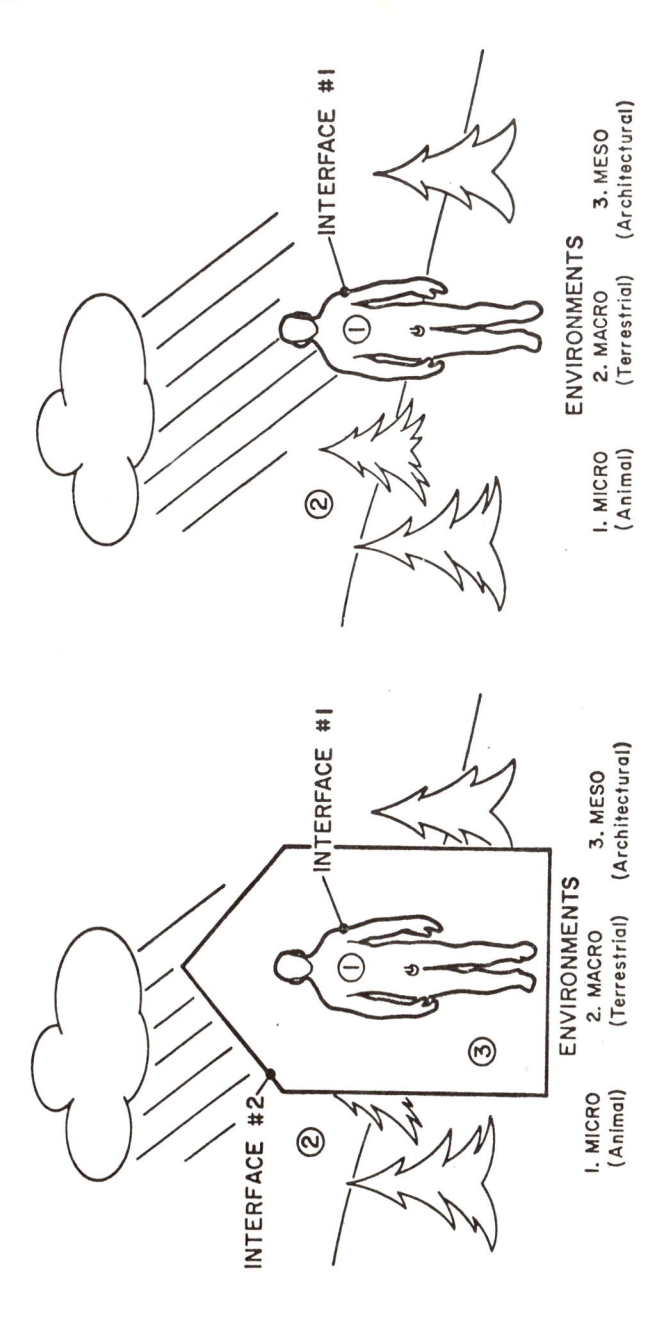

Figure 3. In civilization, relations between man and environment occur across two interfaces. This relationship is manipulated in his favor by clothing along the surface of his epidermis (1) and by architecture as the boundary of his "built" environment (2). By thus lifting gross environmental loads, a greater portion of his energies can be focused on socially-productive work.

References

1. This paper, in somewhat expanded form, will appear in Fitch, *American Building: The Experiential Factors that Shape It* (2nd edition, vol. II) to be published by Houghton Mifflin in 1970.

2. This proposition cannot, of course be refersed—i.e., that perception can exist independently of its metabolic base. The metabolic process constitutes the platform of consciousness. Cf. Fitch, "The Aesthetics of Function", *Annals of N. Y. Academy of Sciences, vol. 128,* art. 2, pp. 706-714.

3. Raymond G. Studer, "The Dynamics of Behavior—Contingent Physical Systems," Proceedings of the Symposium on Design Methods, Portsmouth College of Technology, England, December 1967.

4. James J. Gibson, *The Senses Considered as Perceptual Systems,* Houghton Mifflin (New York & Boston), 1966, p.111.

5. C. E. A. Winslow and L. P. Harington, *Temperature and Human Life,* Princeton University Press (Princton) 1949.

6. *Ibid.,* p. 178.

7. Samuel H. Bartley and Eloise Chute, *Fatigue and Impairment in Man,* McGraw-Hill, (New York), 1947, p. 39.

8. Hans Selye, *The Stress of Life,* McGraw-Hill (New York) 1956.

9. Stanley Schacter, et al., *Social Pressures in Informal Groups,* Berkeley, University of California Press, 1962.

10. Edward T. Hall, *The Hidden Dimension,* Doubleday and Company, (New York) 1966.

11. But these attenuated spatial relationships will be reflected in an analagous diminution in the experiential potency of the contact. No phone conversation can replace the face-to-face talk; no musical form can be amplified without distortion; no cinematic facsimile is a full surrogate for the multi-sensory reality of the live theater. Cf. Fitch, "For the Theatrical Experience, An Architecture of Truth", *Journal of U. S. Institute for Theater Technology,* no. 16, February 1969, pp. 16-20.

12. Robert Sommer, "The Ecology of Privacy," *The Library Quaterly* (Chicago) vol. 36, no.3, July 1966, p. 236.

13. Humphry Osmond, "The Relationship Between Architect and Psychiatrist," in *Psychiatric Architecture,* American Psychiatric Association (Washington, D. C.) 1949.

14. Humphry Osmond, quoted in *Progressive Architecture* (New York) April 1965, vol. 46, p. 160.

Peter Batchelor
School of Design, North Carolina State University

RESIDENTIAL SPACE SYSTEMS: GENERIC AND SPECIFIC CONCEPTS OF URBAN STRUCTURAL ANALYSIS

Urban Form and Residential Structure

With minor exceptions, the city has existed primarily to serve a set of residential functions. While this is growing less true, it is still a fairly safe assumption that the patterns of residential activity of urban populations will continue to be associated with patterns of working, spending and recreation for some time to come. Thus, the structure of cities may be examined by analyzing residential environments, with non-residential activities viewed as complementary and dependent functions.

There are some compelling arguments for analyzing the geometric properties of cities. The first, and most important, is that the social, behavioral and design disciplines have been unable to develop a comprehensive descriptive theory of urban structure which satisfies the multiplicity of analytic approaches of each contributing discipline. The second is that the built environment represents a massive investment in resources and justifies all attempts to rationalize the nature of urban structure. Finally, little progress can be made in solving city problems if the urban form is not rigorously and systematicly defined.

The fact that the city has defied definition except at macro levels of generalization appears to be the result of classification concepts which separate form and space from activity. This, in turn, stems from the apparent difference existing between urban space as a function of behavior and urban space as a function of legislative, administrative and cultural constraints.

Variations in the Topology of Residential Space

There are two surfaces or topological conditions describing the three-dimensional structure of cities. The behavioral surface of a city is the boundary of the sum of all the activities which represent the manifest behavior of its inhabitants. Such a surface is necessarily very complex, possessing many overlapping sets of activities. Its complexity militates against precise administrative definition and it is only during the last decade that serious research has started upon the mapping of surfaces and volumes defined by behavior.

A second surface, called for want of a better definition, the cultural surface, is represented by all the separate volumes created by codes, regulations, functional divisions and land usage, subdivisions of property and culturally inherited or derived standards. This surface is considerably less complex than the behavioral surface and is therefore more susceptible to specification. In addition, it is not necessarily coterminous nor coextensive with the behavioral surface and tends to govern the formation of residential space to a greater degree than behaviorally determined space. Cultural space is relatively static, evolving rather slowly over time to meet the needs of people, while behavioral space can fluctuate with the daily activities of a city's inhabitants.

Residential Space Systems

A model of residential spatial structure requires that the interaction between the two topological conditions previously described be defined. This requires the definition of performance standards based upon both user specifications and cultural standards. Urban residential space defined in this manner requires two models: A generic model based upon a comprehensive set of population characteristics, and a specific model based upon a specific population group. The generic model yields a set of relationships which generalize the composite structure, while the specific model derives a spatial structure based upon a unique group of households according to density and income constraints.

A generic residential space model is based upon the assumption that residential space is composed of specific components or variables with relationships existing between them such that a change in any one component or variable results in changes in one or more of the others. The basic components of the generic model are represented by population groups defined in terms of household composition, size, and age structure. Each population group generates a total space requirement which is subdivided according to ownership, enclosure, and function and use of space. In theory, all households will try to maximize their budgeted quantity of space, though in reality many households fail to do so for a variety of reasons. In order to overcome the problem posed by differential utilization of income available for space, the generic model assumes that income is utilized efficiently for all population groups over a complete range of densities.

Specific residential space models explore the structural characteristics of portions of cities according to unique population, income, and density specification. The relationships thus revealed may exhibit parametric variations within the generic model of the same environment, thus giving rise to a way of measuring deviations in the urban residential structure. At any rate, the concept of residential space systems incorporates the capacity for both general and detailed analysis of urban residential form.

Classification of Theories

Theories of urban structure tend to be classified according to scale of investigation and to location on a structural to non-structural continuum. Some theorists have attempted to include a third category—historic theory—although it is fairly easy to demonstrate that historic concepts are also structural in nature.[1] Using the idea of scale, a classification matrix can be developed showing four approaches to the analysis of form, two of which are of interest here: Structural theory of urban micro-form, and structural theory of urban macro-form.

Urban Micro Form

Most of the numerous contributions in this area come from historical analysis of urban form, and planning and design theory. A considerable portion of historical literature was generated by students of European urban history at the turn of the century, and a great deal of influence was exerted on the early city planning movement. Camillo Sitte's careful analysis of medieval streets and plazas led to a series of principles which generated enthusiasm for design of urban

In particular, a bridge must be made between the characteristics of locators in urban environments and the form and structure of space. Investigations of this kind should ultimately make it possible to develop standards for the specification of urban structure based upon rational rather than arbitrary standards, and to develop predictive models of residential spatial structure.

Factors Which Influence the Quantity of Space

Two of the factors which influence the quantity of space relate to the user rather than to the environment: household composition and income. If a matrix of household types is constructed to determine space requirements, the number of variations becomes too large to manipulate and family types need to be grouped to reduce the number of combinations. (See Figures 1 and 2) On the other hand, by allowing any mix of adults and children to occur, it is possible to identify a complete range of social units, including those which do not qualify as legitimate families at the present moment. If marital status is introduced as a variable, then it is possible to generate an infinite number of family types.

In reality, families of varying size and composition tend to adjust to the amount of space that can be purchased, either as a matter of necessity, or as a matter of luxury. Furthermore, the amount of variation in dwelling unit characteristics is sharply limited by the factors which limit choice, the available supply, the locational preferences of the household, knowledge of market and discriminatory practices, etc. It is anticipated that sufficient generalization of family types and incomes can be made as to permit the creation of a population vector. Such a vector would be least accurate when mixes of family types and incomes occur, as is often the case in transitional residential areas.

Insofar as the environment is concerned, density has the major influence on quantity of space. Increases in density are a function of economic utilization of land and as long as a family can purchase all the space it requires, the volume of enclosed space is not affected by changes in density. This is not normally the case, however, and an increase in residential density for a given family size and income causes decreases in both the amount of enclosed space and the amount of private open space, the latter occurring at a more rapid rate than the former.

Location theory suggests that the amount of space purchased within a specific income represents a trade-off between a composite bundle of goods and services and transportation costs. It has been assumed here that any analysis of existing environments deals with a satisfactory resolution of the options among which income can be distributed.

Factors Which Influence the Relationship of Space

The smallest geometric component of a residential environment which satisfies the variety of functions associated with the activities of households is traditionally referred to as the dwelling unit. It occupies a finite quantity of space in a three-dimensional co-ordinate system in which a whole variety of residential and non-residential uses may occur. The dwelling unit is frequently, though not necessarily, related to a surrounding area of land which may be privately or publicly owned. This land and the dwelling unit within it is subject to a series of land use and structural controls which affect the spacing, position, height, mass, and density of dwelling units as well as some of their internal dimensions. Other controls, such as mortgage company regulations and building codes can also alter the relationship of spaces, although this has its greatest impact on the internal arrangement of dwelling units rather than on the total environmental space of cities.

places.[2] A second and distinctly formal movement was created by American attempts to superimpose a Baroque grandeur on the urban landscape. A third movement originated in Utopian schemas and progressed through Ebenezer Howard, Raymond Unwin, Henry Wright, and Clarence Stein. This movement focused attention on two major form concepts: Garden Cities and neighborhoods.[3] By the early 1950's most of the structural theory of urban micro-form was based on an analysis of cities, historic urban streets and places, and upon planning concepts such as the neighborhood theory and the Radburn concept.

At this point several interesting directions were taken. A serious movement to examine the whole range of visual and other perceptual experiences was initiated by persons such as Rasmussen, Logie, Cullen, Nairn and Lynch. Lynch expanded his concept of urban structure in a classic treatise entitled "The Form of Cities" in which density, grain, shape, and internal patterns were identified as the four major characteristics of urban structure.[4] His attempts to define the urban environment have ranged from perceptual studies of parts of and whole environments to theoretical studies of urban structure.[5] In 1964, Maki and Goldberg published their research on collective form and linkage in which cities were analyzed in terms of structural groupings and separate structural components.[6] Since this time, the Smithsons have attempted to build upon work done by Lynch, Goldberg and Maki, though most of the fundamental concepts have remained unchanged.

Analysis of urban micro-form currently leans toward perceptual mapping and component analysis. New directions indicate that serious attempts to incorporate value structures are causing a trend away from descriptive theory.

Urban Macro Form

Explanations of urban structure tend to be far more numerous at the macro scale, and there is considerable variation in emphasis on the major variables involved. This is a function of the broad range of disciplines contributing to structural theories of urban macro-form

The earliest explanations of urban macro-form came from location theory and urban land economics at the beginning of this century. Richard M. Hurd's *Principles of City Land Values* laid down many of the basic concepts which appeared in the ecological theories of the 1920's.[7] In 1929 Haig established the concept of the "friction of space" in which the costs of overcoming distance gave rise to a series of economic theories of location up to the 1960's.

The first detailed explanation of urban structure involving density gradients and generalized use zones came from Burgess and Park in 1925.[8] Their Concentric Ring Theory was later modified by Hoyt's Sector Theory in 1939, and by Harris and Ullman's Multiple Nuclei Theory in 1945.[9] In the early 1950's, the work of urban geographers began to assist in the delimitation of urban activities, functional areas and classification of cities. Transportation plans for large metropolitan areas gave valuable insights into urban activities and into the relationships between income, density, land use, and transportation patterns. Albert Guttenberg gave one of the clearest explanations of urban structure as a function of size and density changes in 1961,[10] and this was built upon by Webber, Dyckman, Foley, Wheaton and others in *Explorations into Urban Structure*.[11] Most of the theories proposed in the latter half of the 1960's have not progressed beyond the generalizations of Webber, et al, although there is a discernible trend toward the incorporation of growth and decision processes into explanations of urban form.

Current Status of Structural Theory

Analysts of urban environments are faced with proving or disproving most of the speculative theories of earlier authors. The general relationships involving density gradients, aggregate and disaggregate space, spatial configurations, and spatial concentrations have yet to be qualified.

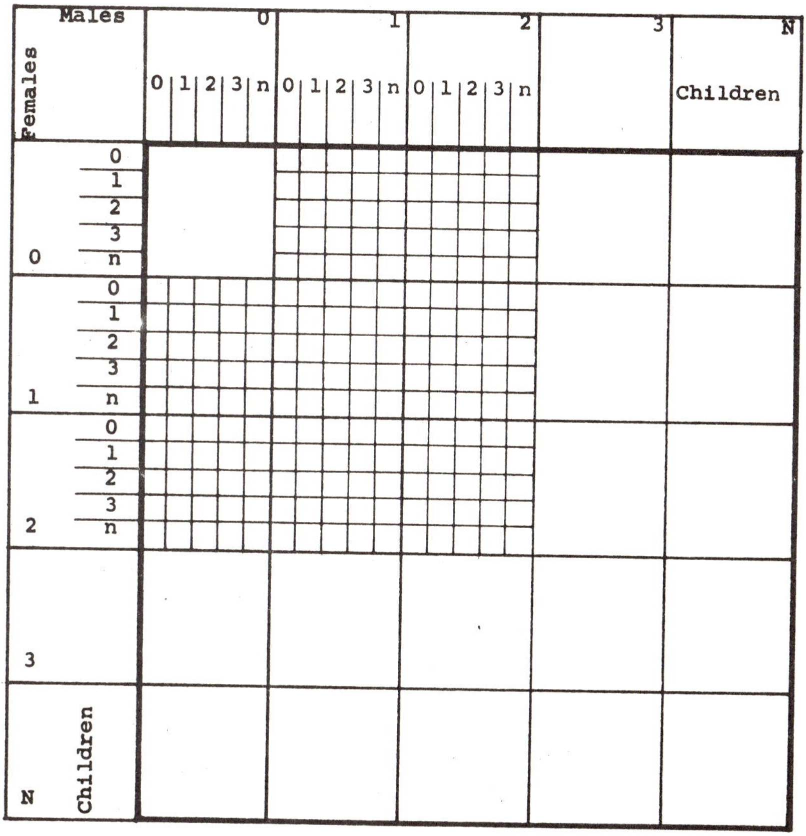

Figure 1. Matrix to Determine Space Requirements for Households According to Size and Composition

38

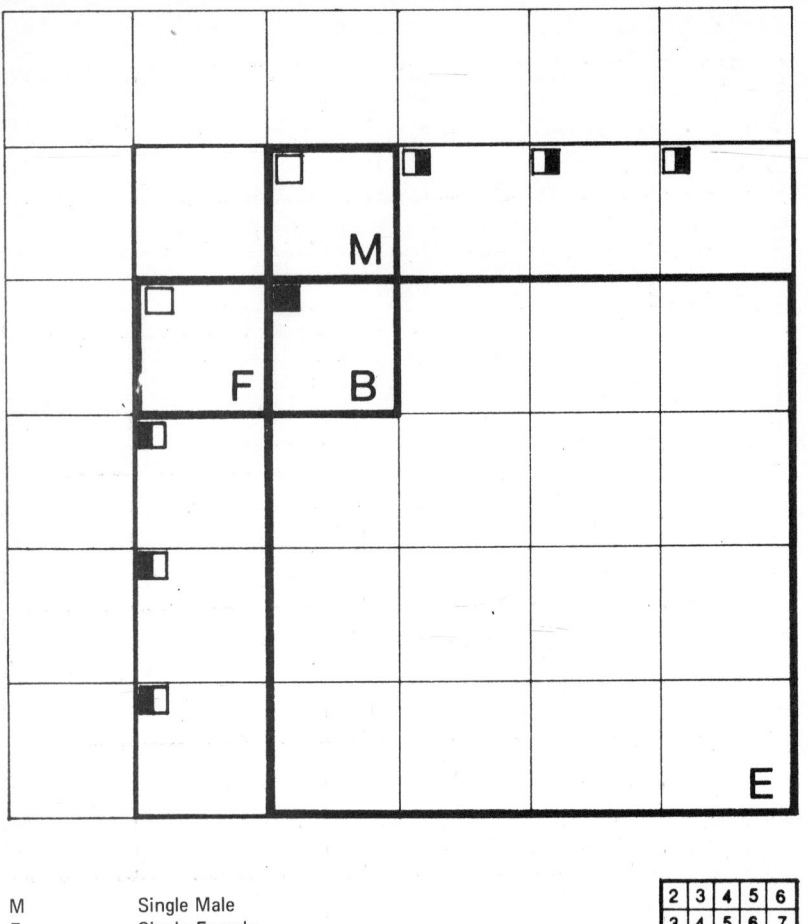

M	Single Male
F	Single Female
B	Bi-nuclear
E	Extended

- ☐ Single Persons
- ◩ All Females
- ◨ All Males
- ■ Couple w/o Children

2	3	4	5	6
3	4	5	6	7
4	5	6	7	8
5	6	7	8	9
6	7	8	9	10

5 x 5 matrix showing household composition and size

Figure 2   Partitioned Matrix Showing General and Special Household Types

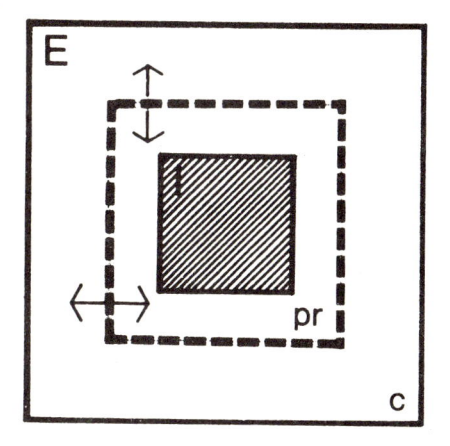

E External Space
I Internal Space
pr Proximal Space
c Community Space
 (c = E-pr)

Figure 3. **Components of Residential Space**

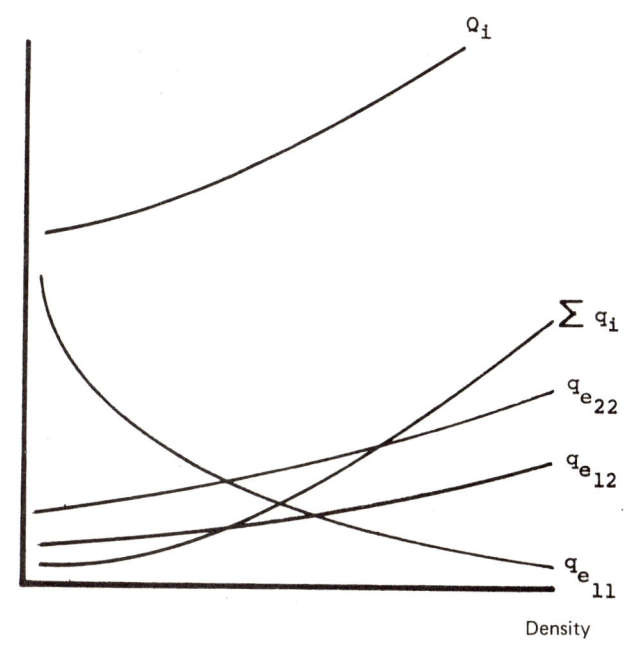

Figure 4. **Specific Model: Effect of Density Changes on q_e and q_i**

If the total quantity of space associated with a dwelling unit is thought of as internal and external space and if external space is characterized in terms of ownership, three space zones may be defined; Internal, proximal, and communal. (See Figure 3) Ownership concepts may apply to all three zones but public use of private property as well as private use of public property creates quasi-functional zones.

Density changes give rise to the most significant variations in relationships existing between the dwelling unit and its associated external spaces. For a given household size and income an increase in density causes an increase in the absolute volume of internal space, and decreases in the absolute amounts of proximal and communal external space. The relative changes which occur as density increases are an increase in the amount of common internal and external space.

Generic Model

The significant elements of the generic model are:

[S] A matrix of space coefficients for a population vector;

[P] A population vector for households with variations based on (a) family size, composition and age, and on (b) income;

[Q] A space vector consisting of the sum of all the spaces required to satisfy the needs of all households in the population vector;

[K] A change of state matrix for all elements of [Q];

[M] A modified change of state matrix for a single element of [Q];

And a set of propositions based on the relationship

[Q] = [S] [P]

[S] is a matrix of coefficients representing the equivalent volumes required to satisfy all household types within a population vector [P]. If P_j is a population group in the range $P_1, P_2, \ldots P_j, \ldots P_n$, then s_j is the volume of space generated by one household in P_j. The matrix [S] has the dimensions n x n, and its trace contains non-negative coefficients $s_1, s_2, \ldots s_n$ as follows:

$$\begin{bmatrix} s_1 & 0 & 0 & * & 0 \\ 0 & s_2 & 0 & * & 0 \\ 0 & 0 & s_3 & * & 0 \\ * & * & * & * & * \\ 0 & 0 & 0 & * & s_n \end{bmatrix} \quad (s \geq 0)$$

Now s_j is the aggreagte of a set of separate volumes ($s_e + s_i$) representing internal and external space. Both s_e and s_i may be disaggregated to indicate separate kinds of internal and external space $(s_{e1}, s_{e2}, \ldots s_{en})$ and $(s_{i1}, s_{i2}, \ldots s_{in})$. However, reduction of spatial coefficients into constants representing different internal and external space characteristics does not affect the accuracy of the model as a whole.

The population vector [P] has the dimensions n x 1 and is a set of elements representing the number of generic households selected from the household composition matrix in combination with a survey of a specific population group. Population groups are differentiated by economic class as well as by size, composition and age category. [P] is therefore a vector of non-negative elements of the following form:

$$\begin{bmatrix} P_1 \\ P_2 \\ P_3 \\ * \\ P_n \end{bmatrix} \quad (P \geq 0)$$

(Note that the "n" dimension of [S] must correspond to the "n" dimension of [P]).

A vector of space [Q] for the entire residential system is the product of [S] and [P]:

[Q] = [S] [P] where [Q] is an nx1 vector containing the quantities of space generated by each population group in the population vector [P].

Operations Which Can Be Performed With the Generic Model

If [K] is a matrix of coefficients for the change of state in s_j, J = 1 to n over a discrete time period $t_m \rightarrow t_{m+1}$, then $S_{t_{m+1}} = S_{t_m} \; K$

where [K] =
$$\begin{matrix} k_{11} & k_{12} & k_{13} & * & k_{1n} \\ k_{21} & k_{22} & k_{23} & * & k_{2n} \\ k_{31} & k_{32} & k_{33} & * & k_{3n} \\ * & * & * & * & * \\ k_{n1} & k_{n2} & k_{n3} & * & k_{nn} \end{matrix}$$

$$(0 \leq k \leq 1)$$

and k_{ij} = rate of change of space from i to j over $t_m \; t_{m+1}$

Reverting to the general form [Q] = [S] [P] it can be shown that $Q_{t_{m+1}} = S_{t_{m+1}} \; P$

If each k in [K] above represents a constant rate of change over discrete time periods $(t_m, t_{m+1}, \ldots t_{m+n})$ then calculation of a new population vector $P_{t_{m+1}}$ would render the general form of this model applicable in the following manner:

$$Q_{t_{m+1}} = S_{t_m} \; K \; P_{t_{m+1}}$$

where

$$P_{t_{m+1}} = P_{t_m} + \triangle P_{t_{m \rightarrow m+1}} \quad (\triangle = \text{-ve, 0, +ve})$$

$\triangle P_t$ is readily calculated using survivorship methods.

The change of state matrix shown above may be modified slightly to yield a discrete time model of evolution. An element of [Q], Q_i (i=1,n) is assumed to pass through m-1 evolutionary phases during the time period $t_0 \rightarrow t_m$. This can be represented as follows:

Time	0	1	2	m-1	m
Phase	A	B	CV	X	
Coefficient	k_0	k_1	k_2	k_{m-1}	k_m

where k_1, k_2 and k_3 correspond to k_{11}, k_{12} and k_{33} in [K] and refer to the amount of space left in its former state during the course of a discrete time period. From this it is seen that 1-k is the coefficient representing the change of state to the next level and that:

$$[M] = \begin{pmatrix} (A & B & C & D & - & X) \\ k_1 & 1-k_1 & 0 & 0 & * & 0 \\ 0 & k_2 & 1-k_2 & 0 & * & 0 \\ 0 & 0 & k_3 & 1-k_3 & * & 0 \\ 0 & 0 & 0 & k_4 & * & 0 \\ * & * & * & * & * & * \\ 0 & 0 & 0 & 0 & * & k_m \end{pmatrix} \begin{matrix} t \\ 1 \\ 2 \\ 3 \\ 4 \\ - \\ m \end{matrix}$$

The calculation of Q_i at time t would be: $\quad Q_i = P_i$ [M]

With P_i being progressively changed over time: (t=1) $Q_{i_1} = P_{i_0}$ [M]

(t=2) $Q_{i_2} = Q_{i_1}$ [M]

(t=3) $Q_{i_3} = Q_{i_2}$ [M] etc.

For example, assume that $Q_i = 1000$ at time t = 0 and that Q_i passes through four phases in which

$$M = \begin{bmatrix} .8 & .2 & 0 & 0 \\ 0 & .8 & .2 & 0 \\ 0 & 0 & .8 & .2 \\ 0 & 0 & 0 & 1.0 \end{bmatrix}$$

Therefore,

$Q_{i_1} = (1000, 0, 0, 0)$ [M] = (800, 200, 0, 0)

$Q_{i_2} = (800, 200, 0, 0)$ [M] = (640, 320, 40, 0)

$Q_{i_3} = (640, 320, 40, 0)$ [M] = (512, 384, 96, 8)

and $Q_{i_4} = (512, 384, 96, 8)$ [M] = (409.6, 409.6, 153.6, 27)

The major property of this discrete time model is that no space is lost in the system, indicating that k_m must equal unity.[12]

Specific Model

The relationships expressed in the previous model are not altered by studying a variety of urban environments. A specific model, however, deals with discrete portions of a single environment. These portions are selected on the basis of population groups whose residential environment exhibits certain common socio-economic characteristics. In a given city population vector [P] is the same for both generic and specific models.

If Q_i is the coefficient for the quantity of space required by a household in population group P_i then: $Q_i = \sum_1^n q$ where $q = q_i + q_e$ (q_i = internal space, q_e = external space)

The major classification of residential space is based upon whether space is related to individual households or shared by others, and whether this space is in the public or private domain. This classification system is as follows:

Internal	Private	Individual	$q_{i_{11}}$	
		Common	$q_{i_{12}}$	
	Public	Individual	$q_{i_{21}}$	q_i
		Common	$q_{i_{22}}$	
External	Private	Individual	$q_{e_{11}}$	
		Common	$q_{e_{12}}$	
	Public	Individual	$q_{e_{21}}$	q_e
		Common	$q_{e_{22}}$	

Preliminary surveys show that $q_{i_{11}}$, $q_{i_{12}}$, $q_{e_{11}}$, $q_{e_{12}}$, $q_{e_{22}}$ are the quantities most likely to form a constant Q_i for a given population group P_i. For a specific urban situation, the relationship which exists between each q_e and q_i can be plotted relative to all Q_i's. (See Figure) The hypotheses currently being tested are:

Q_i	d	$(Q_i = (Kd)^e)$
$\sum q_i$	d	$(\sum q_i = (K_i d)^{ei})$
q_{e_i}	i/d	$(q_{e_{11}} = (K_{11}/d)^{e_{11}})$
q_{e_i}	d	$(q_{e_{12}} = (K_{12} d)^{e_{12}})$
$q_{e_{22}}$	d	$(q_{e_{22}} = (K_{22} d)^{e_{22}})$

where K is a constant and e an exponent for specific environmental situation.

Conclusion

The generic model has two major benefits: It offers a method of allocating space without resorting to arbitrary standards, and allows a predictive dynamic model of residential space to be developed in conjunction with population growth and density assumptions. The specific model offers a means of analyzing urban residential space for the purpose of developing design criteria for partial environments. Application of these models is limited by the definition of a population group, which however, could conceivably be infinitely large. Analysis of the use to which space may be put and of the relationships which exist between internal and external space has been deliberately omitted for the sake of simplicity but it is possible to reconstruct the two models on the basis of activity space allocations.

References

1 Alonso, William, "The Historic and the Structural Theories of Urban Form," *Land Economics,* May 1964, pp 227-231.

2 Sitte, Camillo, *City Planning According to Artistic Principles,* London, Phaidon Press, 1965 (Translated from the original German text *Den Stadtebau* by G. R. and C. C. Collins).

3 Batchelor, Peter, "The Origin and Evolution of the Garden City Concept of Urban Form," University of Pennsylvania, 1966.

4 Lynch, Kevin, "The Nature of Cities," *Scientific American,* April, 1954.

5 See the following works by Lynch: "A Theory of Urban Form," *JAIP,* Vol. 4, 1958; *The Image of the City,* 1960; "The Pattern of the Metropolis," *Daedalus,* Winter 1961; and *The Form of Metropolitan Sectors,* 1962.

6 Maki, F., and Goldberg, J. *Investigations in Collective Form,* Washington University, 1964.

7 Hurd, Richard M., *Principles of City Land Values,* New York, 1903.

8 Park, R. E., and E. W. Burgess, *The City,* Chicago, The University of Chicago Press, 1925.

9 Each of the three theories is compared in "The Nature of Cities" by Chauncy D. Harris and Edward L. Ullman (*Annals of the American Academy of Political and Social Science,* November 1945, pp 7-17).

10 Guttenberg, Albert Z., "Urban Structure and Urban Growth," *JAIP,* May 1960, pp 104-110.

11 Webber, Melvin, et al, *Explorations into Urban Structure,* Philadelphia, University of Pennsylvania Press, 1964.

12 See Wolfe, Harry B., "Models for Condition Aging of Residential Structure," *JAIP,* Vol. 33, No. 3, May 1967, pp 192-196.

Michael Brill and Richard Krauss
Institute for Applied Technology
National Bureau of Standards

PLANNING FOR COMMUNITY MENTAL HEALTH CENTERS:
THE PERFORMANCE APPROACH

History of the Project

    A major purpose of this project has been to find a way to allow local groups to set the standards by which the federal government will administer and fund mental health building programs in their area. In 1963, when Congress made large amounts of money available to build Community Mental Health Centers,[1] the administrators of the CMHC building program proposed that all new mental health centers be exempted from any federal building standards, especially those of hospitals. They felt, first of all, that the CMHC should not be institutional, "hospital-like" buildings; they might in some cases be just a set of services carried out by existing community elements. Second, that they could not anticipate what programs imaginative professionals in the community might want to run. In principle, then, they used few standards, and emphasizing the good judgments of their own review panels and consultant staffs in the hope that local applicants for funds would de-institutionalize their operations and be innovative, though a few applicants did develop community oriented and innovative programs, many did not. In fact, many of the proposed programs and facilities made it clear that NIMH's goals were not understood at the local level even though a mechanism existed for advising and consulting local professionals. It became obvious that this mechanism, the Architectural Consultation Section, could not keep up with the demand for their services with a professional staff of one senior architect and several young designers. In the face of this and other problems such as the stringency of local codes and the difficulty of defining catchment area, the Architectural Consultation Section discussed with members of the National Bureau of Standards a way to set up "standards" and documentation for their advice which could amplify their capability to meet the local level demands.

    In response, the National Bureau of Standards recommended a system for improving the planning process. Instead of proposing a "standard" for buildings, they developed a "standard" for the planning process. The conceptual basis and techniques for doing this stemmed from work being done at the National Bureau of Standards[2] on *performance specification techniques*.

Planning Method

    In the past, people who felt a need for a more suitable environment stated their assumed needs and simply described a building that they thought might meet those needs. As the words "performance specification" imply, environments might be procured by asking for a certain kind of performance from the environment rather than a specific physical configuration or building.

    There are five basic levels of specification: first, specifying building hardware and elements; second, specifying what kind of *human performance* it shall support; fourth specifying what kind of *human problems* the performance will solve (what kind of objectives it shall meet); and fifth, specifying how to *decide what problems* should be solved.

    Diagrammatically, these five levels are:

(Specific set	(Performance of	(Performance of	(Objectives to	(Choosing
1 of hardware)	2 hardware)	3 people)	4 be met)	5 objectives)
Hardware	*Setting*	*Activities*	*Problems*	*Process*

Upper Arrows: Supply Potential of the System
Lower Arrows: Demand Placed on the System

Note that as you move to the right, requesting performance for an environment rather than specifying a building itself, you permit designers/builders to consider a greater number of alternatives and therefore, more freedom to innovate.

At present the National Bureau of Standards has begun to find ways to specify performance for hardwere (see block 2). They have done this for exterior walls of dwelling units for the Federal Housing Administration and for the operating office space for Federal Office Buildings. In addition, studies for the Department of Housing and Urban Development attempted to describe the activities housing should support. The aim of these studies was to encourage the building industry to develop innovative solutions which would be either less expensive and/or give better performance. An important difficulty was encountered in the HUD project. It was not always easy to specify sets of *activities* and even more difficult to specify the *problems* they were attempting to solve. Though many of each could be stated—activities could be identified (like sleeping, dining, entertaining) and problems (privacy vs. group options, identity with "place") could posed— it became clear that these lists were open ended. Whatever list would be compiled would depend on *who* was compiling it and *when* he was doing the compiling. Hence, specifying the performance the environment should accommodate and what problems it would solve is, in fact, a political decision making process.

What we have devised is a prescription *for the planning process* rather than either a specification for a building, a fixed list of activities, or set of objectives. With this, the Architectural Consultation Section will be able to help *improve the quality of local planning process* by requesting planners to take certain steps, by supplying information to them dureing the process, and by providing a formal explicit way for the participants in a mental health planning process in a local area to inform each other. This planning operation is conducted through a set of forms and directions called the Planning Aid Kit.

The Programming/Planning Process Using PAK

The basic intention of PAK is to help articulate the environmental implications of the Mental Health treatment programs in a specific community. The emphasis given by the NIMH to community psychiatry has had a fundamental effect on the character of these treatment programs. The central tenet of community psychiatry is that "the community is ultimately responsible for the mental health of its members." This concept implies an individual's mental health partially depends on his interaction with his sociophysical environment, and that, in effect, describes his community.

Further implications of this concept are:

1. Environmental stress leads to clinically categorized types of mental illness.

2. Removing such stresses results in environment conducive to mental health

3. Treatment programs are no longer solely based on the needs of the identifiable mentally ill but rather on needs related to an individual's interaction with the social relationships and physical attributes of his environment.

4. A beneficial way to treat this individual/environment interaction is by treating the community so it becomes the delivery mechanism for a range of explicit & implicit mental health programs.

In order to improve the understanding of both the social and physical components of the environment and their relationships to community psychiatry, it is necessary to develop a language for communication among the Mental Health professionals and the others involved in the planning process. Since the application of treatment programs has environmental design implications which can only become explicit by means of such communication within the programming/planning process, we have articulated the steps of the process in a way that allows for interdisciplinary communication. These steps are general here, and specific descriptions of the kind of information for each step occur in specified forms.

In the early stages, the type of communication most desirable between people using PAK are statements in terms of goals or objectives, rather than in terms of solutions, buildings, or existing programs. The desired Performance of the proposed mental health "system" in the community should be stated clearly and then solutions can be devised.

Although it is generally easier to think and communicate in terms of past solutions (we use our experience with success and failure as models for future planning). this planning model has three flaws:

1. Problems, people, and times change. What was successful before may not be successful this time.
2. Early decision as to solutions tends to preclude examination of alternatives and acts as an effective barrier to innovation.
3. Often these decisions are made by high echelon professionals, whose tasks place them at some distance from specific and actual patient needs. These professionals possess only "generalized" information about patient success or failure. There should be enough "slack" in upper level decisions to permit satisfying idiosyncratic needs at the patient level.

In addition to overcoming these flaws, the flexibility of PAK also makes it possible to incorporate a greater variety of appropriate resources which at present seem therefore, it is beneficial to mental health programming that performance characteristics be specified rather than specific solutions. If, for example, a treatment needs certain environmental qualitites which are present in a hospital and also in the patient's own home and neighborhood pub, there is no therapeutic reason why the latter should not be considered as legitimate resources

Using the Performance Concept in PAK

Solving a problem through performance techniques entails putting the problem statement through a series of transformations which converts the statement of the Problem to a stated set of Activities (a therapeutically desired act involving the user of Mental Health services) to a description of an "ideal" Setting (an environment that will support the activities). The steps involved in carrying out these translations for community mental health problems are described in the following series of steps and flow chart:

Step 1: Select and Unite Participants

The information needed to carry out these planning operations is collected at a series of meetings on a special set of forms. (See the attached forms). The issues that are brought up and the ones that get acted upon will range in accord with who is participating in the planning process. Hence, there are also rules for who participates and how they may vote.

Step 2: Articulate the Problems

For a mental health problem to be considered a "legitimate" concern of the planning group, it must stem from one of the following sources:

1. Problems implied by the objectives of the services currently provided (e. g., Emergency Service implies the need for unscheduled services at all times)

2. Problems inferred from demographic data about the specific community population (e. g., amount of alcohol per capitol consummed in the community).

3. Problems perceived from recurrent aberrant behavior in the population, informally recorded by the profession or the community.

4. Problems of specific interest to the professional group.

5. NIMH's judgment that local condtions might lead to the local eruption of a recurrent national problem.

These problems are to be recorded at any level at which they are perceived and without implicit priorities. PAK permits the recording of problems, stimulates discussion of their causes and effects, and aids in establishin priorities for the allocation of resources to solve them through various courses of action. It is critical that the planning group be explicit as to whose mental health disfunction is responsible for the Problem's existence (e.g. while juvenile delinquency may be a threat to safety in the streets, the delinquents' mental health disfunction and not that of the public, is of interest.)

Step 3: Select COURSES OF ACTION

A COURSE OF ACTION is essentially the mobilization of resources by a particular people who solve a pro lem. There may be many alternative Courses of Action to solve a particular Problem. The process of selecting Courses of Action has been set up in accord with the following NIMH activity priorities: (1) prevention, (2) intervention, (3) treatment--rehabilitation , and (4) custodial care. These will be listed for each problem, and priorities will be established in order that resources may best be utilized in planning.

Step 4: List ACTIVITIES

Activities are observable acts between people which have a purposive quality within a specific cause of ac and which involve the recipients of Mental Health Services. We are primarily concerned with those Activities de therapeutic. Obviously, this will vary with the nature of the recipient, the staff, the community, the Problem, There may be alternative sets of Activities which constitute the same Course of Action. However stated, the ch sen Activity must be specific enough so that it makes demands that affect the environment in specific ways. H an activity can be described at any scale so long as its environmental implications are clear.

These Activity alternatives are ideally described and selected by the local professional staff, the people mo aware of the possibilities a given community and mental health staff can realize.

Step 5: Prescribe PERFORMANCE CHARACTERISTICS

An Activity will place many diverse demands on the environment in order that the Setting permit it to tak place (e. g., the Activity of counseling will require a quiet, physically comfortable and stable place. We are on concerned here with those requirements that have some direct environmental effect on mental health problems hence, we are only peripherally concerned with such things as circulation effeciency, structural stability, appro priate mechanical services, etc. The statement of a Performance Characteristic (PC) is an attempt to link thera peutic and architectural concerns in one set of statements to be used as the communication mechanism betwee Mental Health professionals and architects. The PC's are not standards to be simply "applied" in all contexts. They are a range or continuum of measurable qualities which can be set at different levels as the situation dem

However, for each Activity deemed therapeutic by the PAK participants, there is a set of "best" or "mos therapeutically desirable" PC's which describe, in PC terms, the "ideal" Setting to support the user in that Act These PC statements, and the "weight" or value given to them by the PAK participants, define the characterist of the Setting. A fuller description of the PC's is given later in PAK. These form the environmental "brief" or program given to design professionals to aid the development of a fit environment for the task. PAK essentiall ends here.

Step 6: Selecting or Designing SETTINGS

The product of this step is the physical description of the actual Setting which fits the description of the "ideal" Setting. This step may be a design or selection process, or be composed of both. Where the participants are intent on new facilities, the Setting may be designed by the Architects using PC's. Where the participants wish to make use of existing community resources as Settings, it is a selection process. Where existing Settings are to be modified through design, it involves both selection and design. The use of PC's permit examination of a wide range of alternative solutions.

Step 7: FEEDBACK

The final step is that of checking the results of the decisions. In this case, the criteria for success is whether or not the originally perceived problems have been solved. This process, called "feedback" because you are feeding information back to evaluate success, leads logically to "feedforward," the feeding of information forward to the next cycle of development. This closes the loop, providing an evaluation of this project and information for other communities engaged in the same tasks. Diagramatically, this operation appears as follows:

CMHC PLANNING PROCESS

Select and Invite
    PARTICIPANTS
List All
    PROBLEMS
List and Give Priorities to
    COURSES OF ACTION

Select "Therapeutically" Important
    ACTIVITIES
Select "Ideal"
    PERFORMANCE CHARACTERISTICS
Find, Modify, or Build New
    SETTINGS
Operate
Evaluate

The Performance Characteristics

Each PC is a continuum with no values ascribed to either end. (For example, two different physical settings may require extreme privacy or open communality and either will be considered a positive value for that setting.

PC 1	Communality	Privacy
PC 2	Sociopetality	Sociofugality
PC 3	Informality	Formality
PC 4	Familiarity	Remoteness
PC 5	Accessibility	Inaccessibility
PC 6	Ambiguity	Legibility
PC 7	Diversity	Homogeneity
PC 8	Adaptability	Fixity
PC 9	Comfort	Discomfort

Measures of Performance Characteristics

If we are to compare and evaluate settings, there must be some way to measure each performance characteristic. For each PC, we will define the continuum, then make a concise statement about one end of the scale and assume the other end is its opposite. Then we will state some measure for the PC. Normally we measure and achieve performance within the context of $Y = f(x)$ and get results such as 3.57. This is not always possible and in some cases ratios (y:x) and size comparisons ($Y \times x$ or $x \times y$) are employed. Ultimately we accept (yes-no) as a measure. It is in this context that we attempt to develop measures.

A Scale for the Performance Characteristics

We wish to have a common scale to compare one proposed or actual setting with another. Assuming a scale of 5 increments for all Measures of Performance Characteristics, for each measure we have a scale of, from left to right:

$$+2.........+1.........0.........-1.........-2$$

The profile of all 9 PC's for an activity setting might look like this:

Continuum	PC 1: *Privacy-Communality*. This is a characteristic of a setting that conveys a message to the user as to the degree to which he will be asked to share it with others. *Privacy* then, is the absence of unwanted human stimuli and surveillance within a setting.

Measure	PRIVACY—is inversely proportional to Exposure and Sensory Impingements:
	
	or PRIVACY= $\dfrac{1}{\text{EXPOSURE, SENSORY IMPINGEMENT}}$
	
	where:
	Exposure is measured by the Number of people (other than Mental Health personnel) who are in a position to actually or potentially survey the user.
	
	Sensory Impingement is a measure of "signal" (not "Noise") the user senses. Includes sight, sound, smell, and vibration signals from other people.
	
	NOTE: Two common mechanisms by which privacy is achieved are:
	   a. Sense of Enclosure or Insulation from other people
	
	   b. Sense of Distance from other people.

Continuum	PC 2: *Sociopetality-Sociofugality*. A setting is sociopetal or sociofugal in that it promotes or discourages social interaction within it. Sociopetality then, is achieved by the presence of elements in a setting which suggest and support the presence and interaction of people.
Measure	*SOCIOPETALITY* *if* a setting is (or is open to) the intersection of paths that connect many activity settings or ones highly and often utilized, and *if* a setting has a large measure of Diversity, [See **PC 7**], and *if* a setting is one in which the "normal" distances between people is greater than 2' o.c., but less than 9' o.c., and they face each other with no barriers to interaction between them, and *if* many elements in the setting are conducive to lingering, and *if* there are fewer formal activities than persons, then the setting is a highly Sociopetal one.

Continuum: PC 3: *Informality-Formality.* This is a characteristic of a setting that conveys a message to the user as to whether or not the use of that setting is governed by prescribed behavioral rules.

                Formality is where activities or behaviors are highly pre-determined by setting.

Measure: *Formality*

           if Furniture arrangement is specific, coercive and relatively immobile,

           and/or

           if Space allocation and shape is a precise "fit" for the activity, the actors, and the furniture,

           and/or

           if the Sensory Environment is exclusively responsive to the particular activity,

           then

           the activity setting is a highly Formal one.

Continuum: PC 4: *Familiarity-Remoteness.* This is a characteristic of a setting exhibit which indicates to the user whether or not it is one of a class within which he has operated before.

                Familiarity then, is achieved through the presence of familiar objects in setting familiar to the user.

Measure: *Familiarity*

                The application of this measure presupposes a knowledge of the past and present physical environments of the relevant user groups.

                If the proposed setting duplicates or approximates the users' past and present settings in many of the following categories, that proposed setting will be a familiar one:

                   a. relationship of building to site and community services

                   b. historic style of building

                   c. materials

                   d. shapes

                   e. sequence of spaces

                   f. type of, and placement of furniture, equipment, and information devices.

Continuum: PC 5: *Accessibility-Inaccessibility.* This is an indication of the relative ease with which a setting is approached or entered.

                Accessibility is achieved through minimizing the expenditure of users' energy in reaching a destination and maximizing his motivation, or:

$$\text{ACCESSIBILITY} = \frac{\text{MOTIVATION TO MAKE CONTACT}}{\text{ENEGRY EXPENDED TO ACHIEVE CONTACT}}$$

Measure   Users' Energy will vary with the distance they msut travel, the time it takes and the physical obstacles in their way. Therefore:

$$\text{ACCESSIBILITY} = \frac{\text{MOTIVATION}}{\text{DISTANCE, TIME, OBSTACLES}}$$

where:

     a. *Motivation* is given weight by *importance to program* to Mental Health professionals. Motivation known only to the user and not related to program cannot be considered.

     b. *Distance* is distance traveled by User from point to point within a single facility, or from his normal habitat to an M. H. facility.

     c. *Time* is the lapse involved in distance above. (Mode of transportation becomes important.)

     d. *Obstacles* are physical ones only, scaled by the evaluator—such as stairs, ramps, no through streets, etc.

Continuum   PC 6: *Ambiguity-Legibility.* This is the characteristic of a setting that makes vague or clear the intended use(s) for which the setting has been provided.

Legibility then, is achieved through a high degree of spatial "cues" or "signals" clarifying the use, direction of movement, and location of the setting in relationship to other settings, geophysical elements, or climate.

Elements to be assessed for Information Content are:

1. Shape of setting

2. Furniture and equipment

3. Graphics

4. Finishes

5. Perception beyond setting

6. Place of setting in a sequence of settings.

Continuum   PC 7: *Diversity-Homogeneity.* This is a characteristic of a setting which suggests the number of activities which may coexist in the setting.

Diversity is considered a characteristic of a setting whch both suggests and supports a large number of mutually compatible specific activities.

Measure   *Diversity*

When a specific activity is both suggested and supported by the setting, we have spoken of these as Formal settings, insofar as suggestion and support "predetermine" the activity by excluding all others. Therefore, the Diverse setting contains a number of Formal settings and any setting composed of a considerable number of Formal activities is a Diverse one.

$$D = f_1 + f_2 + f_3 \ldots \ldots + f_n$$

where D = Diversity
and f = Formal activity setting,
and n = number of f's,

then whichever (n) is higher, represents the setting with the greater diversity.

Continuum   PC 8: *Adaptability-Fixity.* This is a characteristic of a setting which predicts the capacity of the setting to successfully adapt to unforseen change.

Adaptability then, is a measure of the resources which must be expended to adapt a setting to support new activities.

Measure
$$\text{ADAPTABILITY} = \frac{1}{\text{MAN HOURS} \times \text{SKILL}}$$

where *Man Hours* is the amount of time required to make the PHYSICAL changes necessary, and *Skill Level* is a rating applied by the measures...for example, a nurse moving chairs has a skill level of 5 (the maximum).

Continuum   PC 9: *Comfort-Discomfort.* The sum of those physical characteristics of the environment, furniture, and equipment which noticeably intrude upon the users' Physical Performance of an activity due to malfunction.

Comfort then, is defined as the ability of a setting to pass human engineering standards tests which define a "comfort zone" in the following 2 categories,

then

the setting is considered comfortable.

Categories:

    A.   Sensory/adequate supply and control of

         1. thermal qualities
         2. illumination
         3. acoustic qualities
         4. air quality

    B.   Physical/

         1. Equilibrium
         2. Anthropometric "fit"   (shape, size, surfaces, position)

Conclusion

In essence PAK is a technique for describing goals and objectives in terms of desired performance while permitting the generation of many alternative solutions which yield this performance. It is a way of encouraging innovation and of increasing and widening participation in the design and selection of solutions. It is the stating of goals unbiased by the means used to achieve them. This concept, in PAK is applied in defining treatment programs in environmental performance language to describe ACTIVITY SETTINGS, i.e., that part of an individual's sociophysical environment within which the individual acts. All human activities take place within some identifiable spatial boundaries which describe such an ACTIVITY SETTING. Each SETTING exhibits social and physical characteristics which tend to support or constrain one or more activities that may take place within it. PAK is the mechanism by which the treatment programs for Mental Health problems are translated into particular lists of ACTIVITIES for which appropriately supportive ACTIVITY SETTINGS must be provided. In order to plan and design a physical setting which must support a certain ACTIVITY, the specification given for that environment must not predetermine or limit the number of alternative solutions. It is evident, therefore, that such specifications must not be phrased in terms of physical dimensions, or materials, but in terms of the performance qualities desired.

Because the PC's describe the characteristics of settings for human activity in Performance terms, many settings could possibly exhibit these characteristics, including some not traditionally thought of as Mental Health settings. This permits the examination of a wide range of community resources as possible settings for the delivers of Mental Health services. It is also obvious that the PC's, although developed in a Mental Health project, are not specific to such concerns, and are applicable in any design project.

P-Prevention R-Rehabilitation
T-Treatment C.C.-Custodial Care

Meeting **1**   Initials _____ Date _____ Profession _____
Duties in planning or operation of center _____

Problem Statement _____

Description of Ultimate User _____ Physical description _____
Location of residence _____ Sociocultural data _____
Economic resources _____ Attitude toward mental health care _____

	Program alternatives to solve problems	Therapy alternatives to carry out the program	Details or Notes e.g. activities, possible settings, etc.
☐R ☐C.C. ☐P ☐T	Program 1 purpose:	Therapy 1 / 2 / 3	
☐R ☐C.C. ☐P ☐T	Program 2 purpose:	Therapy 1 / 2 / 3	
☐R ☐C.C. ☐P ☐T	Program 3 purpose:	Therapy 1 / 2 / 3	
☐R ☐C.C. ☐P ☐T	Program 4 purpose:	Therapy 1 / 2 / 3	
☐R ☐C.C. ☐P ☐T	Program 5 purpose:	Therapy 1 / 2 / 3	
☐R ☐C.C. ☐P ☐T	Program 6 purpose:	Therapy 1 / 2 / 3	
☐R ☐C.C. ☐P ☐T	Program 7	Therapy 1 / 2	

Meeting **2**

Initials _____ Date _____ Profession _____
Duties in planning or operation of center _____

Problem Statement _____
User Data _____
Therapy Alternative _____

Therapy 1 Discribe in terms of necessary activities	Notes on participants and possible activity

ntact
ing in
agnose
eat
sengage
low-up

Therapy 2 Describe in terms of necessary activities	Notes on participants and possible activity

tact
ag in
gnose
at
engage
ow-up

57

Meeting **3**   Initials   Date   Profession
              Duties in planning or operation of center

Activity Statement (subjects, action, place time, manner)

Participants Involved: Directly
Indirectly:

Supports for activity settings (list separately for each setting if more than one)
Relationship to other settings
Access to other settings
Special conditions or equipment

PERFORMANCE CHARACTERISTICS of setting to support the above activity

Remarks	Continuum:					Rank i pretat
	Communality					Privacy
	Sociopetality					Sociofugality
	Informality					Formality
	Identification					Remoteness
	Accessibility					Inaccessibility
	Ambiguity					Legibility
	Diversity					Homogeneity
	Adaptability					Fixity
	Comfort					Discomfort

Suggested by participant:

DEFINITION OF PERFORMANCE CHARACTERISTICS SUGGESTED BY PARTICIPANT:

Continuum: _____   Measure: _____

In the space below, please name settings presently existing in the service area or potential, which most nearly match the characteristics as scored above.

PAK

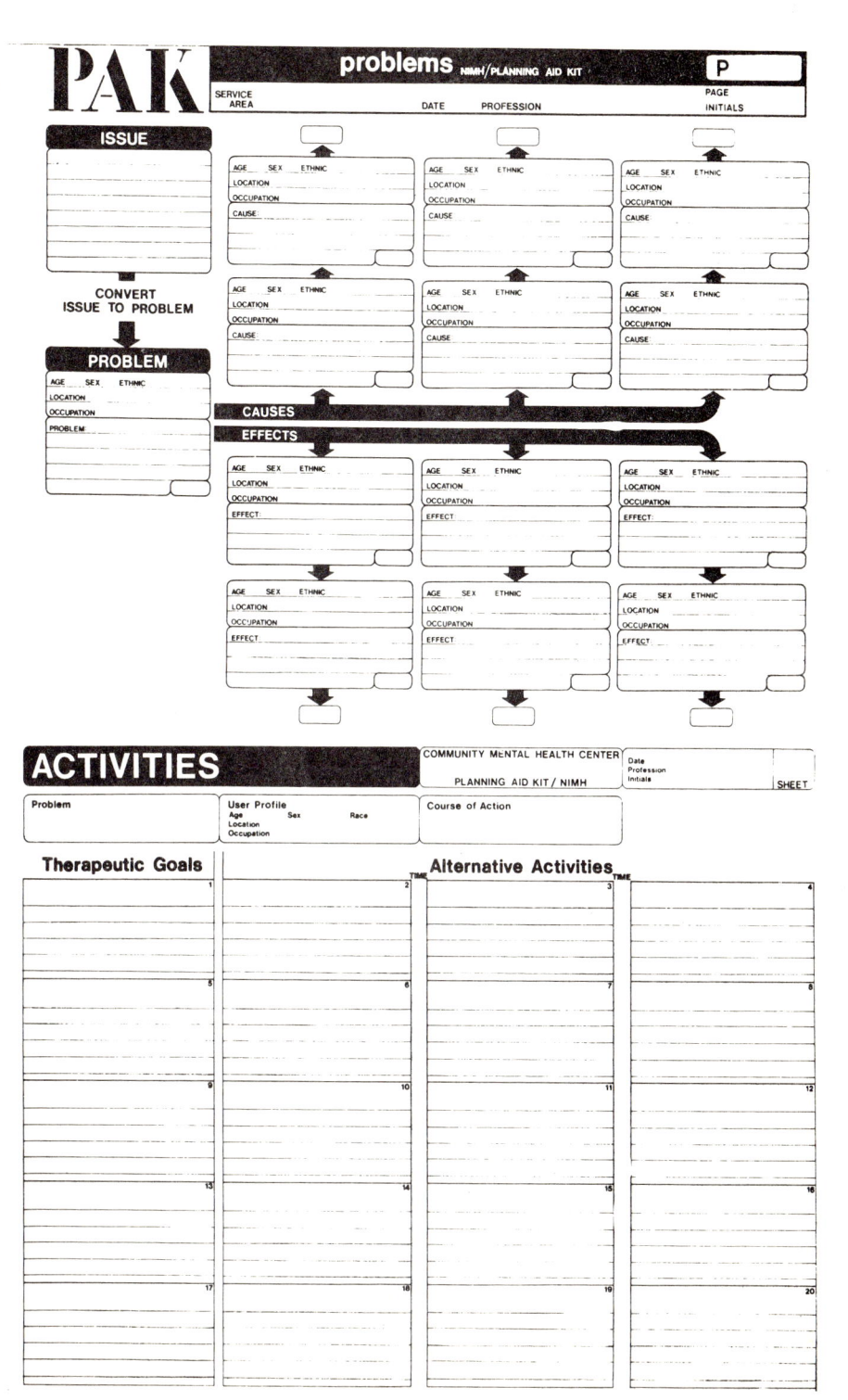

problems NIMH/PLANNING AID KIT

P

SERVICE AREA | DATE | PROFESSION | PAGE | INITIALS

ISSUE

CONVERT ISSUE TO PROBLEM

PROBLEM

AGE SEX ETHNIC
LOCATION
OCCUPATION
PROBLEM:

AGE SEX ETHNIC
LOCATION
OCCUPATION
CAUSE:

AGE SEX ETHNIC
LOCATION
OCCUPATION
CAUSE:

AGE SEX ETHNIC
LOCATION
OCCUPATION
CAUSE:

AGE SEX ETHNIC
LOCATION
OCCUPATION
CAUSE:

AGE SEX ETHNIC
LOCATION
OCCUPATION
CAUSE:

AGE SEX ETHNIC
LOCATION
OCCUPATION
CAUSE:

CAUSES

EFFECTS

AGE SEX ETHNIC
LOCATION
OCCUPATION
EFFECT:

AGE SEX ETHNIC
LOCATION
OCCUPATION
EFFECT:

AGE SEX ETHNIC
LOCATION
OCCUPATION
EFFECT:

AGE SEX ETHNIC
LOCATION
OCCUPATION
EFFECT:

AGE SEX ETHNIC
LOCATION
OCCUPATION
EFFECT:

AGE SEX ETHNIC
LOCATION
OCCUPATION
EFFECT:

ACTIVITIES

COMMUNITY MENTAL HEALTH CENTER
PLANNING AID KIT / NIMH

Date
Profession
Initials

SHEET

Problem

User Profile
Age Sex Race
Location
Occupation

Course of Action

Therapeutic Goals

Alternative Activities

TIME

TIME

1 2 3 4
5 6 7 8
9 10 11 12
13 14 15 16
17 18 19 20

Francis Duffy and James Freedman
School of Architecture, Department of Anthropology
Princeton University

PATTERNS AND SEMIOLOGY

This paper is an attempt to criticize and develop Christopher Alexander's *Pattern Language,* by taking into account a wider range of cultural phenomena than is usual in that design method.

The Pattern Language is practical in intent and has been used in such a way that a close *fit* between built form and programmatic requirements, which are expressed as *conflicts,* has seemed inevitable. However, experience and cross cultural comparison of house forms, for example, suggests that the basic domestic program can be enclosed in widely differing ways. In other words a degree of *slack* may be observed between program and form. It is difficult for the Pattern Language in its present form, to cope with this problem of variation without breaking down under the strain of incorporating more and more subtle and contradictory expressions of human needs.

Is Slack completely arbitrary and subject to endless individual variation? To support the argument that this is not entirely so, we draw on a concept which has influenced the study of other human phenomena in linguistics, in anthropology, and most recently in *Semiology,* the general science of signs. This concept, or rather method, is *Structuralism.* We argue that structuralist methods can also be applied to the study of built form to elucidate unconscious infrastructures, which underpin systems of relations between terms. Ultimately this kind of analysis aims to discover general laws about variation in such systems.

We argue that despite current limitations, patterns are in fact ideal units for the analysis of variations in built form. They are both the relations and the terms of the "language" of built form we are studying.

These arguments are tested by a case study. A detailed comparison is made between Egyptian and Japanese houses, and modern American popular housing.

The possible antagonsim between structuralist thinking and innovation is discussed. Nevertheless we dare make some suggestions about the possible liberating effect of being able to examine the tacit boundaries and silent rules which confine our own culture.

The Problem and the Approach

a) Explaining variations in built form

The main characteristic of rationalizing trend of thought in the moderan movement in architecture is an attempt to fit building forms more precisely to the requirements of the program. It is hardly necessary to remind you that Le Corbusier's[1] interest in the notion of *object-type,* the form so exquisitely moulded to exact technological and programmatic requirements that it becomes a timeless object, is one example of this characteristic, while Gropius's[2] experiments with sunlight as a determinant of the spacing of housing blocks in another. Many others could be cited. The International Style of the twenties and thirties in its austere sufficiency and necessary universality was in itself a manifesto of this preoccupation. Today, formal outward manifestations are less glamorous although perhaps more all pervasive. Nevertheless the same thinking persists. It is very clear, for example, that the increasing interest in the relation between built form and behavior which is reflected in the recent founding of such newsletters as *Architectural Psychology* and *Man and Environment* is at least partly caused by a desire to avoid ill effects on society of thoughtless, unconsidered decisions by architects and other designers. The flowering of serious studies in design methodology, fertilized by strong doses of operations research, and by computer technology, is evidenced by the recent founding of the *DMG Newsletter* and the *DRS Newsletter* on either side of the Atlantic. Both movements in varying degree are seeking to reduce the possibility of arbitrary design decisions, that is, to achieve a more efficient, more exact fit between program and built form.

It is important to ask ourselves what are the rules for inclusion of data in programs, and what rules account for variation in built form. These questions are becoming increasingly difficult to answer. In contrast to the primitive determinism of the twenties, more and more qualifications are made as designers gradually realize the difficulties and ambiguities involved in importing data from the social sciences. *Nihil Humanum alienum puto* and yet man seems to be a very strange and contradictory creature indeed. Of all the social sciences, anthropology, with its wide ranging, comparative methods, makes this point of man's contrariness most strongly.

Anthropology, it is interesting to note, has been torn by a debate about functionalism, similar to that which is familiar to all architects. Malinowski[3] is perhaps the extreme proponent of the functionalist approach. His attempt to explain the complexities of customs and artifacts of primitive peoples in terms of organic needs, i.e. "system(s) of conditions in the human organism, in the cultural setting, and in the relation of both to the natural environment, which are sufficient and necessary for the survival of group and organism," has been strongly attacked as simplistic. Generalizations based on such foundations are likely to be banal. You will see how similar this anthropological functionalism is to the architectural variety which is also characterised by an attempt to reduce the explanation of form to a few very basic needs. Anthropology as well as architecture has experienced a reaction from primitive functionalism, as the subtlety of human motivation becomes apparent.

Anthropology, more than other social sciences conducts a comparative analysis free of cultural favoritism. The application of social sciences to architecture has often been to show how one design is preferred to another. Anthropology does not provide arguments for preferences—rather directs its energies toward extensive bias-free description. In this way, the implicit notion that there is a right form to *fit* a given situation is avoided.

We suggest that the building fabric, especially those buildings reflecting popular taste which are illustrated in *Homes and Gardens* rather than *Architectural Design,* is in fact more varied and subtle than the notion of fit implies. We also suggest that this non-determined variation is not unbounded, that in any one culture, and that even throughout all cultures, less forms are actually exploited than are conceivable.

This is the core of our investigation. We want to refine the concept of "fit" between built form and behavior. Although we accept that "misfits" can occur, we are not focusing our attention upon them but instead concentrating on the variations within fit which we call "slack". Instead of apologizing for such variations, we aim to study them to discover their extent, and the rules by which they are structured and limited.

b) "Slack"

At this point, definitions are in order. "Fit" and "misfit" are of course, terms used by Christopher Alexander[4] in *Notes on the Synthesis of Form.* Although they are no longer prominent in Alexander's later work, there is no doubt about the success of their coinage, since they have not only been widely adopted but in Alexander's own case they represent very clearly a fundamental and continuing distinction between forms which are suitable in a certain context and those which are not. To quote him, "The form is the solution to the problem; the context defines the problem. Good fit is a desired property of this ensemble which relates to some particular division of the ensemble into form and context." In practice since it is impractical to list all properties of form and context, Alexander recommended in 1964 as part of design method that we look for examples of bad fit which are less numerous and more visible "Wherever an instance of misfit occurs in an ensemble, we are able to point specifically to what fails and to describe it."

No doubt this is very sound advice; however some consequences are unfortunate from an analytical point of view. Because Alexander was so concerned—rightly so in a design method—to locate the problems he had to solve, the whole emphasis is taken away from another interesting problem which he recognized but did not begin to investigate. The emphasis, on what is wrong in the fit between context and form, or environment and built form, takes one's attention away from how many diverse ways it may be rightly done It is not Alexander's purpose to show how diverse are the possibilities which may respond to a single problem—rather he wishes to solve the problem as efficiently as possible. We, on the other hand, in taking our inspiration from cross cultural variation, and man's infinite symbolic imagination, chose to show how varied within rules of variation, are the responses to physical problems.

This is a development argument that Rykwert[4a] made so well when he demonstrated that study of the anthropometry of the seating position, however fine and detailed, cannot determine the form of a chair because sitting in chairs is itself a cultural gesture with strong overtones of meaning which is elected in some societies but not in others.

There is no end to the subtlety with which one can dissect human motivation. Practically, especially i one is a design methodologist with insufficient data, there are very real limits and these are reflected in the kind of misfits, or conflicts which are actually recorded.

It will be apparent that our definition of need takes in a lot more than is usually presumed by the wor It subsumes physical requirements as well as those requirements which are associated with cultural institutio such as religion and kinship obligations. This very large area—which is indeed the totality of culture—into v we extend the concept of man's needs, is what we will call in architectural terms, slack. We define "slack" finally as follows: *For any design in a given context, in which a number of requirements have to be met, o conflicts solved, "slack" is the sum of all possible alternative physical ways of meeting those requirements a solving those conflicts fully.*

Perhaps this is opening Pandora's box. To increase the number of possible alternatives is no way to sp the design process. Nevertheless, the point cannot be ignored. Anthropology reveals the extent of the prob by its surveys and comparative methods. Anthropology, can also be relied upon to suggest some possible es cape routes.

c) The limits and rules of variation

So far we have observed that there is a great deal of variation in the built forms of houses, for exampl which enclose the little bundle of requirements and conflicts which make up a family. The question is whether there are any limits to these variations in house form, and also whether there are rules which one can use to generalize about the kinds of variation that can take place. If such rules can be determined, the certain consequences may follow. Firstly we shall understand a good deal more about the loose relation of built form to behavior. Secondly we shall always be aware of the permutations that are possible in built fo in any given situation. Thirdly we may be able to enrich our design vocabulary by constantly making ourselves aware of forgotten or latent possibilities. Our hypotheses are these:

1. that "slack" does exist, that in any design a number of alternative solutions may be freely chosen, all of which satisfy the programmatic requirements however minutely these are expressed.

2. that "slack" is not random, that principles can be found which explain both the limits of the var ation of slack in any type of built form, and also why slack is articulated in certain characteristic ways.

3. that knowledge of these principles is liberating rather than inhibiting, that they can be used to remind designers of possibilities they are not aware of.

Elements of Structural Anthropology

What is a need? Our contention is that a need cannot be identified solely as a physical necessity. There are cultural requirements which every built form must fulfill in order to be comfortable and comprehensible. Just as physical requirements are ways of putting together available material in the most convenient manner, so cultural requirements are ways of building which seem the least uncomfortable to our normal patterns of thinking. The constraints in satisfying these requirements we call conventions—patterns of thinking which condition how we understand and make ourselves understood.

Discovering these conventions is important for two reasons; they reveal our standards of comprehensibility, as well as, and it naturally follows, gives the terms to which even creativity must adhere. To know the rationales behind these conventions is to have access to instruments by which innovation is facilitated.

Finding out what these cultural conventions are, seems, on first view, too easy to merit an elaborate procedure. But this is not entirely the case—for most such conventions have rationales which are not immediately apparent. There are some few assumptions and techniques, utilized in a large part of anthropological literature, which are helpful in this regard. These will be the subject of my part of this presentation.

a) Assumptions

How we walk and how we eat, like language, are habits which people seldom challenge. We are reluctant to intrude on behavior which so little intrudes on us. To search for meaning, other than of a quaint or superficial sort, in these practices, is considered profane if not taboo.

All of which is naive. To think that they do not intrude on us the less simply because they are invisible, subsumed, taken for granted, is to ignore that their grasp is simply the more subtle. The words we use, like table manners, increments of time, customary spatial adjustments—all of these have a vernacular meaning as rich as the elite meanings of our monuments of creativity and innovation.

Yet there is a certain convenience to this rampant renunciation that we practice on our routine world. The very existence of habit makes our lives ever so much easier. How inconvenient it would be if every time we sat down at table, we rehashed all of the rationales for our manners; why the fork goes on the left and why we cut with the right then change the fork back to use as a scoop instead of a stabilizer. To be so critical of our instincts would slow down life unbearably and make of us awkward fumblers.

Further, if every time we wished to say the word good, we had rather to gesticulate or demonstrate this sentiment instead of enunciating the convenient four letter sound sequence, we would be sorely limited in our range of expression. Language is a blessing.

But it is one in disguise. For what a symbol, such as the sound sequence gains in convenience, it loses in meaning. What it gains, to use a well-worked vocabulary, in *medium* power, it loses in *message* power. The more denuded and empty of sense, the more agile it is in conveying sense; and this is its greatest deception. The more habitual we animals become, the more we take for granted, the more senseless we all become.

The process of dehabituating these various adjustments, is our goal. We want to find a way to work a critical dissection on myth, manner and building conventions, so that they are not obstacles to creative thinking, but sources of information of how people build up their perceived world. The following notions are keys for dislodging this information. The first three are assumptions about these conventions. The fourth and fifth describe what to look for in investigation.

1. *Society as a whole is qualitatively greater than the sum of the individuals that make it up.*
The gist of this axiom is that cultural notions which the individual receives from his fellow members, past and present, will influence him more than his individual decisions will effect the social norms. The society is self-justifying and self-perpetuating beyond the individual's capacity to arrest or in large part comprehend it.

2. *Man is a creature of habit.*
   This premise merely aggravates the effect of the first. Not only does society gain control over individual wills, but the individual *wills* this submission. He invites it by slipping easily into a routine, even when it is oppressive and harmful.

3. *Man is a symbol using animal.*
   Symbols and symbol systems are simply one of the habits of man. To picture a symbol thus, is not to ignore the innovative power of metaphor and adventurous representation. Rather it is to stress a more prevalent aspect of the process of representation which can be explained as follows: the process of symbolization, like abbreviation or any other kind of substitution, places meaning under a sign to facilitate storage and communication. In doing so, the original meaning encroaches less and less on our attention, so that the reason for using a particular symbol with a particular idea becomes automatic to our senses. Just as with habit, we tend to ignore the rationales for our action. The rationale for a symbol, that is, why we represent a certain meaning in a certain way, is equally lost with constant use.

4. *Man's unconscious is a composite of the renounced meanings and rationales of his accepted symbols and habits. It is structured. Its structure is the object of structural research.*
   The use of the word "unconscious" to represent the determinants of behavior of which we are unaware, may be objected to, but since it is the terminology of the masters, we use it here. Claude Levi-Strauss, much spokesman for structuralist method is largely responsible, having asserted that an analysis of unconscious structures gives a more objective appraisal of the universal characteristics of man. The conscious structures, usually known as norms, he said, "are, by definition, very poor ones since they are not intended to explain the phenomena, but to perpetuate them."[6]

5. *Structure is defined by the way information is decoded and encoded in it.*
   The concept of structure referred to here is not identifiable by any positive, inherent principle of human behavior, which constitutes man's essence, nor is it a certain quality. It is not what there is to know about man that concerns us, but the manner by which we know and perceive. it is the form by which man understands and transmits information, whose content may vary, but whose form works always by the same rules.

We describe this form as follows: One expresses structure and describes culture systems through significant units, which are minimal units for conveying information. Each of these units has three parts: two terms and ther relation. The two terms are different in some finite regard and are similar in another. By virtue of their difference they are often known as oppositions. By virtue of what they have in common, they may be spoken of as relations.

And now the explanation for this form: why are there two terms and a relation? Why not one term, or three? What is so essential about this form?

Let us retrace a few steps. We have said that symbols and man's committment to them may be seen as habits. A word is a sign much like a gesture at the table, and much like an architectural convention—such as defining house boundaries with fences, hedges, and walls as is universal in England, but not in the U.S.A. But what is unique about these conventions is that we really know very little about why we employ them... neither their history nor often their utility. Just as with symbols, why we represent a certain meaning in a certain way—the rationale for habit is lost with constant use. As we become more routinized in our conventions, the history and utility of our symbolic behavior of which built form is a prime example, is discarded.

Now, if there is no intrinsic meaning in a sign to recommend it to some idea, or if this meaning is lost as is often the case, where does the sign get its meaning? Certainly not from any positive definition. There is no absolute source of meaning as in a god-head or universal mandate. There is no essential necessity between the shape of sound or house and what it means. All depends on what social custom ordains. The only way we can see meaning or value in a concept, then is in comparison with another concept by contrast, or in similarity. Saussure explained it thus: "Language is a system of interdependent terms in which the value of each term results solely from the simultaneous presence of the others."[7]

Money works on the same principle. We have long since forgotten the intrinsic value of a coin; it is what one gets in exchange for it that determines its value. To determine what a piece of money is worth, one must know what it can be exchanged for, and how it can be compared with a similar coin in the same system. In other words, like language, money is a system based on critical differences.

If a symbol system is made up of critical differences rather than absolute qualities, it is obvious that definitions will not suffice to represent the system What is needed rather is a form that takes account of *meaning by contrast.* The minimal shape of such a form is a set of two terms with a specified relation. This is what we call a significant unit.

Regardless of what one wants to describe—a myth, a house or a language—these minimal units will have the same form. The myth studies of Levi-Strauss contain sets of two terms with a relation. The Art History of Heinrich Wolfflin[7a] in his study of the passage from classical to baroque styles contains five sets of two terms apiece. Christopher Alexander's elementary unit in a pattern problem consists likewise of a single conflict between a minimum of two tendencies for which there is a single resolution.

We will see later how this significant unit, used by anthropology as a purely analytical tool, can be converted into an aid to design.

The Pattern Language and Variation

The origin of patterns as used by Alexander is in the disatisfaction he felt with the misfit variables we described earlier. Although in theory misfits constitute a diagram of stress to which a formal solution corresponded, they were too numerous and the links between them too complicated to be a working design method. To cope with this methodological problem, Alexander hit upon the idea of isolating *minimum units* of built form which can be said to solve particular problems. Once such units are established, the study of linkages between them becomes much more possible. The question is, or course, how they can be established.

Alexander's answer is simple.[8] A problem occurs if people are thwarted in trying to do something. What people want to do can in the long run be discovered from their actions; each move towards some goal can can be called a *tendency.* Tendencies are concrete and observable; they are the operational equivalent of needs or requirements. Sometimes two tendencies come into conflict—for example when a man in an office organization wants to talk to a client and yet at the same time be free to answer confidential phone calls. It is when this happens, that some kind of solution is required to allow both tendencies free play, to prevent them getting in each other's way. If the solution can be achieved by physical means, for example in the office by positioning the telephone sufficiently far away from where intimate conversations take place, then this is a *pattern,* a generic physical solution to an insoluable conflict problem. A number of conditions must be fulfilled, however. First, the context in which the tendencies come into conflict must be described accurately. Secondly, not only must the solution be described precisely in all relevant details, but this must be done in such a way that the description covers all possible solutions of that problem. This is a problem of course, of topological complexity and very difficult to solve. Thirdly, it must be clear that the pattern solves the problem completely, and not only partially. Fourthly, the pattern must solve one and only one problem and not two or three.

So we can define a pattern as follows: *the arrangement in space within a defined context of the minimum number of physical objects, which permit two tendencies to coexist without coming into conflict.*

The pattern language is the total of individual patterns and the rules or "grammar" which regulate how they are combined into designs for individual buildings on particular sites. Unfortunately little is known about these rules although a considerable amount of time has been spent trying to discover what they are and how they work. It seems clear for example that in the design of most kinds of buildings, some patterns, particularly those to do with circulation affect the rest of the design powerfully while other patterns are practically independent and have very little influence on what is around them. The closest investi- of this problem is in *A Pattern Language which Generates Multi-Service Centers,* [9] in which the concept of "links" between individual patterns is put forward, and actually investigated in design proposals for eight such centers. However, this is as far as investigation has gone, and it would be wrong to claim that a grammar of the Pattern Language has been fully described.

The success of the pattern language is threefold: first, it allows simple programmatic tasks as well as the more subtle social considerations to be expressed under the same rubric; second, it provides a format in which the material facts of built form and the whole range of social considerations which bear upon them to be brought together without undue stress on either; third, and above all, patterns are the nearest thing we know to *minimum units* in design. The tendencies can be considered as two conflicting terms—the generic physical solution can be thought of as the relation between the terms. In this way a pattern is analagous to the significant units of structural linguistics and structural anthropology.

After having identified this pattern format, Alexander carries his design procedure one step further: the conflicting terms of a pattern, he would say, may be said to have a single and best solution. This best solution, he argues, is that which provides for the absence of conflict between the terms. Simple programmatic needs, which can be expressed as performance specifications may have a single and best solution, but this is not the case with social and cultural problems. And here is where we find fault with Alexander, for he assumes a mechanical unilateral solution to all problems, tangible and intangible alike.

In this light we particularly appreciate Amos Rapoport's new book because it indicates that such a unilateral approach is not practicable. Rapoport shows with profuse examples that man responds to physical necessity, such as technology, climate, site, economics and the like, with the most varied and contradictory solutions—contradictory to each other and contradictory to the physical problem. He shows by inference that Alexander, though he may be utopian, does not render a complete account of what appears to be the more complex nature of man.

A study of the Winnebago,[11] North American Indians, provides another excellent example of this complexity: the Winnebago lived in small circular villages, the shape of one village not varying very much from another. But within a single village depending on whom one asked, the spatial arrangement was said to differ enormously. Depending on which half of the village a member of the village belonged to, he held his own distinct conception of the village. So, in this example, members of a single, well-knit social group respond to a single cultural problem with distinctly different design solutions.

According to the legends of the Winnebago tribe, each village was made up of two halves, one of which was superior—the upper moiety—and the other half inferior—the lower moiety. If one asked a member of the upper moiety what his village looked like, he would draw a picture which would accentuate the divisions between the two moieties, and by implication underline the superiority of his own, the upper moiety. It was very simple: he would draw a circle divided by a diameter line, which sharply showed that the spatial structure articulated the division between the two halves.

If one addressed a member of the lower moiety, he would present a different picture. He would deny that the myth about moiety division had any relevance for village structure implying that there was no spatial division based on moiety division. He thus virtually denied the legends, the hierarchy established between the two moieties. The village plan he would draw was of two concentric circles, one inside the other. Both moieties had their ceremonial huts scattered within the inner circle, which was the holy area. The outside circle stood in distinction to the inner, to distinguish profane from sacred, but there was no mention of a social division.

So each moiety gave its own interpretations of the myth, conceiving of the village in its own particular way. One would be hard put to find a better example of how, within a single well-knit social group, one problem gave rise to two distinctly different concepts of design solutions.

That there are diverse design responses, by virtue of cultural and social differences, is not all we want to show. Our more fundamental aim is to find a method which accounts for the broad range of cultural variation, be it within a single, or across many cultures. In order to achieve this aim, we return to minimal units as used in the pattern language as a means of description, and do not worry about the operational task of solving immediate design problems in our own society today.

Levi-Strauss[12] has applied methods of structural linguistics to the study of various of the subject matters of anthropology, for example kinship and myth. Encouraged by this example we are attempting to use these same simple methods—the study of unconscious infrastructures as well as conscious and accepted phenomena; the elucidation and demonstration of systems of relationships; the search for general laws—in our study of variation in built form.

An analysis of myth in the structuralist manner is a better introduction to these methods than more abstract studies of phonetic structure.[13] Myths like buildings make sense at more than one level; like buildings they are somewhat concrete and are easily comprehended if not always fully; like buildings their overall form is rich and confused, their parts hard to define, and yet everyone is aware of fague correspondences between myths of widely varying cultures.

> Myth like the rest of language, is made up of constituent units...These constituent units presuppose the constituent unit present in language when analyzed on other levels—namely, phonemas, morphenes, and sememes—but they, nevertheless differ from the latter in the same way as the latter differ among themselves; they belong to a higher and more complex order. For this reason, we shall call them gross constituent units.[14]

Having made this vitally important assertion, Levi-Strauss takled the practical problems of myth analysis by breaking down each myth into the shortest possible sentences, and writing down each sentence as an index card bearing a number corresponding to the unfolding of the myth story. "Practically each card will thus show that a certain function is, at a given time, linked to a given subject. Or to put it otherwise, each gross constituent unit will consist of a *relation.*"[15] The unit of myth analysis is then an actor and an action, a subject and a verb. The unit of building analysis, we argue, is a problem expressed by two conflicting tendencies, and the successful relation achieved between them by built form, in other words a pattern.

The constituent unit of myth analysis, the *mytheme*, is such a simple relation as "Salvation of the tribe", "food value of game". "brother and sister sacrificed," "Oedipus kills the Sphinx". The constituent unit of built form, the pattern, or to be fanciful the *archeme* is also such a simple relation of subject and verb, or conflict to be solved and the physical means by which it is resolved. For example, the tendencies when entering a house to seek either kitchen or living space are resolved by "one entrance 'which') connects the car to both kitchen and living room." It must be admitted that patterns, as they are currently expressed, seem somewhat "programmatic" and narrow. Nevertheless, there is no limit to the kind of data which can be incorporated in the pattern format, and no end to the subtlety of problem conflict described.

But this is only the beginning of the analysis. In his work, Levi-Strauss argues that bundles of elementary relations make up myths, and he shows how various seemingly dissimilar myths are in fact comparable because the similar myths is finite. If the elements are recognised, the rules and limits of variation can be explicated and light can be thrown upon the structuring principles of man's mind.

We do not see any reason why buildings regarded as bundles of patterns or archemes, cannot be analyzed in the same way. The case study which follows is a step towards this kind of analysis.

The Case Study

Let us take a problem of the entrance to houses as an example of a minimum unit. This minimum unit is characterized by two terms, inside and outside. Their relation is described by the threshold.

Alexander[16] describes a pattern for the entrance to a house which is intended to resolve the discontinuity between the public atmosphere of the street and the private atmosphere of the home. He phrases the problem in terms of their conflicting tendencies:

a. On the street people adopt a mask of 'street behavior;' the momentum of this mask tends to persist until wiped clean.
b. Arriving home, people search for the security of an inner sanctum, where they can relax completely.
c. Those arriving at a house seek as much 'contact' with the people inside as possible.

The first of these tendencies conflicts with the other two. People who are still in a street mood, maintain the general level of 'closedness', tension, and distance which is appropriate to public encounters; and they are thereby prevented from relaxing or from opening up sufficiently to meet people with maximum contact."

The solution to this conflict is expressed in the form of a preferred set of relations between physical objects: Erect "as many discontinuities as possible between the street and the inside of the house; changes of surface, changes of level, changes of view." Various kinds of suitable and unsuitable solutions are described.

Here we have an excellent example of a pattern, a minimum unit in design. What we want to point out is first that although this problem of the antinomy between inside and outside occurs in *all* societies, there is considerable variation in the way physical solutions are achieved. Secondly we want to show that this variation is not random but is articulated in a number of characteristic and one might almost say, logical ways.

Japanese, American and Egyptian houses are so different that it is daring to compare them on the same basis, especially when the examples we use are drawn from different sources and different times. We cannot pretend that our examples are even typical in any representative sense. However, the descriptions of both Egyptian and Japanese houses are only available to us because they seemed to the sensitive Western Observers,[17] who had experienced the pleasures and discomforts of living in them, to reveal something about those cultures.

When we study the Egyptian plan and particularly the relation of outside to inside, we observe that there is in fact only one entrance into a protected court which is an internal space where much circulation takes place on a different domestic level. The major opposition between public and private worlds is stressed to an almost incredible extent. The plan shows that the main entrance is not only narrow; but it is also tortu. Moreover, within the house each progression toward a more private space is marked by a twist, a turn, a stair or some kind of log leg arrangement which obstructs the distant view, and above all protects and encloses. Obsession with privacy is a major theme in this plan and is made apparent not only in the arrangement of doors and circulation, but also in the very windows which are small, latticed, and shuttered, more loopholes than outlooks.

Can we express this Egyptian "pattern" in terms of the same conflicting tendencies? Certainly street behavior is very clearly distinguished from domestic behavior. The mask is wiped clean. It seems too, from this description that the Egyptians searched for the security of an inner sanctum, where they felt able to relax completely. However, the third tendency which Alexander mentions as characteristic of American, or rather Californian society, that those arriving at a house seek as much contact as possible with those inside is contraverted. The Egyptians fought hard in their designs to segregate groups, classes, and sexes within the house almost as rigourously as they segregated their households from the public realm.

The Japanese solution is very different although the familiar basic conflict between inside and outside tendencies in behavior is again resolved. The plan has one major entrance. However, the relation between the outside and inside worlds is extremely diffuse. It is, in fact, possible to enter the house in many places through sliding partitions. The veranda wrapping round two sides of the house makes easy the passage from within to without, and indeed creates an important feature out of this in between area. However, the house is like a flimsy box and can be completely shut off by shutters if necessary and in fact this is done every night—to the extreme discomfort of the sleeping occupants—although Taut describes driving past and catching glimpses of families inside lighted interiors at night, through open panels, before all lights are extinguished, and all shutters drawn.

In Japan as in other cases street behavior is sharply distinguished from domestic behavior. All evidence supports the assertion that Japanese too search for security and relaxation when they return home. Unlike the Egyptians, and even more so than Alexander's Californians, they seem to seek as much contact with the people inside as possible. Indeed it might even be said that families and even aquaintances crowd together as much as possible in one space without compunction, enjoying the security and intimacy of actual bodily contact. It is not surprising that the theme of flexible openness characterizes the interior arrangement too and that the relation of space to space is by internal sliding panels. Clearly such an arrangement is very different from the Egyptian solution which is rigidly demarcated, and quite different from the American solution described by Alexander[18] which lies somewhere in between.

What this case study reveals is that conflicting tendencies which are the locus of a pattern, can be resolved differently in different cultures. It also shows incidentally that we need not always expect the same tendencies to come into the same kind of conflict.

Variation around the solution of the same conflicting tendencies is what we have called *slack*. Let us now examine how this slack is articulated.

This conflict between inside and outside tendencies interests us chiefly because of its physical solutions. The first point to make is that physical solutions may not be necessary since it is conceivable that they may be displaced by manners and customs, for example, which are designed to achieve the same end. We have, therefore, a scale from an extreme dependence on the physical to an extreme lack of dependence. In the first case the opposition between outside and inside is emphasized in order to overcome it; in the second, the opposition is deemphasized also in order to overcome it. On this scale, the Egyptian house represents one extreme, and the American house with its picture windows but guarded door represents a middle position, while the flexible Japanese house, which is only shut up at certain times, lies toward the de-emphasis end of the opposition scale. Perhaps somewhere in the world an example could be found of a dwelling in which no physical barriers mediate between outside and in.

This is only one example of a variation along a simple scale in one pattern. In the case of the Egyptian and Japanese houses other loci of investigation immediately present themselves in the vivid contrast between those patterns which are associated with them:

segregation and non-segregation of internal functions,
permanence and flexibility of building elements,
regularity and irregularity of building units,
decoration and plainness of surfaces
family togetherness and individual or group privacy
and also in the similar patterns which are associated with the physical phenomena of:

little moveable furniture
fixed seating
hierarchical principles of space sharing.

Each of these contrast sets can be manipulated in various ways. Instead of expecting one design solution to answer a particular problem, we anticipate a range. But this range will not be limitless; the variation will take place along some kind of scale and will be measured by some kind of contrast between opposite possibilities.

So far we have discussed variation within each elemental unit. We argue that similar structuring principles will apply to the combination of elemental units, or patterns, into whole designs. Certain juxtapositions will be mandatory; others possible; some forbidden.

A limited number of such dimensions of opposition and relatively few rules may control what appears to be the bewildering variety of building forms displayed, for example, in Rapoport's[19] book.

Conclusions

Our principle conclusions are these:

1. The proper unraveling of man's most habitual actions (language, manner, and myth) will reveal an underlying structure. The structure thus derived is neither intrinsic nor subjective.
2. An underlying structure is best seen as consisting of minimal units which explain its internal consistency
3. Even though these units do not necessarily present themselves to our awareness, they condition to a large extent our conscious behavior as well as our receptivity to new information, since they impose con straints on what is comprehensible. They are like the rationales for popular metaphors which, because of constant use and the fact that one no longer needs them to make the metaphor make sense, have slipped away from conscious appreciation. Because we are generally not aware of these units and the connections between them, does not mean that we are not sensitive to them. The connection which these units maintain, condition our existence and do so more than any other principles of our conscious perceptive tools. They are what make us happy or sad, pleased or displeased, without our knowing exactly why.
4. As much as these units are the agents of our habitual and highly conditioned actions, that is of how we remain the same, so they must be equally important in how we change. No innovation can be possible without taking into account these loci of sensitivity.

Since all culture can only be described as communication, the one thing that all aspects of it must have in common is an underlying code. The code is a formal organization which does not vary. It is the invarian underlying structure. The content it can receive may be of any sort. In the context of language such a code guides our speech—sound production—as much as our hearing—sound perception. In the context of architect such a code conditions the forms which we impose upon the environment as much as they control our passive understanding of built form.

Moreover, the shape of the communication unit, two terms and a relation, is a ready made format for the expression of a design problem and its solution. Our task has been to show that it is possible to use such units to uncover the underlying code in built form.

Amos Rapoport[20] envisaged the possibility of converting cultural information, the verbal and pictorial messages of a social group, into information readily usable in design. One way of doing this is what we have described: to express such messages in minimum units. These minimum units are similar in shape to the minimum units which transmit and store speech and other kinds of information. They are also coincident with Alexander's patterns. As such they form a potential conceptual bridge between observation and design.

The novelty of our presentation does not lie in the discussion of the pattern language which is already well known as a design method; nor in introducing structuralist thought into architecture.[21] What we do claim to be original about our work is that we have begun to put structuralist ideas to work in tis field by proposing an analytical method, allied to a design method, which can be used to investigate the rules and lim of variation in built form. In this way we are merely advancing the general study of signs, now known as semiology,[22] one step further from its origins in structural linguistics.

References

1 Le Corbusier, L'Art Decoratif d'Aujourdhi. Paris. Les Editions G. Gres et Cie. 1925. p.79.

2 Gropius. *Scope of Total Architecture.* New York: Harper & Row, 1955. Chapter 11, "Houses, Walk-ups or High Rise Apartment Blocks?".

3 B. Malinowski. *A Scientific Theory of Culture.* New York: Oxford, 1960. p. 90.

4 C. Alexander. *Notes on the Synthesis of Form,* Cambridge, Harvard University Press, 1964.

4a J. Rykwert, "The Seating Position—a Question of Method," *Arena, Architectural Association Journal,* June 1967.

6 C. Levi-Strauss, "Social Structure", *Structural Anthropology,* New York, Anchor Books, 1963, p. 274.

7 F. De Sausseure. *Course In General Linguistics,* New York, McGraw-Hill, 1959, p. 73.

7a H. Wolfflin. *Principles of Art History: The Problems of the Development of Style in Later Art,* New York; Dover, 1950.

8 C. Alexander, *The Atoms of Environmental Structure.* London; Min. of Public Building and Works, Directorate of Development, R. & D. Paper, 1968.

9 C. Alexander. *A Pattern Language which Generates Multi-Service Centers,* Berkeley; Center for Environmental Structure, 1968.

10 A. Rapoport. *House Form and Culture,* Englewood Cliffs, N. J.: Prentice Hall, 1969.

11 C. Levi-Strauss, "Do Dual Organizations Exist?" *Structural Anthropology.* New York, Anchor. 1963, pp. 128-159.

12 C. Levi-Strauss, "The Structural Study of Myth", *Structural Anthropology,* New York, Anchor, 1963, pp. 202-228.

13 R. Jakobson and M. Halle. *Fundamentals of Language.* The Hague; Mouton, 1956.

14 C. Levi-Strauss, *op. cit.* p. 208

15 C. Levi-Strauss, *op. cit.* p. 207.

16 C. Alexander. *The Atoms of Environmental Structure,* London, MoPBW, Directorate of Development, R. & D. Paper, 1968, p. 45.

17 Egyptian House. E. W. Lane. *Manners and Customs of the Modern Egyptians,* London; Charles Knight & Co. 1936.

   Japanese House. B. Taut. *Houses and People of Japan,* London, Gifford, 1938.

18 C. Alexander, *op. cit.*

19   A. Rapoport, *op. cit.*

20   A. Rapoport, address at EDRA Conference, Chapel Hill, June 1969.

21   See George Baird, "La Dimension Amoureuse in Architecture", *Arena, Architectural Association Journal,* June 1967.

22   R. Barthes, *Elements of Semiology,* New York; Hill and Wang, 1969.

Stuart Silverstone
Department of Architecture
Massachusetts Institute of Technology

WORKSHOP IN PERFORMANCE SYSTEMS

During our workshop discussion it was suggested that there are two distinct types of research now being conducted. One is operational—something that can be done now—and the other is more philosophical, with longer range value. This an unfortunate, but accurate distinction. Unfortunately, those research approaches which seem to have the greatest potential do not receive much attention, because they are not presently operational.

A similar distinction was made by John Archea who suggests that current research can be classified along two exclusive spectrums. The first is whether the research is good or bad, the second is whether it is right or wrong. From criticisms which have been made on this panel, some of the papers presented at this conference would seem to be both bad and wrong. That is, they exhibit a low measure according to what might be called scientific approach, and they do not contribute positively to what is referred to as individual, humanistic, or societal values.

It is probably premature at this time to expect any signigicant research which is good and right. The problem seems to be that too much time and resources are spent on work which is good and wrong, and not enough on that which is bad and right. The assumption is always implicit that what starts out in a small way can eventually become substantial. But to go from wrong to right content is not logical. To move from bad to good technique is much easier, and even natural.

There are examples of the good/wrong trend where certain information is measured, collected, and stored in a computer or where a process is quantified, made deterministic, and computerized as a matter of operational convenience. It is too easy and dangerous to store that information which is easiest to measure and collect, or to automate that process which is possible to quantify and define, just because it fits into the technology. Often these are not the most significant information or process for a particular problem, but they are depended upon because they are available.

From our workshop on performance systems, there emerged a realization that product oriented goals in environmental design are very limiting. To define a problem in terms of preconceived solution. It is preferable to have a process oriented approach to defining a problem. I think this is a very valid realization. The thing that upsets me is I don't think we can afford to attack the problem with one realization at a time. I am very encouraged that the people who presented in the workshop were criticized on this point by the participants. Too much attention is paid to good and bad research techniques and not enough to research content and its implications. But, the two should not be separated.

In my own work I see a similar dilemma in trying to study communication systems. There is a tremendous bias that the tools whith which we work have on the problems that we can solve. If we use numbers or if we use pictures, there is a tremendous influence on what we're able to do. There are popular myths about being able to go out and collect raw data or being able to make objective decisions. But, the information which you can use, and thus, the decisions which you are able to make are limited by the languages which are available. It's a vicious circle, because representation techniques are developed in response to new information on the basis of numerical representations, than they might not be able to deal with the relevant variables.

It's possible that the most important variables cannot be predicted, and that the most important information cannot be replicated. Themodels which try to perfect predictions for these variables are going to be, if they are successful, very good at predicting what is going to happen withing the model. As Jim Fitch suggested this morning, the environment is very, very difficult to abstract. Some of the discussion in the workshop strongly suggested that we shouldn't even try, that we should work directly in the environment with the people who are going to live there.

It might be that such social and political actions are needed to be taken directly in the real-world arena. Surely, some bigger and differently directed steps are needed, if this organization and the interests which it represents are to be influential in shaping the environment. Maybe such shaping requires more than creating shapes and words on paper or computer displays. Such an issue deals with the role of the professional and the question of who he serves. I think this conference should discuss such issues and the methods by which changes can be achieved most effectively, as well as reviewing currently operational research.

It is, of course, easier to criticize what is being done than to propose a substitute. But that fact should not impede criticism—it should stimulate discussion of goals and values, and hopefully result in understanding the kind of new context which is needed for urban research and might require the relinquishing of current vested interests and of past investments. The bad/right kind of research might take longer to become operational, but it will never be able to compete with the good/wrong kind if the allocation of resources continues to prejudge the issues.

Raymond M. Craun Jr.
Department of Architecture
North Carolina State University, Raleigh

VISUAL DETERMINANTS OF PREFERENCE FOR DWELLING ENVIRONS

There are many problems associated with the design and planning of public housing. Quite often, so much stress is placed on economic feasibility, durability and ease of maintenance that public housing developments seem to lack visual appeal. Peter Rossi, in a study entitled *Why Families Move,* indicates that some very basic spatial and utilitarian needs rank before the social and physical environment as impetus for complaint.[1]

To be sure, the aesthetic attractiveness of a dwelling complex does not rank as the primary problem with most low-income families. However, when other more functional criteria have been satisfied, the appearance of the immediate physical environment does affect preferential choices.

Many of the problems of visual appearance seem inherent to mass public housing developments. Catherine Wurster states:

> Public housing projects tend to bery largely and highly standardized in their design. Visually they may be no more monotonous than a typical suburban tract, but their density makes them seem much more institutional.[2]

Architects and planners often strive for solutions to such problems without any knowledge of the values of the occupants. Practically speaking, designers often lack any direct contact with the future occupants of such developments and, consequently must rely on planning procedures not appropriate for the social groups involved. In actuality most low-income persons do aspire to look and live like the middle classes.[3] If this is true, the problem of designing more satisfying visual environments becomes simpler. However, there is little evidence from research findings that visual satisfaction can be measured or predicted. This lack of understanding would apply to any development which affected the visual environment, regardless of the income or social group.

There has been speculation that the visual image may be subject to measures of satisfaction based on inherent psychological and perceptual phenomena. Rapoport, Kanter, Miller and Vigier have all indicated that satisfaction with the perceived environment may depend upon the amount and organization of the information presented the observer.[4] Fiske and Maddi assert that it can be demonstrated in more basic situations that "there is a level of activation which is necessary for maximally effective performance".[5]

The Problem

The purpose of this investigation is to survey the preferences of people for various public housing environments and to isolate some of the aesthetic criteria which are instrumental in the choice of a satisfying residential environment. The housing environment in this study refers to the general exterior appearance of the public housing dwelling group, as it may appear on its particular site. Criteria under investigation are limited to the qualities of visual appearance which may influence the aesthetic satisfaction of a prospective resident.

The scope of the study is limited to include only multi-family dwelling structures, one to three floors in height with emphasis on the relationship of the individual dwelling to the site and to surrounding dwellings. The housing environments are selected to represent a range of aesthetic types normally associated with public housing projects.

Research Design

The research design follows a single-cell model of investigation commonly used in exploratory surveys. The subjects are asked to select those residential environments which they would prefer if they were actually selecting a dwelling. Each of the situations are represented by photos of the unit and its site. (See Appendix A) Since the design of a successful instrument depends on the quality of the simulation, care was taken to provide an accurate and descriptive photographic representation of each housing group. Eighteen different situations were chosen to offer a variety of dwelling and site conditions. Each could be classified as a multi-family structure and none was over three floors in height. A typical dwelling-site relationship was represented for each housing environment.

The subjects, as the sample group, are treated as a control variable. The sample was chosen primarily on the basis of common age, educational training and social standards. The survey results which are reported were obtained from a sample of twenty-five fourth and fifth year students studying architecture, landscape architecture and product design. The judgments were obviously made on the basis of limited information. In each situation there may have been factors present in the everyday situation which would have affected judgments. However, this is a recognized possibility for each of the photographs. None of the observers had the opportunity of actually experiencing the dwellings.

The preference of the sample group is considered to be dependent on three independent evaluations of the housing environments. These are the degree of visual complexity, the perceived cost of the dwelling and the amount of privacy offered by the dwelling environment. The data collection device is designed to provide an indication of the respondent's experience in housing types and living patterns. Because the sample group was limited to students of the School of Design, whose educational and social background was known, no other general questions were included. Each subject was asked to indicate his impression of the characteristics of an imagined *good residential environment.* This provided a basis for comparison with the characteristics of the preferred housing environments.

The remaining three resorts required categorical judgments of expensive/economical, private/public, and simple/complex. Each subject was presented a deck of eighteen cards which had been randomly identified by letters of the alphabet. Each card was a photo-simulate of a housing type. Four separate evaluations were required and each involved sorting the card deck. The forst required grouping according to preference, in five categories ranging from "like very much" to "dislike very much."

Hypotheses

The everyday environment is much more complicated than isolated and controlled laboratory experiments simulating the phenomenon. For this reason, it is difficult to relate the specific measurements and findings of this survey to the theoretical framework. There are many variables which may influence the choice of the observer. Those which seem most amenable to the judgments sought and to the means of simulation used are included in the conceptual framework.
1. People prefer residential environments which to them seem expensive. When money is not a factor, and when the observer is given a free choice of many residential environments, he will not choose economical or inexpensive examples.
2. People prefer those housing environments which they believe offer a high degree of family privacy. When there is a choice of dwellings ranging in the degree of privacy they afford, people will not choose those which seem public.

3. People will prefer housing environments which are highly complex visually. If there is a range of dwelling and site conditions which offers a corresponding choice of visually simple and visually complex situations, people will not select the extremely simple situations.[6]
4. As far as privacy can be valued in monetary terms, the private dwellings will be positively associated with the expensive dwellings. This hypothesis is quite dependent on the nature of the dwellings offered. Single-family detached dwellings are not included in the housing environments presented in the sample.
5. In general, there will be a correlation between those dwelling environments which are least preferred and those which are judged to be simple, economical and public.

Verification of the hypotheses is limited by the inability to control the amount and kind of information each observer will receive from the stimuli situation. There will be absolute values obtained. The results depend upon the subjective judgments made by the sample group.

Analysis

The data were collected by means of a two page questionnaire, which took an average of twenty minutes for each individual sampled. (See Appendix B)

The first card sort required categorization of each housing environment according to a five part preference scale ranging from judgments of "like very much" to "dislike very much." The frequency of response for the entire sample group is represented in Table 1. Numerical values were assigned to the categories ranging from one to five and combined with the frequency, they were used to obtain mean preference ratings for each of the housing environments. These are arranged in Table 2. The mean score for the entire sample of housing environments was 3.42.

Table 3 indicates the correlation of preference with those housing environments judged to be private, complex and expensive. The only housing environments judged to be all three—private, complex and expensive—were ranked one, two and three in terms of preference. Conversely, those which were judged to be public, simple and economical were least preferred and ranked at the bottom. Table 4 summarizes the results by comparing the categories of preference with the judgments made concerning the housing sample.

Summary

The analysis indicates the relationships do exist, however, it should not be assumed that the preferences were based entirely upon the hypothetical relationships. Other factors may have influenced the preferences. Factors which were considered important by the sample group are presented in Table 6. Some of these relationships cannot be determined from the visual information presented in the photographs. These would have to be applied only in the actual housing environment. Others are social relationships which are limited to the experience of neighbor relations.

Much more should be done to indicate the relative values assigned to the individual requirements. The subject, as an individual and as a member of a particular socio-economic group, will bring to the situation many feelings and experiences which determine his perception of the housing situation. This will determine the relevancy of many of the concepts which are considered important in this study. For example, even though economy was a stated requirement for many in the sample group, houses which were preferred were usually judged to be expensive. This does not mean, however, that the housing preferred was actually expensive. No monetary values were available. Expensiveness was based entirely on the visual appearance.

Better operational definitions must be used. The concept of complexity is measureable only in a nominal sense. In order to relate the findings to theory, the concept of complexity should be more precisely defined. Objective measurement is difficult to obtain when the amount and variety of perceptual information is so vast.

Table 1: *Frequency distribution of response to like-dislike categories*

	Like very much 1	Like a little 2	Neutral 3	Dislike a little 4	Dislike very much 5
A	-	3	9	9	4
B	-	2	1	4	18
C	-	2	1	8	14
D	1	8	7	7	2
E	3	8	7	4	3
F	14	8	2	-	1
G	-	-	1	6	18
H	-	-	4	10	11
I	1	7	14	2	1
J	4	6	8	3	4
K	-	4	6	8	7
L	-	-	-	4	21
M	5	8	9	2	1
N	-	3	7	8	7
O	1	6	7	8	3
P	-	3	4	15	3
R	4	8	10	3	-

Table 2: *Housing environments ranked according to mean preference scores*

1	F	1.64	
2	M	2.44	
3	R	2.48	
4	Q	2.76	
5	E	2.84	
6	J	2.88	
7	D		3.04
8	I		3.20
9	O		3.24
10	K		3.32
11	A		3.56
12	P		3.72
13	N		3.76
14	H		4.28
15	C		4.36
16	B		4.52
17	G		4.68
18	L		4.84

Table 3: *Housing environments ranked according to mean scores with corresponding evaluations of private-public, simple-complex, expensive-economical*

	private / public		simple / complex		expensive / economical	
f	17	8	6	19	24	1
m	18	7	-	25	23	2
r	16	9	6	19	21	4
q	7	18	15	10	21	4
e	21	4	18	7	21	4
j	11	14	21	4	3	22
d	4	21	16	9	15	10
i	10	15	16	9	11	14
o	12	13	15	10	21	4
k	6	19	19	6	4	21
a	5	20	20	5	9	16
p	7	18	23	2	6	19
n	2	23	15	10	8	17
h	3	22	23	2	2	23
c	4	21	7	18	6	19
b	5	20	21	4	1	24
g	-	25	24	1	-	25
l	-	25	22	3	-	25

Table 4: *Categorical groups according to preference correlated with number of housing environments evaluated as private-public, simple-complex, expensive-economical*

	private/public		simple/complex		expensive/economical	
1.0-1.5 (like very much)	0	0	0	0	0	0
1.5-2.5 (like a little)	3	0	0	3	3	0
2.5-3.5 (neutral)	1	6	7	0	4	3
3.5-4.5 (dislike a little)	0	5	4	1	0	5
4.5-5.0 (dislike very much)	0	3	3	0	0	3
totals	4	14	14	4	7	11

Table 5: *Impression of good housing areas: frequency of response*
(As a reference for the results of the preference testing, an impression of "good housing" was asked of the sample group. The response was in the form of a set of semantic differential scales.)

	-3	-2	-1	0	+1	+2	+3	
urban	1	4	6	(8)	3	1	2	rural
simple	-	3	4	(6)	(6)	4	2	comp
public	-	1	-	3	5	(9)	7	privat
quiet	8	(11)	3	1	1	1	-	noisy
expensive	-	5	3	(8)	3	4	2	econc
spacious	2	(6)	5	(7)	4	-	1	comp

( ) indicates modal frequency

Table 6: *Attributes*
(One question was asked to indicate the features or requirements people desired in a dwelling. In response to the question: What are some qualities you would look for in a house if you were seeking a place to live?)

Dwelling attributes
    Spatial conditions
        spaciousness   16%
        variety   12
    Design requirements
        efficiency   16
        flexibility   8
        privacy   40
        environmental control   8
    Costs
        economical   16
        quality   12

Neighborhood attributes
    Site conditions
        wooded site   28
        quiet location   12
        open view   16
    Location
        convenient   40
    Neighbors
        friendly   24

Appendix A

82

83

Appendix B

This questionnaire has been designed to evaluate and compare your response to various housing situations.

1. Would you say your background is mainly: rural ( ), suburban ( ), or urban ( )?
2. In which types of house has your family lived: single family ( ), two-family duplex ( ), multi-family apartments ( ).
3. By using the following set of scales describe your impression of what constitutes *good housing areas*.

    urban____:____:____:____:____:____:____rural
    simple____:____:____:____:____:____:____complex
    public____:____:____:____:____:____:____private
    quiet____:____:____:____:____:____:____noisy
    expensive____:____:____:____:____:____:____economical
    spacious____:____:____:____:____:____:____compact

4. What are some qualities you would look for in a house if you were seeking a place to live?
    _____
    _____

5. Turn to the second page and complete the questionnaire using the group of photographs.
6. Sort the cards into five piles according to how much you would like living in them. After you have grouped them, list the letter of each photograph in the order of your preference in the correct blocks below.

like very much	like a little	neutral	dislike a little	dislike very much

7. This time sort the cards into two piles according to those which you feel are *expensive* and those which are economical. Record the letters in the proper blocks.

expensive	
economical	

8. Resort the cards into two piles: those which you feel are *simple* and those which you feel are *complex*.

simple	
complex	

9. Finally, sort the cards into two groups: those which you feel offer *privacy* to the resident and those which you feel would be quite *public*.

private	
public	

References

1. Peter H. Rossi, *Why Families Move* (Glencoe, Illinois, 1955) p. 82.

2. Catherine B. Wurster, " The Dreary Deadlock of Public Housing," *Architectural Forum,* 106, No. 5 (May, 1957), pp. 140.

3. Clare C. Cooper, "Some Social Implications of House and Site Plan Design at Easter Hill Village: A Case Study," Institute of Urban and Regional Development (Berkeley, 1965), p. 278.

4. See bibliography for sources cited.

5. Donald W. Fiske and Salvatore R. Maddi, *Functions of Varied Experience* (Homewood, Illinois, 1961), p. 31.

6. A practical measurement for the degree of visual complexity is difficult to determine. For the purpose of this survey, complexity will be relative to large scale physical form changes, tonal and textural variety and the absence of repetition.

Robert G. Hershberger
Institute for Environmental Studies
University of Pennsylvania

A STUDY OF MEANING AND ARCHITECTURE

There are few forms in architecture to which men do not attach some meaning either by way of convention, use, purpose, or value. This includes the very mundane realization that a wood panel approximately three feet wide by seven feet high is a *door* (object which one opens to pass through), the more subtle feelings of warmth and protection at the entrance of some buildings and some of the most profound experiences of beauty and art. Indeed, the transmission of meaning through the architectural medium is essential to both the use and enjoyment of architecture. Meaning is of considerable importance in perception,[1] "one of the most important determinants of human behavior,"[2] and unquestionably involved with human feelings. Furthermore, it has been argued by architects and planners alike that in an increasing number of situations, the underload, overload, or confusion of meaning in architecture are seriously jeopardized.[3,4] In consequence, it would seem appropriate at this time to undertake serious studies of the nature of architectural meaning to learn what is needed to create physical environments which can be satisfactorily perceived, felt, and used.

The research reported here was addressed to this problem, taking the point of view that the forms, colors, spaces, etc. of architecture are media through which architects communicate to the users of their buildings, and focusing on the "fidelity" of this communication. Is there a close correspondence between the meanings architects intend for buildings and the meanings which laymen attribute to them? Do architects and laymen share similar representations when they experience architecture? Are they affected in the same way by their representations? Are their resultant evaluations and behavior similar?

It was decided that the most straightforward and effective way to approach this problem was to conduct an experiment in which architects and laymen would be directly compared in their attribution of meaning to buildings. The primary objectives of the experiment were: (1) to determine if the physical attributes of buildings can be considered to constitute a "code" capable of communicating an architect's "intentions" to the users of his buildings; and (2) to determine if the areas of disagreement, if any, should be attributed to the professional education of the architects. These objectives were translated into three experimental hypotheses: (1) architects and laymen will not differ greatly in the underlying dimensions of meaning used to judge architecture; (2) significant differences is specific judgements of meaning will occur most often on the affective and evaluative dimensions and least often on the representational dimensions;[6] and (3) the differences which are found (dimensional and specific) will be attributable to the professional education of the architects.

Method and Procedure

The experimental design required each of four respondent groups to rate the connotative meanings of twenty-five building aspects (represented by colored slides) on thirty semantic scales.

Respondent Groups

The experiment utilized three groups of twenty-six students each from the University of Pennsylvania as respondents: (1) the graduating thesis students in architecture, (2) a group of pre-architects, and (3) a

simple	_	:	_	:	_	:	_	:	_	:	_	:	_	complex
beautiful								ugly
passive								active
strong								weak
rational								intuitive
unique								common
ambiguous								clear
exciting								calming
plain								ornate
interesting								boring
generalized								specialized
confined								spacious
delicate								rugged
good								bad
accidental								controlled
open								closed
gloomy								cheerful
superficial								profound
permanent								temporary
welcoming								forbidding
chaotic								ordered
uncomfortable								comfortable
bold								timid
continuous								broken
revolutionary								reactionary
delightful								dreadful
considered								arbitrary
tight								loose
annoying								pleasing
straightforward								contradictory

Figure 1. **Semantic Scale**

sample of non-architects. It also utilized a group of twenty-one architectural students from Drexel Institute of Technology. The choice of three student groups from the same institution allowed the testing of the hypothesis that differences which might be found between architects and non-architects could be attributed largely to the professional education of the architects. The pre-architect group, thus, served as a control group whose ratings, if similar to the non-architects, would tend to confirm the hypothesis, or, if similar the the architects, would tend to descredit the hypothesis. The architectural students from Drexel Institute of Technology, on the other hand, were included in an attempt to determine if differences between the Penn Architects and the other two groups should be attributed specifically to the education of architects at the University of Pennsylvania or, rather, if they could be attributed to more general characteristics of the architectural education.

Building Aspects

A wide range of building types, sizes, styles, aspects and qualities would be represented. Members of the respondent groups would have seen or used a majority of the buildings prior to the experiment (to insure that the dimensionality of meaning uncovered would relate to the everyday use and experience of architecture). The majority of the twenty-five building aspects selected for the experiment were from the campus of the University of Pennsylvania.

The Measuring Instrument

The "semantic differential," a general measuring technique developed by Osgood, Suci, and Tannenbaum to measure connotative meaning was selected to obtain judgements of meaning from the various respondent groups on the architectural material.[7] Briefly, the "semantic differential" utilizes a number of scales consisting of polar adjectives such as "good-bad," "strong-weak" and "active-passive" to differentiate the meaning of concepts (in this case aspects of buildings). The scales are divided into seven steps as follows:

strong____:____:____:____:____:____:____weak

Each subject is asked to consider the concept (building aspect) and place a check in the blank in which he feels the meaning of the concept (building aspect) lies. From left to right on the above example a check in the blank would indicate: extremely strong, quite strong, slightly strong, neither strong nor weak (or not applicable), slightly weak, quite weak, extremely weak. Each concept is judged on several such scales, the scales being varied to suit the subject material.

Semantic Scales

For this experiment the attempt was to generate a "logically comprehensive" list of adjective pairs applicable to architecture since an incomplete list will necessarily yield an incomplete list of factors.[8] To accomplish this, reference was made to the extensive lists of adjective pairs generated for previous studies utilizing the semantic differential,[9] an assortment of architectural books and periodicals, Roget's Thesaurus, an adjective check-list developed by Kenneth Craik,[10] and adjective lists developed in studies by David Canter[11] and John B. Collins.[12] The extensive list of adjective pairs generated were then reduced to the total of thirty used in the experiment by grouping them into ten major categories of meaning and selecting the three most diverse scales from each of the categories. In this way enough scales were provided from each major category to create a separate dimension in the factor analytic study *if* the meaning of the three adjective pairs selected from that category were, in fact, used by the respondent groups to have nearly the same meaning for architectural subject material. At the same time, the selection of the adjective pairs with the most diverse meanings from each category tended to insure enough dissimilarity of meaning to prevent the automatic occurrence of all of the ten categories as dimensions of meaning.

The Experiment

The judgements of meaning over all aspects were obtained from the three respondent groups from the University of Pennsylvania during a one and one-half hour session in a large lecture room. The judgements of the Drexel Architects were obtained in a similar manner during two scheduled design studios.[13]

Phase I: Dimensionality of Judgements

The first concern in comparing the four respondent groups was with respect to similarities or differences in the dimensions used to rate the architectural subject material. The basic procedure here was to subject the data obtained from each respondent group to a separate factor analysis, and then to compare the four sets of factors obtained.[14] A factor analysis, in effect, "groups" scales which are consistently used in the same way by members of a respondent group into "factors" or "dimensions." For example, if building aspects rated "complex" were consistently rated "interesting," these two scales would be highly correlated. If several such scales were highly correlated, factor analysis would group them around a common factor or dimension.

The comparison of the four respondent groups at this level was of fundamental importance because differences in the dimensions of meaning used by architects and laymen would involve total disagreement; the lack of a "common frame of reference" with respect to architectural subject material. In this event, further comparisons between the groups would become spurious since the groups, in effect would not share a common language.

Comparison of Factors

Three common factors, together accounting for approximately 50% of the total variance, were obtained for each respondent group. Factor loadings over .40, by factor and respondent group over all aspects, are shown in Table A. The first factor for each respondent group was clearly a Space-Evaluation factor, displaying high and generally restricted loadings on such space scales as spacious-confined, open-closed, and loose-tight, and such evaluative scales as cheerful-gloomy, pleasing-annoying, and welcoming-forbidding. The second factor was considered to be an Organization dimension for the three groups from the University of Pennsylvania, because it had high and restricted loadings on such scales as ordered-chaotic, clear-ambiguous, and rational-intuitive. The same group of scales also loaded highly on the second factor for the Drexel Architects along with such scales as beautiful-ugly, comfortable-uncomfortable, and good-bad; indicating that the Drexel students consistently attributed beauty, comfort, and goodness to those buildings which they considered to be rational and clearly organized.

The third factor for all four respondent groups was composed of scales indicating "potency": strong-weak, bold-timid, permanent-temporary. However, beyond this common characteristic, there was considerable difference between the groups. The Non-Architects showed no other type of scale loadings, making this a "pure" Potency dimension. The Pre-Architects, however had moderately high, but not restricted, loadings of such scales as unique-common, interesting-boring, and exciting-calming. For the two groups of architects the scales indicating novelty and excitement loaded highly and exclusively on this factor. In addition, there were moderately high, although secondary loadings of such scales as good-bad, pleasing-annoying, and delightful-dreadful. It appeared, therefore, for these two groups, that this factor could correctly be labeled a Potency—Aesthetic dimension. The architects were "pleasurably moved" by buildings which they considered to be potent and unique.

The results of the comparisons of factor structure, thus, seemed to indicate that the spaciousness, organization, and potency parts of the factors were stable and orthogonal for all four respondent groups.

Table A

Varimax Factor Loadings Over .40 by Factors and Respondent Groups Over All Aspects

PENN ARCHITECTS		PRE-ARCHITECTS		NON-ARCHITECTS		DREXEL ARCHITECTS	
I *Space-Eval*		*Space-Eval*		*Space-Eval*		*Space-Eval*	
cheerful	.80	cheerful	.82	cheerful	.85	open	.73
welcoming	.69	delightful	.82	pleasing	.82	cheerful	.71
spacious	.69	pleasing	.79	delightful	.81	spacious	.68
delightful	.68	good	.77	good	.80	welcoming	.57
open	.67	welcoming	.76	interesting	.78	delightful	.56
good	.64	beautiful	.75	beautiful	.75	loose	.55
beautiful	.64	open	.69	welcoming	.73	pleasing	.48*
pleasing	.64	spacious	.68	comfortable	.67	comfortable	.47*
comfortable	.64	interesting	.67	open	.66	beautiful	.41*
interesting	.53*	comfortable	.63	spacious	.66		
revolutionary	.44	loose	.58	exciting	.54		
profound	.43*	revolutionary	.56	unique (c)	.54		
loose	.40	exciting	.52	loose	.54		
		unique	.50	revolutionary	.51		
		active	.42*	active	.49		
II *Organization*		*Organization*		*Organization*		*Organ-Eval*	
ordered	.76	ordered	.85	ordered	.72	ordered	.74
straightfwd	.65	clear	.69	simple	.69	straightfwd	.64
clear	.62	simple	.69	clear	.65	simple	.63
rational	.57	straightfwd	.68	straightfwd	.64	considered	.61
controlled	.56	controlled	.67	rational	.53	beautiful	.60
considered	.48	rational	.65	controlled	.51	clear	.55
simple (c)	.43*	considered	.58	continuous	.50	rational	.52
		continuous	.57	common (u)	.47*	comfortable	.51
		plain	.55	considered	.43	continuous	.51
				plain	.41	controlled	.50
						good	.49*
III *Poten-Aesth*		*Poten-Excite*		*Potency*		*Poten-Aesth*	
bold	.74	strong	.69	strong	.63	interesting	.73
strong	.70	bold	.67	permanent	.61	unique	.72
interesting	.68	permanent	.57	profound	.59	bold	.69
unique	.66	profound	.54	rugged	.48	strong	.67
activity	.64	rugged	.49	bold	.41	good	.62
exciting	.62	active	.48	controlled	.40*	exciting	.62
profound	.61	unique	.48*			pleasing	.56
rugged	.58	interesting	.48*			active	.55
specific	.58	exciting	.46*			profound	.52
complex (s)	.57					specialized	.51
good	.56*					delightful	.48*
permanent	.54						
pleasing	.51*						
ornate	.48						
delightful	.42*						

*These scales load more highly on one of the other factors.

( ) Parentheses indicate that the loadings on the opposite pole of the scale are indicated under another factor.

The affective and evaluational portions of the factors, on the other hand, appeared unstable and non-orthogonal for all four respondent groups. The affective and evaluational portions of the factors, on the other hand, appeared unstable and non-orthogonal, loading on one factor for one respondent group and on another factor for another respondent group, and in some cases loading on two or even all of the three factors at one time. Indeed, it appeared that the affective and evaluative aspects of connotative meaning in architecture were not "independent" dimensions of judgement at all, but "dependent" on one or more of the three stable and orthogonal subfactors: Spaciousness, Organization, and Potency.[15]

Relative Factor Salience

It has already been seen from inspection of the factor matrices that some differences existed in the relative dominance of the three factors obtained for each of the four respondent groups. Comparisons at this level are rather important because strong emphasis on different dimensions by architects and laymen could cause disruption in communication fidelity. That is, if architects place primary importance on the potency attributes of a building while laymen place primary importance on the spatial attributes, then even if both attributes were seen to be approximately the same by the two groups, there would be a dissimilar overall impression of the building. Accordingly, the groups were compared in this respect using the percent of common variance accounted for by each factor as the index of factor salience.

Table B shows the results of comparisons of the common variance accounted for by the three factors. It is apparent that some differences existed here between the four respondent groups. The Potency-(Aesthetic) dimension was of primary importance for the two groups of architects, but was least dominant for the Non-Architects. Conversely, the Space-Evaluation dimension was dominant for the Non-Architects and Pre-Architects, but of only secondary importance for the Penn Architects, and of least importance for the Drexel Architects.

Having already observed the shifting of the affective and evaluational scales from one to another of the more stable and independent, perceptually related portions of the factors, it seems probable that it was these shifts which accounted for the changing dominance of the overall factors. Nevertheless, it is significant that the two groups of architects were more concerned with the aesthetic nature of the buildings viewed: their potency, interest, excitement, etc; while the Non-Architects and Pre-Architects were more concerned with the pleasantness of the buildings: their spaciousness, comfort, cheerfulness, etc.

Homogeneity of Judgements

Another important comparison between the four respondent groups concerned their internal agreement, or homogeneity of judgements. Ordinarily it has been found that members of an organization of profession are more homogeneous in their outlook than the population in general. Likewise, it has been found that homogeneity of judgement increases during the college years for gorups following the same course of study.[16] Therefore, it would be expected that the two architect groups would display the most homogeneity of judgements, that the Pre-Architects would be next because of their similarity of interests, and the Non-Architects would be last because of their diversity of interests and education. Table C presents the mean standard deviations for each respondent group over all building aspects as indices of their homogeneity of judgement. The table also presents the results of comparisons between the respondent groups utilizing the Mann-Whitney U-test. Scores for this comparison were the mean standard deviations computed over all thirty scales for each building aspect, with the twenty-five building aspects as replications.

The results tended to confirm expectations, showing the Drexel Architects and Penn Architects to be significantly more homogeneous in their judgements than the Pre-Architects and Non-Architects.

Similarly, the Pre-Architects were significantly more homogeneous in their judgements than the Non-Architects. The exceptionally high significant differences between the Architects and Non-Architects can perhaps be accounted for by a combination of the professional education of the architects and exceptional diversity among the Non-Architects, whose random selection provided twenty-three major study areas out of twenty-six respondents.

Meaningfulness of Judgements

Another characteristic on which the four respondent groups were compared was their extremity or meaningfulness of judgements. Since the mid-point of semantic differential scales is taken as an index of meaninglessness, departure from the mid-point may be referred to as meaningfulness. The appropriate measure here is the generalized distance function, D, computed as the multidimensional distance of each concept from the origin.[17] In this study, mean scale values for each respondent group were used in computing the D (or distance) between the groups. The four groups were compared in pairs utilizing the Mann-Whitney U-test with the twenty-five building aspects serving as replications. It was believed, although not hypothesized, that the Penn Architects and Drexel Architects would find the architectural material to be most meaningful; that they would be followed by the Pre-Architects; and that the Non-Architects would find the architectural material to be least meaningful.

Table D presents the mean D (or distance) from the origin of the "semantic space" for each respondent group taken over all twenty-five building aspects, and the results of the comparisons utilizing the Mann-Whitney U-test. As was expected, the Penn Architects found the building aspects to be most meaningful. They were followed by the Pre-Architects and Non-Architects from the University of Pennsylvania. Surprisingly, the Drexel Architects found the building aspects presented to be even less meaningful than did the Non-Architects. In fact, the only difference approaching significance was that between the Penn Architects and the Drexel Architects. It was significant at the .02 level when considered alone; however, when Ryan's correction for experimentwise error rate was applied even this comparison fell below the .05 level of significance.[18]

The lack of a significant difference in the meaningfulness of the architectural material for the architects and laymen was quite surprising, since it was believed that the architects would consider architectural material to be much more meaningful than would the non-architects. The result, on the other hand, can be considered encouraging since the differences were probably not great enough to create a serious problem of communication between any of the respondent groups. The even more surprising difference between the Penn Architects and the Drexel Architects can, perhaps, be accounted for by the different experience with architecture of the two groups. Most of the Drexel Architects had been employed full-time in architects' offices. In this capacity they were more likely to have considered buildings primarily as objects to be designed, detailed, and constructed, rather than as a potential method of communication with the users of the buildings. The Penn Architects and Pre-Architects, on the other hand, owing to their educational and social backgrounds as well as to their lack of practical office experience, might have tended to be more concerned about how buildings are perceived and experienced by their human occupants, that is, with their communication possibilities.

Phase II: Judgements On Specific Building Aspects

The objectives of this phase of the analysis were essentially the same as for the first phase: (1) to determine if there is a sufficient basis to assume that buildings can be used effectively by architects to communicate their "intentions" to laymen, and (2) to determine if the areas of difference can be attributed to the professional education of the architect. In this regard, the difference between this phase

Table B

Percentage of Common Variance Accounted for by Each Factor
For Each Respondent Group, Over All Aspect Groups

	Respondent Groups	I	Factors II	III
1.	Penn Arch	38.3%	22.2%	(39.4%)
2.	Pre-Arch	(47.6)	30.9	21.4
3.	Non-Arch	(53.8)	29.1	17.1
4.	Drexel Arch	28.6	34.2	(37.2)

NOTE: The dominant factor for each group is indicated by parentheses. The factors were labled as follows:

Factor I   "Space-Evaluation" for all four groups
Factor II "Organization-Evaluation" for the Drexel Architects, "Organization" for the other three groups
Factor III "Potency" for the Non-Architects, "Potency-Excitement" for the Pre-Architects,
           "Potency-Aesthetic" for the Penn and Drexel Architects

Table C

Homogeneity of Judgements

Respondent Group	Mean Std. Dev.	Penn Arch	Significance of Difference Pre-Arch	Non-Arch	Drex Arch
Penn Architect	1.333	----------------	.001	.000001	.27
Pre-Architect	1.415		-------------	.001	.000001
Non-Architect	1.509			---------------	.000001
Drex Architect	1.303				----------------

Table D

Meaningfulness of Judgements

Respondent Group	Mean "D" Score	Penn Arch	Significance of Difference Pre-Arch	Non-Arch	Drex Arch
Penn Architect	6.523	----------------	.48	.21	.08
Pre-Architect	5.892		-------------	.98	.22
Non-Architect	5.744			---------------	.21
Drex Architect	5.296				----------------

and the first phase was not in objectives or hypotheses, but in the level and methods of comparison. It was an important phase of the analysis, however, because respondent groups which use the same dimensions of meaning might, nevertheless, be quite different in their judgements of meaning for specific buildings. The converse could also be true; groups which use different dimensions of meaning might be quite similar in their judgements of meaning for specific buildings. In any event, it was only by considering these two levels of comparison in relation to each other that a comprehensive picture could be obtained of the similarities and differences in the judgements of meaning of the four respondent groups.

The basic procedure here was to use one-way analysis of variance and Duncan multiple range tests to compare the factor scores of the four respondent groups on each building aspect. A separate analysis was conducted for each of the five common factors extracted in the oblique factor analysis: Pleasantness, Organization, Potency, Novelty-Excitement, and Spaciousness.[19] The results of these analyses over all factors, and for each factor, are shown on Table E. Looking first at the summary over all buildings and all factors at the top of the table, it can be seen that the greatest number of significantly different factor scores (at the .05 level) were between the Penn Architects and the Non-Architects with fifty-three significant differences out of 125 comparisons, and with twenty-one of the significant differences in opposite directions. The smallest number of significantly different factor scores, on the other hand, were between the Penn Architects and Drexel Architects, with twenty-one significant differences out of 125 comparisons, and with only three in opposite directions. It should be emphasized that both the number and direction of significant differences are extremely useful in revealing the nature of the differences between the specific judgements of meaning of the respondent groups. Considering, for example, the Penn Architects and the Drexel Architects it can be seen that in only three instances on the five salient dimensions of meaning were the judgements of the two groups on opposite sides of the neutral point of the "semantic differential." They disagreed once as to whether a building was basically organized or unorganized, once as to whether a building was spacious or unspacious. Most of their comparatively small number of differences were, therefore, only in magnitude, not in direction. Indeed, it is likely that most of the differences were due to the general tendency of the Penn Architects to consider architectural material to be more meaningful than did the Drexel Architects.

When the Penn Architects and the Non-Architects were compared with respect to the sign of differences, a rather different picture emerged. In this case almost half (21) of the sizeable number of overall differences (53) were in opposite directions. This pointed to a rather fundamental difference between the two groups in their attribution of meaning to specific building aspects. The differences between the Drexel Architects and the Non-Architects were similar in this respect, although the number of significant differences were not as pronounced, with a total of thirty-five significant differences and with fourteen in opposite directions. The Penn Architects and Pre-Architects, while second in the overall number of significant differences with a total of thirty-eight, had a somewhat lower proportion of differences in sign with twelve. Most of the differences in this case were only differences in intensity of meaning and, therefore, did not indicate a communication difficulty of the magnitude of that between the Penn Architects and the Non-Architects.

Comparisons Between Factors

An advantage of the comparisons utilizing analysis of variance and Duncan multiple range tests was the ability to determine on which of the dimensions of meaning the greatest differences between the architects and laymen would occur. In this regard, it was hypothesized that very few differences, particularly in the direction of judgement, would be found for what were considered to be the perceptually related dimensions of meaning: Organization, Potency, and Spaciousness. Conversely, it was hypothesized that there would be a substantial number of differences in both magnitude and direction on the affective and evaluative dimensions of meaning: Novelty-Excitement and Pleasantness. As usual, it was hypothesized that the professional education of the architects would account for the differences.

Table E

Significant Differences Between Respondent Groups
Over All Buildings and Factors
(Duncan Multiple Range Tests: .05 Level of Significance)

Over All Factors   Total: 193 (67)

	Penn Arch	Pre-Arch	Non-Arch	Drexel Arch
Penn Arch	---------	38 (12)	53 (21)	21 (3)
Pre-Arch		--------	30 (9)	23 (8)
Non-Arch			--------	35 (14)
Drexel Arch				-----------

Pleasantness   Total: 54 (28)

	Penn Arch	Pre-Arch	Non-Arch	Drexel Arch
Penn Arch	---------	14 (6)	14 (7)	6
Pre-Arch		--------	9 (4)	5 (5)
Non-Arch			--------	11 (6)
Drexel Arch				-----------

Organization   Total: 40 (12)

	Penn Arch	Pre-Arch	Non-Arch	Drexel Arch
Penn Arch	---------	11 (5)	8 (2)	4 (1)
Pre-Arch		--------	2 (1)	8 (1)
Non-Arch			--------	7 (2)
Drexel Arch				-----------

Potency   Total: 33 (3)

	Penn Arch	Pre-Arch	Non-Arch	Drexel Arch
Penn Arch	---------	5	11 (2)	3
Pre-Arch		--------	6	2
Non-Arch			--------	6 (1)
Drexel Arch				-----------

Novelty-Excitement   Total: 40 (14)

	Penn Arch	Pre-Arch	Non-Arch	Drexel Arch
Penn Arch	---------	5	11 (5)	5 (1)
Pre-Arch		--------	8 (2)	4 (2)
Non-Arch			--------	9 (4)
Drexel Arch				-----------

Spaciousness   Total: 26 (10)

	Penn Arch	Pre-Arch	Non-Arch	Drexel Arch
Penn Arch	---------	3 (1)	9 (5)	3 (1)
Pre-Arch		--------	5 (2)	4
Non-Arch			--------	2 (1)
Drexel Arch				-----------

NOTE: The first value in the matrix indicates the number of significant differences. The second value, in parentheses, indicates differences in opposite directions.

By inspecting the remainder of the matrices on Table E, it can be seen that the expectations were generally confirmed. The greatest number of differences between the architects and laymen were on the Pleasantness dimension. The second greatest number were on the Novelty-Excitement dimension. Differences on the Organization, Potency, and Spaciousness dimensions followed in that order.

On the Pleasantness dimension, there were only six significant differences between the Penn and Drexel Architects and none in opposite directions. In comparing the Penn Architect with the Non-Architect, however, the results were quite different; there were fourteen significant differences out of a possible twenty-five (over half) and seven (nearly one-third) were in opposite directions. Furthermore, an almost identical relationship held with the Pre-Architects with fourteen significant differences and six in opposite directions. Similar, but less pronounced results can be seen comparing the Drexel Architects with the Pre-Architects and Non-Architects

On the Novelty-Excitement dimension, the differences were similar, with only five significant differences between the Penn and Drexel Architects and one in opposite directions, but eleven significant differences and five in opposite directions between the Penn Architects and Non-Architects. On this dimension, the judgements of the Pre-Architects were more like those of the Architects than those of the Non-Architects.

The most surprising finding was that the differences between the Architects, Pre-Architects, and Non-Architects on the Organization dimension were nearly as great as those on the Novelty-Excitement dimension in both magnitude and direction. The differences between the Penn Architects and the Pre-Architects were particularly puzzling. The greatest differences in both magnitude and direction on the Organization dimension were between the Penn Architects and the Pre-Architects at eleven and five rather than between the Penn Architects and the Penn Architects and the Non-Architects at eight and two. This can, perhaps, be accounted for by the fact that several of the scales making up the Organization dimension really did not relate closely to the perceptual qualities of the building, but rather to inferences regarding the processes used in planning the building: rational-intuitive, considered-arbitrary, controlled-accidental. The Pre-Architects being visually oriented might have made their inferences solely on the basis of the perceived surface organization, while the architects, knowing more about design, could have made their judgements on functional criteria. In any event, it appeared that the professional education of the architect actually caused his judgements of meaning relative to organization to move from one side of the Non-Architect to the other.

The judgements on the Potency and Spaciousness dimensions were as expected with a relatively small number of differences in magnitude and direction between the groups. Even where there were a relatively large number of differences, as between the Penn Architects and Non-Architects on the Potency dimension at eleven and two, they were primarily with regard to magnitude and not direction.

Conclusions

Architectural Communication

Since the same three independent (orthogonal) and two dependent (non-orthogoanl) dimensions of connotative meaning were extracted for all four respondent groups, it is felt that there is no fundamental reason why the physical attributes of buildings should not be considered to constitute a "code" capable of use to communicate an architect's "intentions" to the users of his buildings (laymen and other architects). This would seem to be especially true, even now, for the Potency and Spaciousness dimensions of meaning. On the Organization dimension, however, the relation to non-perceptual aspects of architecture (and the Drexel Architects' exclusive tendency to attribute goodness and beauty to buildings considered well organized) creates some doubt if the architect's intended meaning would coincide with the user's attributed meaning—at least in any uniform way.

The gravest doubts concerning communication fidelity, however, related to the affective (Novelty-Excitement) and evaluative (Pleasantness) dimensions of meaning. Here there were rather substantial differences in the scalar composition of the dimensions, in their loadings with the three independent dimensions, and in the judgements of meaning on specific building aspects. First, the Novelty-Excitement dimension of the Non-Architects was generally composed of a minimum number of scales and exhibited low factor salience. The equivalent dimension of the Penn Architects and Drexel Architects included a greater number of scales—several of an evaluative nature—to form a strong Potency-Aesthetic dimension. Since the Non-Architects appeared not to consistently value (find good and pleasing) architectural objects which they found interesting, exciting, unique, etc., there is a strong possibility that forms intended by the architects to be both exciting and pleasant would not be interpreted in this way by laymen. Similarly, since the Aesthetic sub-factor of the architects loaded with the Potency dimension, while the equivalent sub-factors of the Non Architecs loaded with the Spaciousness dimension, it would seem very likely that here, too, some communication difficulties would arise. The most serious difficulties on these dimensions, however, appeared to relate to judgements on specific building aspects.. On both the Pleasantness and Novelty-Excitement dimensions approximately half of the judgements by the Penn Architects were significantly different from those of the Non-Architects, and half of these differences were in opposite directions. The combination of significant and non-significant differences in opposite directions were, of course, even more numerous to the point that it could be expected that approximately thirty percent of the time when the Penn Architects would judge a building to be good, pleasing, beautiful, interesting, exciting, and unique; the Non-Architects would judge it to be bad, annoying, ugly, boring, calming, and common. Such a large number of differences between the two groups would, of course, seriously affect the success of the architect in communicating his intentions to laymen.

The Architect's Professional Education

The professional education of the architects in this study appeared not to have a great effect on the basic dimensions of meaning used to describe or judge architecture. The Penn Architects, Drexel Architects, Pre-Architects, and Non-Architects all employed the same basic dimensions or factors.

The professional education did, however, appear to have minor effects on the scales loading on some of the dimensions, how the dependent dimensions loaded with the independent dimensions, and the salience of the dimensions. With regard to factor salience the effect was quite substantial, the professional education of the architects causing them to attach greater affective and evaluative importance to building aspects perceived to be potent. The Pre-Architects and Non-Architects, on the other hand, attached their judgements in the affective and evaluative areas primarily to the perceived spaciousness of the building.

The meaningfulness of architectural material appeared not to be related at all to the architects professional education. Rather it appeared to relate either to personality type or previous experience with the buildings judged. The homogeneity of judgements, on the other hand, appeared to be influenced quite markedly by the professional education of the architects, the differences between the architects and the other tow groups being highly significant. Again, it was on judgements of the meaning of specific building aspects that the professional education of the architects appeared to have the greatest influence. While the two groups of architects were almost identical in the direction of their judgements on all dimensions, they were quite different from both the Pre-Architects and Non-Architects in their judgements on the affective, evaluative, and organization dimensions.

Implications

Assuming that the results of the experiment could be generalized beyond the immediate respondent groups, the following implications would be apparent. If architects hope to utilize their media (architecture) to communicate intentions to laymen, they must either (1) re-orient the architectural education such that it does not change the architects' way of experiencing architecture from that which he had as a pre-architect, (2) re-orient the architectural education such that the architect is taught how forms, spaces, and the like are interpreted by laymen, as well as by architects, so that he can consciously manipulate them in such a way as to successfully communicate with both groups, or (3) make greater efforts to educate the layman (general public) to see and appreciate architecture in the same way as architects. It is felt that the first alternative is neither desirable nor possible without abandoning the architectural education almost in its entirety. A combination of the second and third alternatives would seem appropriate along with greater efforts to teach the architect to empathize with what is important to those who use his buildings.

Some will argue, of course, that most schools of architecture have already come to this conclusion. Nevertheless, the fact remains that there was a wide disparity in the specific judgements by the architects and laymen in the present study, such that if this course of action is now being followed, it is not being done successfully. More important, however, is the fact that some of the specific problems of communication have been isolated in this study. We need not be concerned about bringing the basic dimensions of meaning closer together—they are already similar. We need not be concerned about the perceived spaciousness of a building—there are few disagreements in this regard. We do need to concentrate on what it is that makes laymen fail to appreciate some "potent" buildings which architects appreciate. We whould also try to determine which characteristics of buildings cause them to appear organized to laymen and pre-architects but not to architects and vice versa. Similar efforts should be directed toward the extreme differences between architects and laymen with respect to the affective and evaluative areas of meaning. Indeed, to discover what it is in architecture that "pleasurably moves" some men but not others should be at the very core of our concern as architects. If we hope to communicate our intentions to those who use our buildings, we must know the meanings they will attribute to the forms, spaces, colors, etc. which we choose to employ. The curtains are hardly drawn on the vast amount of research which could be done in this area.

References

*   This paper is based on the experimental portion of research conducted by the author in partial fulfillment of the requirements for a Ph.D. in architecture at the University of Pennsylvania. The dissertation, by the same title, includes detailed descriptions and explanations of all aspects of the experiment as well as an extensive theoretical study into the nature, types, and levels of architectural meaning. The dissertation study was supported in part by grants from the University of Pennsylvania's Computer Center and Institute for Environmental Studies. Copies of the dissertation can be obtained from University Microfilms, 300 N. Zeeb Road, Ann Arbor, Michigan, 48106.

1   M. B. Creelman, *The Experimental Investigation of Meaning: A Review of the Literature* (New York: Springer Publishing Co., 1966), p. 15.

2   Charles E. Osgood, George J. Suci, and Percy H. Tannenbaum, *The Measurement of Meaning* (Urbana: University of Illinois Press, 1967), p. 10.

3   Denise Scott Brown, "The Meaningful City," *Civic Design Miscellany,* University of Pennsylvania, pp. 53-58.

4   Ian McHarg, "The Ecology of the City," *AIA Journal* (Nov., 1962), pp. 101-103.

5   Implicit in research that takes such a focus is the belief that meaning is not *contained* in the elements of architecture, but rather something which is intended for or attributed to them by human beings; that such meanings may or may not be held in common by those who experience architecture; indeed, that fundamental differences in human experience will cause fundamental differences in the meanings people attribute to their environments.

6   In the theoretical portion of the dissertation a model of meaning was formulated which indicated that there are essentially two kinds of meaning related to architecture. The first kind, representational or objective meaning, was condidered to embrace all meaning dealing with objects, events, ideas and the like which are external to the person having the meaning. This meaning dealt with such phenomena as percepts, concepts, and thoughts. The second kind of meaning, responsive or subjective meaning, embraced all meaning relating solely to the people having the meaning. This meaning, thus, dealt with such phenomena as feelings, emotions, attitudes, and evaluations.

7   Osgood *et al., The Measurement of Meaning,* pp. 125-128.

8   See Joyce Ardell Vielhauer, *The Development of a Semantic Scale for the Description of the Physical Environment,* Ann Arbor: University Microfilms, 1965, for an effort to generate an empirically satisfying set of semantic scales.

9   Osgood *et al., The Measurement of Meaning,* pp. 37, 53-69, 67, 69.

10 Kenneth H. Craik, "Environmental Display Adjective Checklist" (unpublished, Institute of Personality Assessment and Research, University of California, Berkeley, Spring, 1966).

11 David Canter, "The Study of Meaning in Architecture" (unpublished paper, Building Performance Research Unit, University of Strathclyde, Glasgow, Scotland, May, 1968).

12 John B. Collins, "Some Verbal Dimensions of Architectural Space Perception" (dissertation in progress in Architectural Psychology, University of Utah, S.L.C., 1966).

13 The students from the University of Pennsylvania were each paid $2.00 for thier efforts from a grant provided by the Institute for Environmental Studies.

14 Factor analysis was by the principal factor method using the Pearson product-moment correlation matrices as input. The diagonal elements, initially the squared multiple correlations, were iterated by refactoring until the maximum change in communality estimates was less than .001. Rotations were based on the oblimin criteria utilizing Kaiser normalization. In the first series of analyses the factors were kept orthogonal to yield Varimax rotations. The minimum eigenvalue for this series of analyses was unity. This value determined both the number of principal components computed as well as the number of rotations. In the second series of analyses the oblimin criteria was applied to the primary factor loadings, allowing the factors to be oblique and yielding simple loading rotations. For this series extraction and rotation of six factors was required. Both series of calculations were performed utilizing the BMD X72 program on the IBM computer at the University of Pennsylvania Computer Center.

15 Subsequent factor analysis utilizing oblique rotational criteria and requiring extraction of six factors confirmed the existence of the evaluative and affective subfactors. It also provided a quantitative comparison of the correlation of the various factors which revealed that the Organization, Potency, and Spaciousness dimensions were essentially uncorrelated (orthogonal), but that the affective (novelty-excitement) and evaluative (pleasantness) dimensions, while strong, generally correlated with one or more of the "independent" dimensions as well as with each other. The factors obtained in the oblique analyses were Organization, Potency, Spaciousness, Pleasantness (evaluative), and Novelty-Excitement (affective).

16 P. H. Tannenbaum and J. M. McLeod, "On the Measurement of Socialization," *The Public Opinion Quarterly,* XXXI (Spring, 1967), pp. 27-37.

17 Osgood et al., *The Measurement of Meaning,* pp. 91-92.

18 T. A. Ryan, "Significance Tests for Multiple Comparisons of Proportions, Variances, and Other Statistics," *Psychological Bulletin,* LVII (1960), pp. 318-328.

19 See footnote No. 15.

George L. Peterson, Robert L. Bishop, and Edward S. Neumann
The Technological Institute
Department of Civil Engineering
Northwestern University

THE QUALITY OF VISUAL RESIDENTIAL ENVIRONMENTS

Introduction

   A major part of the interaction between man and his surroundings is the reception and processing of sensory information. The physical environment should be controlled so as to harmonize with human requirements. Frequently, this means managing our surroundings so that they evoke pleasant and satisfying response at the human scale. However, in the complex urban society, unaided human response to the environment sometimes falls short of maximizing individual or collective welfare. If this is true, information, education, or control services may be desirable. In either case, before human response to the environment can be served effectively, implementable criteria are needed. The proper description of such criteria requires an understanding of the process by which man experiences and responds to the world around him. The complexity of the process will preclude useful understanding unless it can be organized. The organization should occur both conceptually and methodologically. Conceptually, a model is required which organizes the processes and components into a simplified structure which can be understood and utilized. Such a conceptual model is of limited usefulness, however, if it cannot be related to the real world of man and his environment through measurement and quantitative analysis.
   The purpose of this paper is thus to inquire into the way people receive sensory information and respond with preferences. What are the objects, conditions, or circumstances to which preferences are sensitive? What are the differences between people? How can the complexity of these questions be organized conceptually and methodologically so that they become manageable?

The Hypothesized Nature of Preference

   The human being can be thought of cautiously as a discriminating process which receives information about its surroundings and computes satisfaction. If the situation requires a decision among several alternatives, then the computation is of expected satisfaction, and the decision is an expression of preference. Preference, then, is used here to refer to the fact that something is desired and that it is more desirable than other things with which it competes. It is assumed that behavior is an expression of preference and that preference has as its source the satisfactions that are received or expected.
   The reasons for the satisfaction might be moral or aesthetic.[1] This might be based on complete or incomplete information or correct or incorrect expectation. What we are talking about is the observation of operative value,[2] whatever the reasons for why it operates the way it does. The essential elements are (1) objects, conditions or alternatives which are encountered (environment), (2) information about the environment which is received by the human discriminating process (perception or cognizance), (3) interpretation of the information and calculation of satisfaction or desire, and (4) expressions of preference through decision-making or discriminating behavior. Hopefully, there is some systematic operation of the discriminating process which can be described, perhaps stochastically, by simple rules of preference (preference func

tions). It is hypothesized here that in any given context, the environmental alternatives can be described as a vector of measurable components or characteristics to which preferences are sensitive. It is hypothesized further that measurable characteristics of individuals determine or influence strongly the way these environmental attributes stimulate preferences and that a model of preference which includes individual differences is therefore plausible.

It would be convenient if such rules or laws for predicting human response to the environment could be developed with the degree of success enjoyed by the physical sciences. This does not seem likely, either because of fundamental differences in kind between human and physical phenomena or simply because human beings are so profoundly complex. Even if the human being is not more than a physical device for processing and responding to information, the problem is perhaps unmanageable.

If human response to the environment is indeed determined or influenced strongly by identifiable environmental and human characteristics and if there is any systematic communality across individuals in the human characteristics or the processes by which they influence response, then, with a lot of luck and hard work, it may be possible to identify some of the significant characteristics of man and gross environment, and perhaps, even a process or two. Approached empirically through observations of behavior, results are likely to be crude and simple but hopefully of use to those who are faced every day with controlling the environment or aiding or controlling the people who use it.

It is apparent that people have much in common; otherwise, they would not be classified together as human beings. They have a reasonably uniform physical structure handed down through a long history of survival und fairly common conditions, although minor variations are observable. These physical characteristics which peop share in common are partly responsible for the way they experience and interpret their surroundings. Human preferences certainly are part of a system for guiding behavior and behavior is the way man achieves survival. Successful behavior requires successful preferences.

Some aspects of the preference process consist of genetically-controlled physical mechanisms or "hardware to use computer jargon, which is part of man's "standard" physical structure. It is possible that this "hardware may be partially pre-programmed with some very fundamental preferences or preference tendencies or that the hardware itself limits or biases the way preferences can be developed.[3] Apparently physical pain does not have to be physically damaging to cause pain. Conscious awareness of the conditions has to be learned, but the pain does not, and it seems impossible to learn not to feel pain when such conditions are encountered. Are there more subtle genetically-programmed dimsinions of pain and pleasure unalterabley linked with environmental conditions which, when encountered, have tended to damage us physically, mentally or socially? If so, then they are predictable to the extent that it is possible to predict what will happen when someone steps on a tack or burns his finger. It is only necessary to identify the conditions which evoke the pleasure or pain. These may be called protective or defensive responses which the species has learned through genetic trial and error.

Other environmental circumstances amy cause unalterable pleasure or aversion because of quirks in the human system and not as a reward for good behavior or as a warning against damage.

When observed as the scale of human diversity, however, it is obvious that any pre-programming of preferences or preference tendencies is only a minor part of human response tb the environment. Man is a self-organizing system which organized preferences largely in response to his own individual struggle for survival and satisfaction. He develops and refines his preferences through experimentation and selective retention of whatev tends to be re-enforced by external approval, enchanced survival, or pleasurable sensation. This process is goverened to a large degree by unique conditions of environment and culture and by unique sequence of events. This, it seems, is responsible for the great variety of human preferences, and is the source of perhaps unmanagea complexity.

Nevertheless, it seems that this variety, as rich as it is, may be interconnected with some common threads. Through observations of behavior, an empirical study of preferences cannot hope to reduce the human being to a predictable physical device, but it might uncover some of the manageable threads of communality, thus contributing to an improved ability to design satisfying and benevolent environments or to inform, educate, or control individuals. It may not be possible to maximize the satisfaction of everyone, but it should be possible to avoid needless sources of pain, stress, and frustration.

A Conceptual Model of Preference

The purpose of this section is to suggest a strategy for organizing conceptually man's very complicated feelings about his visual environment. The strategy consists of a simple model of the individual preference process and a model of interpersonal differences. The model of the individual preference process defines a system of measurable components. The model of interpersonal differences allows the richness of human variation to be included. These models are a simple pragmatic hypotheses which lend themselves to analysis by available methods.

The Individual Preference Process

Figure 1 shows schematically a suggested model of individual perception and evaluation of the visual environment. The component of primary interest here is the preference function. It is defined as a process which receives information about the environment after a filtering by the perception process. It is assumed that the perception process simplifies environments into a vector of properties or conditions. The preference function receives this simplified description of the environment and evaluates it by referring to "context" which is a bundle of stored information, programmed responses, and physical mechanisms.[4] A more detailed model might subdivide context accordingly into hardware, software, and stored data, and it might also define functional interactions among these components.

For present purposes, the preference function can be thought of as a programmed response which transforms stored data and information about the environment inot preferences. The preferences define what is desired and would thus determine behavior except that they must pass through an implementor. The implementor considers constraints, competing desires, and other factors before generating behavior. In order to avoid assumptions of determinism, the preference function and the perception filter can be assumed to be probabilistic estimators. That is, the outputs are not deterministic functions of the inputs necessarily, but might be generated stochastically.

The Structure of Interpersonal Differences

It is assumed that the preference function, the perception filter, and the context are unique to each individual. However, it is assumed that the uniqueness in these components is determined or influenced strongly by physical, cultural, social, and experiential characteristics of the individual. Let $(S_1, S_2, \ldots, S_h, \ldots)$ be a vector of such characteristics. The index "h" identifies the characteristic, and S measures its magnitude. Figure 2 shows two extremes in the way individuals might be scattered with respect to these characteristics. In the illustration, the S vector is collapsed to two dimensions for simplicity. These are only two of several conceivable cluster patterns.[5]

If it is assumed that the individual characteristics determine (probabilistically) the strength of preference stimulated by visual messages from the environment, then the preference function can be written as:

$$P_{ij} = f(_i x_{kj}, S_{ih}) \qquad (1)$$

where
P_{ij} = the amount of prefernece demonstrated by individual i for environment j,
f = the preference function,
$_i x_{kj}$ = the quantity of condition or factor k perceived by individual i in visual image j, and
S_{ih} = the quantity of characteristic h possessed by individual i.

If it is assumed that all individuals receive identical information about the visual environment through the perception filter, the general preference function can be simplified to:

$$P_{ij} = f(x_{kj}, S_{ih}) \qquad (2)$$

As in Figure 2b, if individuals tend to be grouped naturally with respect to S, the model can be aggregated and written as

$$P_{gj} = f_g(_ix_{kj}) \qquad (3)$$

or $\qquad P_{gj} = f_g(_gx_{kj}) \qquad (4)$

or $\qquad P_{gj} = f_g(x_{kj}) \qquad (5)$

where g is an index which identifies the group, and f_g is the preference function for that group. These aggregate functions are more useful empirically than (1) or (2) because they do not require the invention of some universal preference function which operates on both x and S. Equations (3), (4), and (5) aggregate individuals and disaggregate the functions.

A simple form of the model has been tested empirically, as will be explained in a later section. It is the following linear function:

$$P_{gj} = \ _ka_{gk}x_{kj} \qquad (6)$$

where a_{gk} is the value of each unit of perceived factor k to group g.

The perception filter should be studied because it seems unreasonable to assume that all individuals receive the same messages from the environment. This is implied crudely in Figure 1 by the feedback from the context and the preference function to the perception filter. However, the methodology which will be discussed later allows perceptions to be measured directly, so it becomes an empirical question for purposes of th study. It will be left to psychologists to unravel the phenomenology of perception.

The Structure of the Preference Function

The simple linear model expressed by equation (6) is only one form that the model might take. It is convenient to work with empirically because of the abundance of linear statistical methods. It leaves many questions unanswered, however. Such forms of the function require the assumption that the perception processes reduce the visual environment to a vector of independent attributes, each of which enjoys linear marginal utility, and all of which are linearly additive. In general, this does not seem reasonable but serves as a first approximation. Many alternative functional forms seem equally plausible but are not as simple to work with. One might justify the linear form by hypothesizing that the human preference process is a set of independent evaluators, each linearly sensitive to a perceived environmental variable. This would be convenient, but data are not available to test the assumption.

An alternative and equally convenient form is the logarithmic linear form of (6) given by:

$$P_{gj} = \ _ka_{gk} \text{Log}(x_{kj}) \qquad (7)$$

or $\qquad P_{gj} = \ _k (x_{kj})^{a_{gk}} \ 6 \qquad (8)$

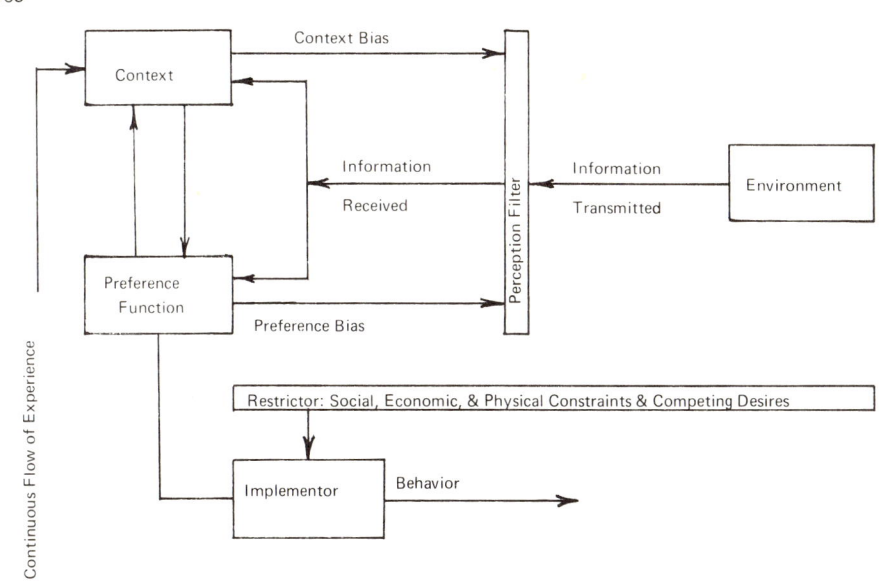

Figure 1. Conceptual Model of Individual Preference Process

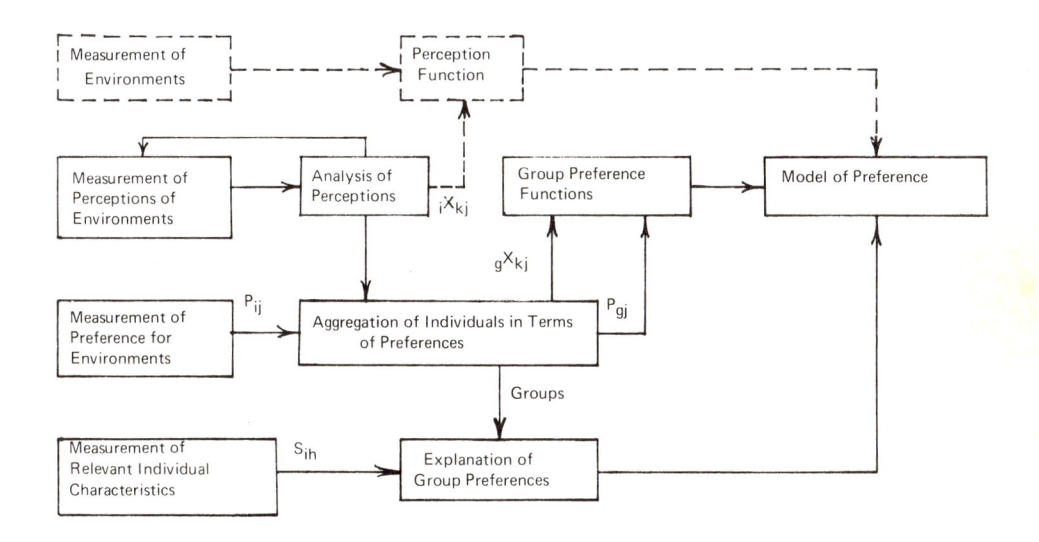

Figure 3. Methodology for Grouped Model of Preference

Figure 2a
No Natural Grouping of Individuals

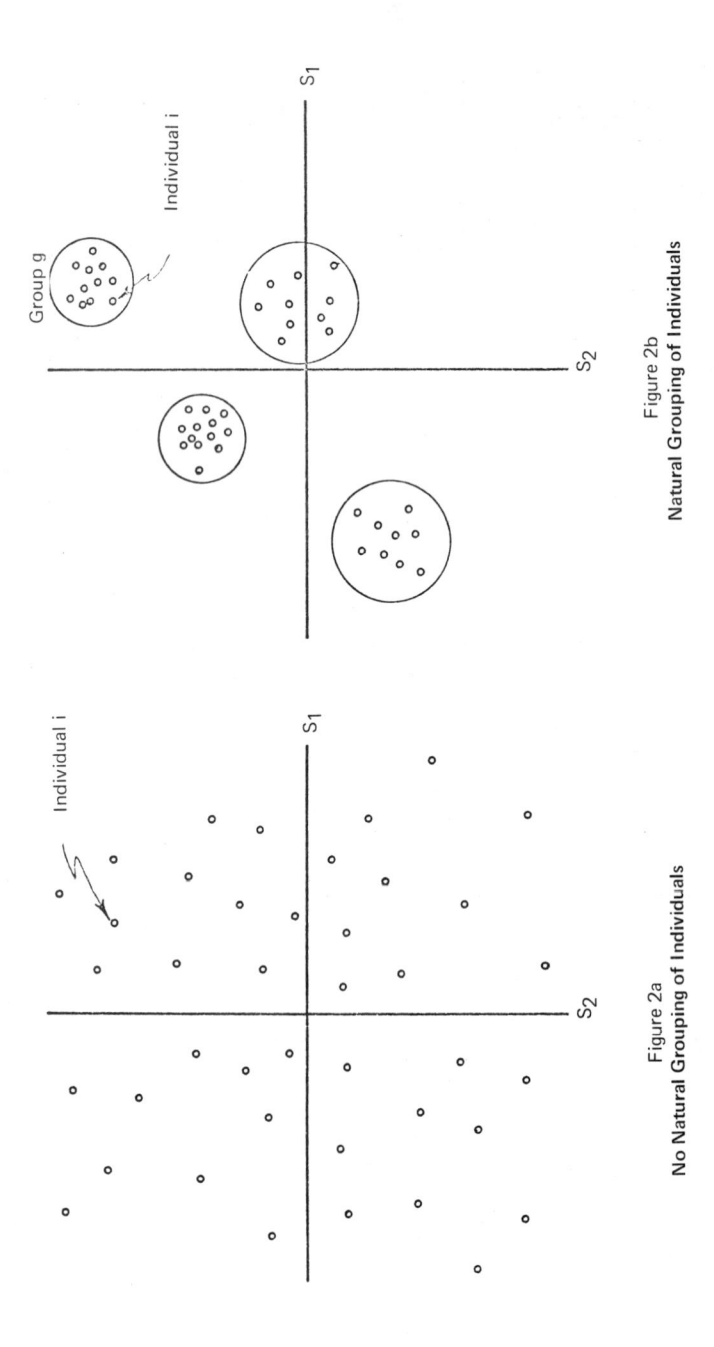

Figure 2b
Natural Grouping of Individuals

However, if the preference function is to be more than empirical fiat, it should be structured according to knowledge of the processes of perception and preference. Such knowledge might come from various sources, including the study of psychological processes or micro-analysis of the nervous system. Such knowledge does not seem to be available at present. Presumably, if the mind of man is a physical device consisting only of physical mechanisms, energy relationships, and information codes, then, like a computer, its behavior is subject to description and prediction. Complexity, by itself, is not capable of defying description forever, no matter how intricate its devices. On the other hand, if the mind is capable of arbitrary action (free will) or if there is more to the mind than physical systems, the problem becomes infinitely more difficult.

A Methodological Model

Methodology for the form of the preference function given by equation (6) is relatively simple. The method consists of a set of activities employing available tools. These activities interact to produce the model of preference. They can be classified as activities which concern
1. definition and measurement,
2. analysis and refinement of variables,
3. aggregation of individuals, and
4. estimation of functions.

The general process is outlined schematically in Figure 3. Space limitations preclude a detailed discussion of methodology.

A Study of Visual Residential Environments

An aggregate linear model of preference in the form of equation (6) was developed for the visual characteristics of residential environments. This model was reported in the *Journal of Regional Science*.[7] Briefly, it was noted that the average preference response of a group of diverse individuals to alternative neighborhoods (represented photographically) could be explained as a response to three independent environmental attributes:

x_1 = sound physical quality as evidenced by an appearance of newness and expensiveness
x_2 = harmony with nature as indicated by an appearance of greenery, open space, privacy, and "naturalness," and
x_3 = variety and rihcness of appearance as opposed to uniformity and monotony

The preference function reported in that study is

$$\text{Preference} = 0.72\, x_1 + 0.60\, x_2 + 0.26\, x_3 \tag{9}$$

This relationship explains 95 percent of the variance of the means for a sample of 140 individuals and 23 photographs. Black and white versions of the color-photographs are shown in Figure 4.

A subsequent study of the same data demonstrated that the alternative environments could be described quantitatively in terms of the three-dimensional preference space used in the preference function (equation 9).[8] The photographs tend to form natural groupings in this space, and the analysis provides useful insights into the perception of environmental similarities and differences. Selection of the initial attributes considered in the study was the author's, however, and there is no assurance that all significant dimensions of preference were included.

Figures 5 and 6 show the dendograms that result from analysis of euclidean distances in the three dimensional preference space. In Figure 5, the three dimensions have been considered to be of equal importance. In Figure 6 they have been weighted according to the strength with which they enter the linear preference function.

Figure 4. Photographs in Rank Order of Mean Preference (decending from left to right, by rows from top to bottom)

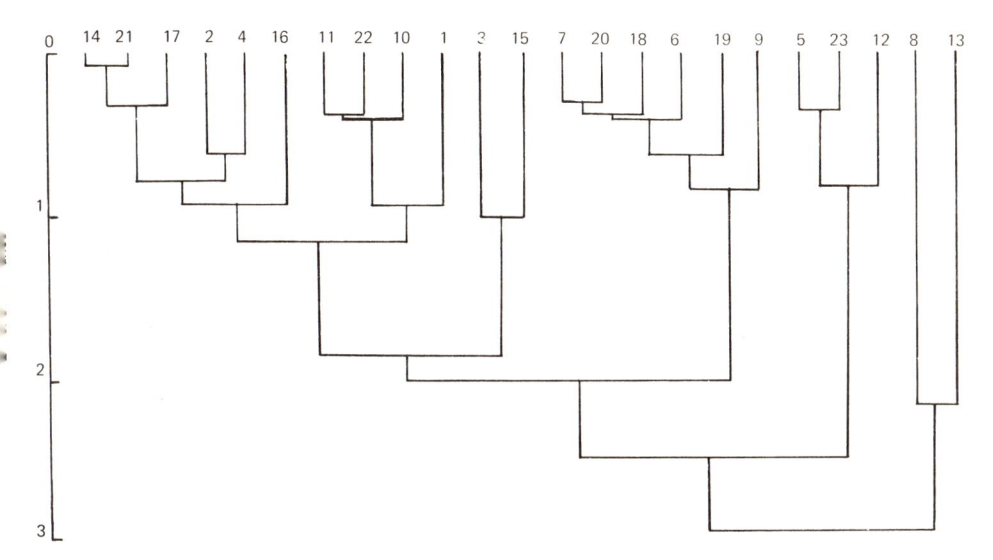

Figure 6. Classification of Appearances by Analysis of Three-Dimensional Distances

Physical Quality
Harmony With Nature
Variety

Dimensions Weighted According to Their Correlation With Preference

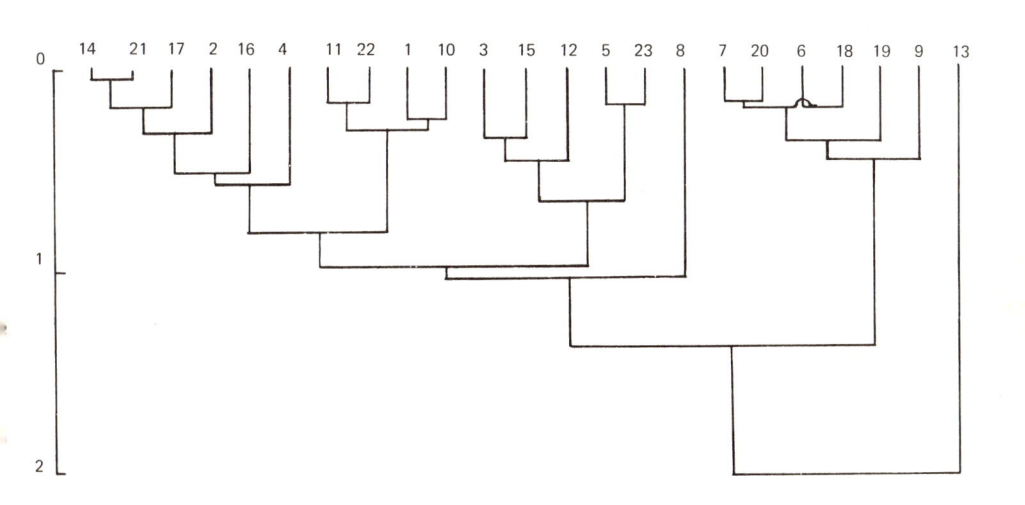

Figure 5. Classification of Appearances by Analysis of Three-Dimensional Distances

Physical Quality
Harmony With Nature
Variety

Dimensions Weighted According to Their Correlation With Preference

Figure 7 shows the weighted taxonomy graphically, and Figure 8 is a photograph of a physical model that was constructed of the the three dimensional weighted preference space. In the weighted space, the scales are in units of preference rather than in units of the attribute in question.

Disaggregation of the preference function into the form of equations (3), (4), or (5) is only partially complete. Fitzgerald attempted to aggregate individuals into groups using the method of nearest neighbor analysis.[9] His results were not definitive, however, because the method was not sensitive at the scale of individual differences encountered.

Further work by Bishop[10] used discriminant iterations to refine preliminary groups which were obtained through elementary linkage analysis produced five preliminary groups.

Using these five groups as a starting point, discriminant iteration was used to refine the classification scheme. Only six of the twenty-three photos were significant in discriminating among the five groups. The results of the analysis are abbreviated in Tables 1 and 2.

Mean preferences of each of the five groups for the six photographs which entered the discriminant analysis are shown in Figure 9 as group profiles. The six photographs responsible for the differences among the five groups are of three quite distinct types, as can be seen in Figures 4 and 7. Photos 3 and 12 are multiple family and distinctly urban in character, whereas photos 8 and 13 are rural. Photos 14 and 17 are suburban and were the two most desired environments on the average.

The profiles of Figure 9 show that each of the five groups has responded uniquely to the six discriminating photographs. Group E, for example, is particularly unique because it shows significantly less preference for the "ideal" suburban environments and is attracted strongly by rural environments. Group B is unique in its pronounced disdain for the rural photos while group D is definitely not attracted by multiple family dwellings. Groups A and C are very similar in everything except their response to photo 3, which portrays an urban high-rise environment.

The next step in the model of preference is to explain the group difference in response to alternative environments by means of socio-economic or other characteristics of the groups. This part of the analysis has proven to be difficult indeed. Thus far, only weak relationships have been observed in the data. The perference groups tend to be weakly different with respect to years of residence in the Chicago area, years at present place of residence and life objective as measured in terms of dedication to career vs. dedication to family where a choice must be made. Life objective is the most strongly related variable.

Limitations in the data, particularly control problems, prevent rigorous interpretation of the socio-economic characteristics of the groups. Before any valid conclusions can be drawn about determinants of preferences, more work must be done.

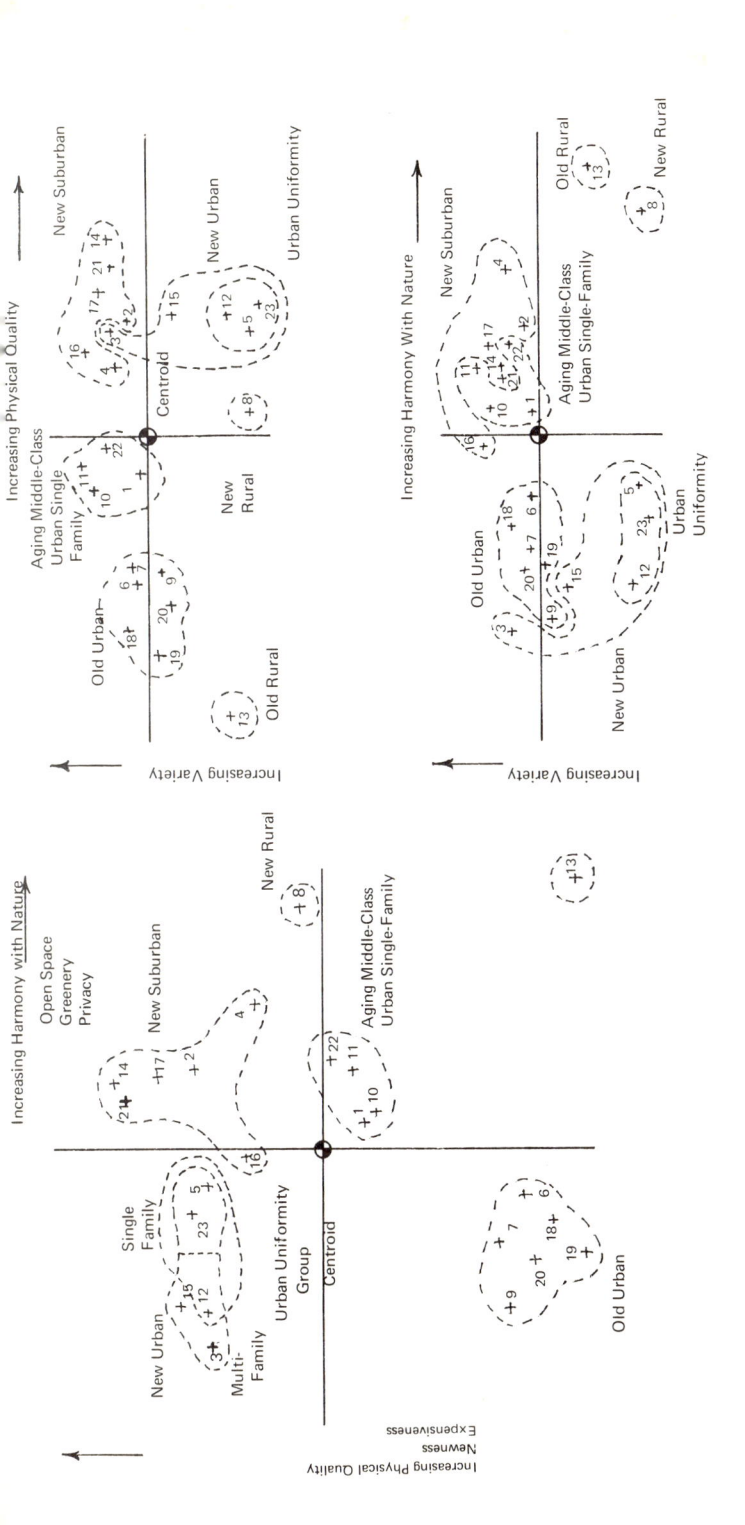

Figure 7. Two-Dimensional Projections of Interpreted Classification System

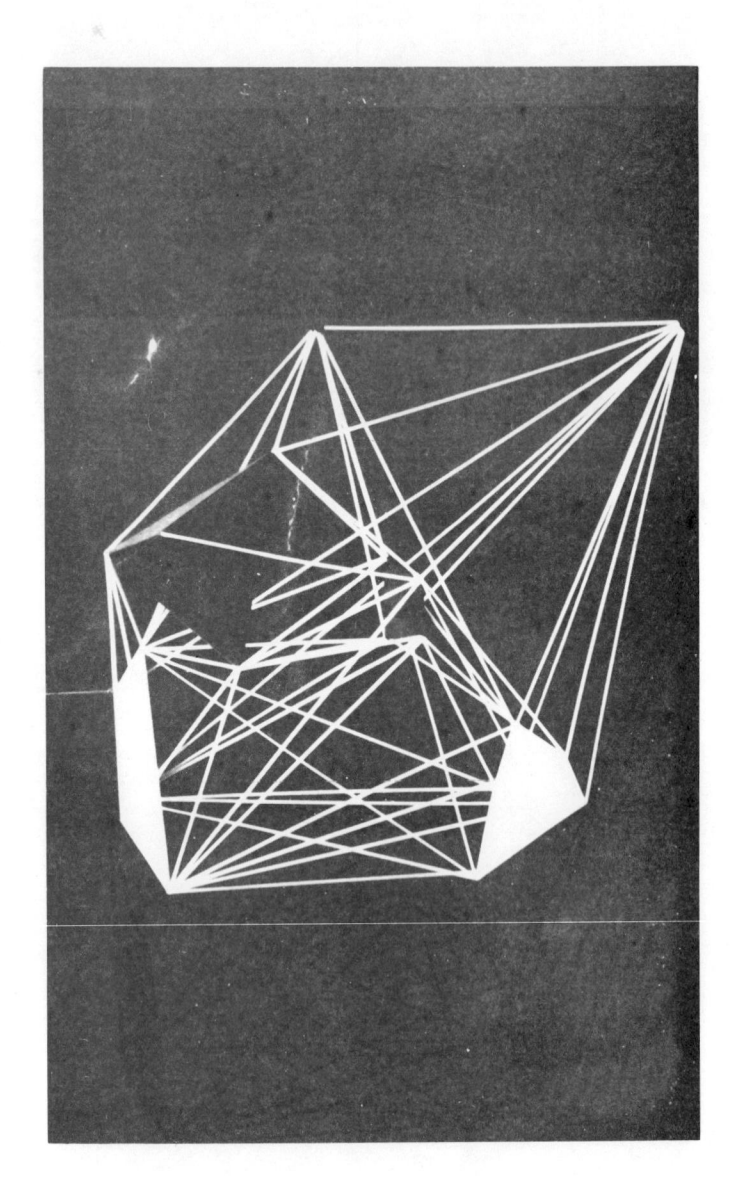

Figure 8. Photograph of Three Dimensional Model of Preference Space

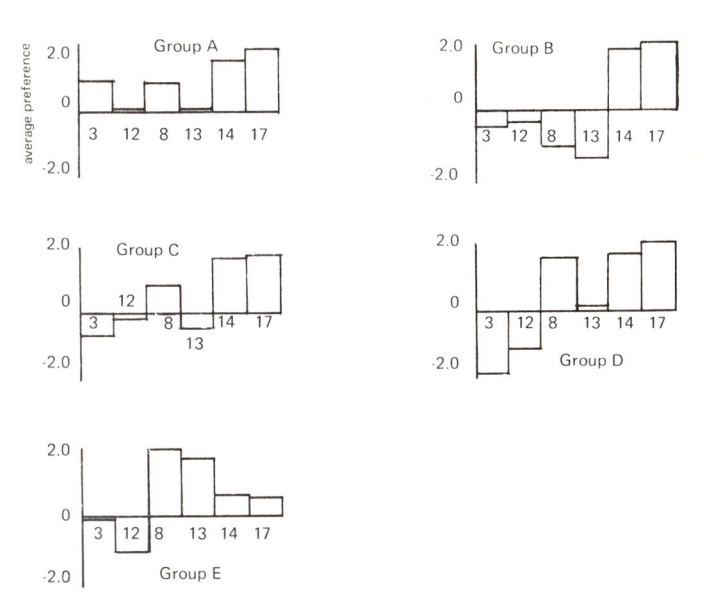

Figure 9. Average Response of the Five Preference Groups to the Six Discriminant Photos

Table 1: Summary of Group Differences

Group	Number of Members	Centroid Self-membership Probability	
A	24	.99	= 0.0626
B	47	1.00	F = 22.9
C	31	.95	df_1 = 24
D	23	.99	df_2 = 455
E	15	1.00	

Table 2: Discriminant Functions

% Explained	Photo 3	Photo 8	Photo 12	Photo 13	Photo 14	Photo 17
82.2	-.112	.755	-.127	.556	-.278	-.122
13.2	-.919	.122	-.218	-.204	.226	.005
3.4	-.072	-.621	-.515	.431	-.069	-.391
1.2	-.200	.149	.459	-.331	.029	-.785

References

1. The words "moral" and "aesthetic" are used here in the same general way that they were used in Santayana's *The Sense of Beauty* (New York: Dover, 1955, Part I). In short, if a thing is of "aesthetic" value, it is desired for its own sake for the pleasure it provides, without being required to serve a purpose other than giving pleasure. If a thing is of moral value, it is desired not for its own sake, but because it is the key or path to something of value.

2. For example, Charles Morris trisects value into operative, conceived, and object values (*Varities of Human Value,* Chicago: University of Chicago Press, 1956, pp. 9-12). Operative value is "the actual direction of preferential behavior, toward one kind of object rather than another."

3. See "Blind Variation and Selective Retention in Creative Thought as in Other Knowledge Processes," by Donald T. Campbell, *Psychological Review,* 1960, Vol. 67, No. 6, pp. 380-400. See also "Kant's Doctrine of the A Priori in the Light of Contemporary Biology," by Konrad Lorenz, *General Systems* Vol. VII, 1962, pp32-35.

4. I have borrowed the concept of "context" and "information" from Edward T. Hall and have adapted the idea for my own purposes.

5. See Figure 7-1 in *Principles of Numerical Taxonomy* by R. R. Sokal and P. H. A. Sneath, San Francisco: W. H. Freeman and Co., 1963, p. 172.

6. Edward S. Neumann discusses this form of preference function briefly in "Evaluating Subjective Response to the Recreation Environment—A Quantitative Analysis of Dissimilar Preferences for the Visual Characteristics of Beaches," Ph.D. dissertation, Department of Civil Engineering, Northwestern University, Evanston, Ill., 1969, p. 85.

7. Peterson, George L., "A Model of Preference: Quantitative Analysis of the Perception of the Visual Appearance of Residential Neighborhoods," *Journal of Regional Science,* Vol. 7, No. 1, 1967, pp. 19-31.

8. Peterson, George L., "Measuring Visual Preferences of Residential Neighborhoods," *Ekistics,* Vol. 23, No. 136, March, 1967, pp. 169-173.

9. Fitzgerald, Robert W., "Perception of the Environment," M. S. Thesis, Department of Civil Engineering, Northwestern University, Evanston, Ill., 1967.

10. Bishop, Robert L., "The Quality of the Environment: A Classification of Individual Preferences for Neighborhood Appearance," M. S. Thesis, Department of Civil Engineering, Northwestern University, Evanston, Ill., 1968.

Dan Carson
Man and Environment Division
Pennsylvania State University

WORKSHOP ON ENVIRONMENTAL ATTRIBUTES

Several things are common to each of these studies which are signal in any discussion of environment attributes. All the studies used judgemental behavior, a type of verbal response and two employed preferences; all used simulated situations which essentially limited judgements to visual attributes as represented by photographs; all were concerned with attributes which were generated by the experimenter; all used designed environments as stimuli, two limiting themselves to residential environments; and all used multivariate design. These points stimulated a lively discussion about the use of judgmental behavior and the problems of methodology.

Regarding the use of judgmental behavior, it was felt that although such experiments serve pedagogy they have severe limitations and low productivity when applied to specific design problems. Relating the one kind of behavior which is judgmental to other kinds of behavior *vis a vis* real environments was lacking and sorely needed to evaluate the *sui generis* nature of the experiments. These questions of relevance revolved around the limitations placed upon actual application and the energy spent on internal consistency of the studies. A different allocation of our intellectual resources was called for; one which may be more productive in the use of other kinds of behavior, such as migration, choices based on race, etc.

More preferences were questioned as an adequate basis for decisions in design or policy. Their use was defonede by asking what other single scale could be used as a criterion. It was felt that multivariate criteria may be necessary, but serious methodological problems arise here. The old devil causation crept into the discussion and reasons for preferences were requested with no clear cut answers offered. Further, the materials covered were considered to be somewhat outdated and a more judicious use of the library would have enhanced the overall contributions. The studies were admittedly exploratory and somewhat preliminary, providing more questions than answered and were lauded for their attempts at precision, irrespective of how misplaced the audience may have felt it was. This point leads us into questions of methodology.

Although some sophisticated multivariate methods of analysis were employed, there were still nits to be picked as well as telling substantive criticism. The limitation to visual attributes by the use of photographs was discussed and rejected as unnecessary, although the pragmatic defense was reasonable: what can be used experimentally and how can we allocate usually small resources to get the greatest amount of realism? The question of validity regarding cost, complexity, and other variables arose in connection with specific limitations placed on the experimental materials by the experimenter. Although the reliability of measurements and the polishing of measuring instruments was correctly defended, it was felt that internal consistency may be at least overconstraining and at worst misleading. Other nits picked covered the standard questions about representativeness of sample, adequacy of sample size and sufficiency of statistical tests, which in some cases were lacking and in others quite appropriate.

These papers stimulated discussion which asked mostly relevant and sophisticated questions, dealt with important issues specific to the topic and produced some heated tempers that permitted further pursuit of certain points. In short, the session was successful in its aims to explore the nature of environmental attributes.

William J. Mitchell
Yale University

SWITCHING ON THE SEVEN LAMPS

Introduction

Discussions of approaches to design normally assume either implicitly or explicitly a normative or descriptive model of the design process. Generally, it is suggested that the designer passes through a succession of stages, from receiving his initial directions to the production of a set of executable instructions for the manufacturer or builder. Very often this progression is represented by a flow-chart.

This "staged process" model type can be traced back as to the method of project organization evolved in the ateliers of the Parisian Ecole des Beaux Arts during the early decades of last century. A rigidly formalized staging of work was imposed on the students by their Masters. In a typical exercise, a student would be presented with a short written program and asked to develop an esquisse solution in a few hours. Upon faculty approval, a second stage was entered . . . of elaboration and perfection over a period of weeks. Finally, an elaborate graphic presentation of the solution was produced for formal presentation to a jury of experts.[1]

In a design text based on the Beaux Arts approach (published in 1927), the advantages of a logical progression of steps within each successive phase were stressed. The student was exhorted to commence an esquisse by making a careful analysis of the written program in order to extract the most salient points. (See Fig. 1) Following this, a large variety of possible solutions should be generated, and each evaluated on the basis of the program requirements. Those judged most satisfactory were to be elaborated (by means of successive overlays on tracing paper), as a basis for evolving a final solution to the esquisse scheme. Time spent in development of the final scheme was to be carefully scheduled, and much of it spent in an iterative process of perfection and refinement.[2]

Recently, interest has developed in attempting to externalize the structure of the design process as a basis for developing more rationally organized and effective design strategies. One of the earliest and simplest models to appear was evolved about ten years ago by Denis Thornley and his colleagues at Manchester University. As did the Masters of the Beaux Arts ateliers, they divided design projects into four stages in order to make visible the progress of students during the development of the project. These stages were:
  (1) Programming (brief, draft program, program);
  (2) General study (meaning, form-finding, evaluation);
  (3) Development;
  (4) Refinement.[3]

Since then, more complex and sophisticated models have been proposed, and empirical studies of the behavior of designers have been carried out. The concepts of other disciplines (particularly information-processing, cybernetics, decision-theory, and psychology) have often been utilized with varying degrees of success.[4]

I will not discuss all these models in detail but suggest that two very severe limitations generally apply to the whole tradition. First, this approach tells us very little about the complex ways in which design activities are embedded in the social, political, and economic structures of society. Second, - and these points overlap to a degree - a very oversimplified concept of goal-direction and motivation is implied . . . that the design goals are quite clear, consistent, and definable, and held in common by all concerned.

Operational games related to area of urban design illuminate the importance of these considerations. The Cornell Land Use Game, CLUG, is a well-known example. Articulation of the rules of such a game requires precise understanding of the kinds of players which may be involved and how they may be interrelated. The development of a game is analgous to the development of a model of the design process.[5] In seeking to understand the economic determinants of a complex process of environmental change such as housing development, other investigators have also constructed quite sophisticated models relating the principal factors involved. The Urban Analysis Corporation's housing model is an excellent example.[6]

An approach which overcomes these falling between two limitations can be developed. This approach would identify the various groups and individuals related to a design project and focus closely on the precise character of the communications channels existing between these entities.   Although this approach is tied to describing rather specific situations and even then must be something of an oversimplification, it can lead to a particularly useful overview of the context and of the significance of the current revolution in orientation to design.

Applying these concepts to the design activity of the architect in private practice, a model is produced which strikingly illustrates the rigidly-defined, almost ritualistic nature of the character and sequence of information-exchanges taking place. Due to all kinds of economic, social, organizational, and even legal constraints, it is usually not easy to step outside this pattern. Since alternate structures are readily imaginable why has this particular structure evolved and been so tenaciously maintained? An answer emerges, if we trace the early evolution of professional organization of architects, commencing in England early in the last century.

Evolution of the Architect's Role and Function

It has been convincingly argued that in the laissez-faire economy of the mid 19th century differences between professional and commercial behavior derived primarily from the inapplicability of the principle of caveat emptor to the professional market. Acceptance of caveat emptor in the commercial market was based on the assumptions that a buyer would know his wants and would recognize those items which fulfulled them. Neither assumption could be made with respect to professional services. Thus, to prevent chaos in the professional market, voluntary professional associations arose with the primary objectives of guaranteeing both the integrity and the competence of their members. Between 1818 and 1870 in England organizations of civil engineers, architects, pharmacists, chemists, mechanical engineers, actuaries, accountants, dentists, surveyors, and teachers were added to those already existing in medicine and law. These developments were paralleled elsewhere.[7]

The Royal Institute of British Architects was founded in 1834 and rapidly set about developing an effective guarantee mechanism. Its first step was to distinguish between the "profession" of design and the "trade" of building, as the opening address of the first meeting proclaimed:

> The Honorary Members shall consist of noblemen, who shall contribute a sum at one time of not less than twenty-five guineas, and of gentlemen unconnected with any branch of building as a trade or business, who shall contribute a like sum.[8]

This separation has been maintained until the present in most Western countries, although it has not always been effectively enforced.

In 1877, the Council of the RIBA imposed a requirement that:
All gentlemen engaged in the study or practice of civil architecture, before presenting themselves for election as Associates, shall, after May 1882, be required to pass an examination before their election, according to a standard to be fixed from time to time by the Council.[9]

This action derived from the need to provide a guarantee of competence as well as integrity, and it marked the beginning of the close connection between professional associations and architectural education systems.

As in many other professions, the establishment of voluntary professional organizations was followed by a movement to establish statutory registration to further strengthen the guarantee structure. International Congresses of Architects in 1900 (Paris), 1906 (London), 1908 (Vienna), and 1911 (Rome) all passed resolutions supporting this practice and legislation was effected in many western countries in the early decades of this century.[10]

It seems unarguable that in relation to market stability, architectural professional organizations have been quite successful; but the price is high. In order to preserve the credibility of the guarantee system, the possible roles of an architect have been very closely defined, restricted and delimited by codes of practice. He is discouraged from becoming too directly involved with the actual business of building and, in many situations, his contact with the actual users of an environment (as distinct from the formal client) may be very tenuous and circuitous. The kinds of legal and organizational relationships which exist in a typical situation between architect and builder, architect and client, etc., often influence the forms of communication to the direct detriment of information-flow efficiency. The effort which goes into making working drawings and specifications legally watertight contract documents is a striking illustration.

Architectural education, which experienced a related process of evolution, has in some ways operated within even more rigidly delimiting frames of reference. From the earliest establishment of schools of architecture[11] up to the present, the core of curricula has normally focused on "design". In general, it meant little more than a series of studio exercises in formal manipulation based on the "requirements" of an anonymous written "program" and evaluated in terms of the current aesthetic value. Despite the Bauhaus leaders' rhetoric of social consciousness and process orientation, their major educational effort now seems to have been the successful substitution of a new aesthetic orthodoxy within this tradition.

When a profession rigidly and narrowly defines its role and accepts an inadequate basis for the education of its prospective members, it risks intellectual sterility and social irrelevance. What is surprising is that this took so long to become obvious; the first widespread signs of a realization amongst architects that "architecture" was failing did not appear until the fifties. Boyd, though a traditionalist (he was neither perceptive nor did he summarize the reaction - he was only on the stylistic plane) summarized one aspect of the crisis which developed:

Not only the International Style, but the whole movement towards a functional, realistic architecture was left without progressive or vigorous leadership. In the latter half of the 1950 decade, the brighter young architects wandered far away, and in many different directions. Then the flood of new styles came flowing back, one after another, from numerous small avant-garde cults. Visual excitement returned to architecture, and in indigestible quantities. Eventually all ranks of the once optimistic and cooperative march to the future broke up; and dispersed.[12]

Boyd described a stylistic crisis and reaction, a familiar situation to art historians and critics; the history of the visual arts in the West over the last few centuries has been a succession of such developments, so it is easy to interpret it this way.

But the more important reaction which took place was far removed from matters of style. There also occurred a major concern for the role and function of architecture. This concern led to an accelerating movement, primarily among students, which rejects the organizational structure, traditionally perceived social roles and methodological paradigms of the architectural profession as a whole.

This development had two components. First, and of overwhelming importance, the movement needs to be discussed within a much broader frame of reference than architecture; it must be seen within the broad context of student political revolt. In broad terms, Vincent Sculley summarized some of the implications of the underlying problem:

> ...architecture is power, architecture grows from power, architecture is the embodiment of power, only power makes architecture, and architecture faithfully reflects the character of the power which creates it...the architect has no power. He has no power, he is a pawn in the manipulation of those who make architecture. Of all those forces which make architecture at the present time, the architect is of least consequence...A generation ago, the architect told himself and told us all that he was a social engineer with high ideals, who constructed society according to various values, and so on; that might have been a good idea or a bad idea, but it's not a fact in practice. He is a servant. He is a servant of forces, in the face of which, because he is not organized, himself, in terms of a union or an effective institute, or whatever it may be which will represent him as a social force, he does in fact have no power, and in my view very little to say. And I say this about the greatest architects in America; I've seen it happen to them, and not a hundred miles from here...[13]

Many students believe that they must learn to swim upstream to the sources of power. But Animal Farm is just nearby; Snowball lives! Unless new methodological paradigms are evolved by designers to expand the intellectual constriction I have described, playing the power game ultimately avails very little.

The second component of the revolt was the emergence, primarily amongst architecture students and researchers, of highly self-conscious programmatic eclecticism. These architects attempt to integrate contributions from a wide variety of fields into a general reconstruction of the theoretical framework within which design takes place, and to expand the spectrum of operational techniques available to designers.[14]

The Product of the Revolt: Current Trends in Architecture

This new consciousness has been the result of the complex evolution of dialogue and experiment occurring in many directions simultaneously. The best way to gain some comprehension of the general implications of the movement is to relate to the model postulated earlier. From this comparison, four overlapping categories of activity can be distinguished:

(1) A search for techniques to establish more meaningful contact with, and sophisticated understanding of, the needs and reactions of users of environments;
(2) A search for ways to introduce effective feedback loops into the process to encourage what Jacob Bronowsky has termed "the habit of truth", i.e. to develop some mechanism analagous to classical scientific method;
(3) Attempts to develop clearer and more efficient techniques of information storage and transmission;
(4) Attempts to escape from some of the constraints imposed by the limitations of the unaided human mind and the models and media traditionally used by architects.

Although this categorization appears somewhat arbitrary, it is useful for clearly discussing current development in architecture.

In considering the attempts to gain a better understanding of user needs, we immediately encounter the sound and fury of the behaviourist—existentialist debate. Although it is not possible to discuss (in this brief paper) this issue in its entirety, I must emphasize that a carefully considered position must be taken if this development is to be evaluated.[15]

It appears that the great bulk of the significant work in this field is firmly within the behaviourist tradition, and is derived from related work in the social and behavioural sciences... sharing the same strengths and weaknesses. It is here that the eclecticism which I noted earlier is most strikingly apparent; methods and data have been directly drawn from psychology, sociology, anthropology, administrative science, and economics. In a less direct way, the whole stream of development related to the work of Christopher Alexander is just as firmly committed in this direction.[16]

An assumption implicit in most of this work is that having gained some understanding of the relationships between environment and behaviour, we should be able to structure physical environments to allow, facilitate, or even produce particular behaviour patterns. The idea is not new; a naive sort of environmental determinism has been implicit in many utopian urban schemes for centuries past as for example, those of Ebenezer Howard. Although few are now satisfied with such a simplistic concept, architects pervasively accept a connection between environment and behavior.

This belief raises a crucial question about the relationship between the designer and those he serves. Enmeshed within the communications net which I have described, it is difficult for the best-intentioned archtiect to escape from what Pat Goeters has called "The Patrician Hangup",[17] i.e., a paternalistic "I-know-what's-good-for-you" attitude toward the environment's user needs, especially if these are not identical with those of his formal client.

This has several important implications. First, there are other ways for designers to relate to people, e.g. advocacy planning, but the problem is not entirely a political and organizational one.

Second, the complexity of the environment often vitiates the goal of achieving simple design objectives; often objectives can only be defined as a complex set of interrelated priorities. We do not, at this time, have an adequate conceptual framework or language for defining such complex objectives, although work by Chermayeff and the Center for Environmental Structure are making contributions in this area.[18]

Feedback

The problems of design—user relationships and objective specification are related to the second category, the problem of feedback. Many writers have pointed out that one of the major strengths of the scientific method is the feedback from experimental results, but that no effective analogous mechanism exists in the traditional architectural design process. (See Figures 1 and 2)

Appropriate organizational structures and explicit objectives are prerequisites for developing such a mechanism; however, effective evaluative techniques are also required. Testing procedures for assessing technical performance have been well-known to building researchers for a long time, but the traditional skills of architects include little more than unsystematic personal observations as means of comparing and evaluating the environment from the users' viewpoint. Once more, techniques developed in other disciplines have been drawn upon in an attempt to fill the gap. In particular, many statistical techniques in use among psychologists are being adopted, adapted and tested, perhaps the most influential being the semantic differential technique of C. W. Osgood.[20]

121

Figure 2. **Information Flow System**

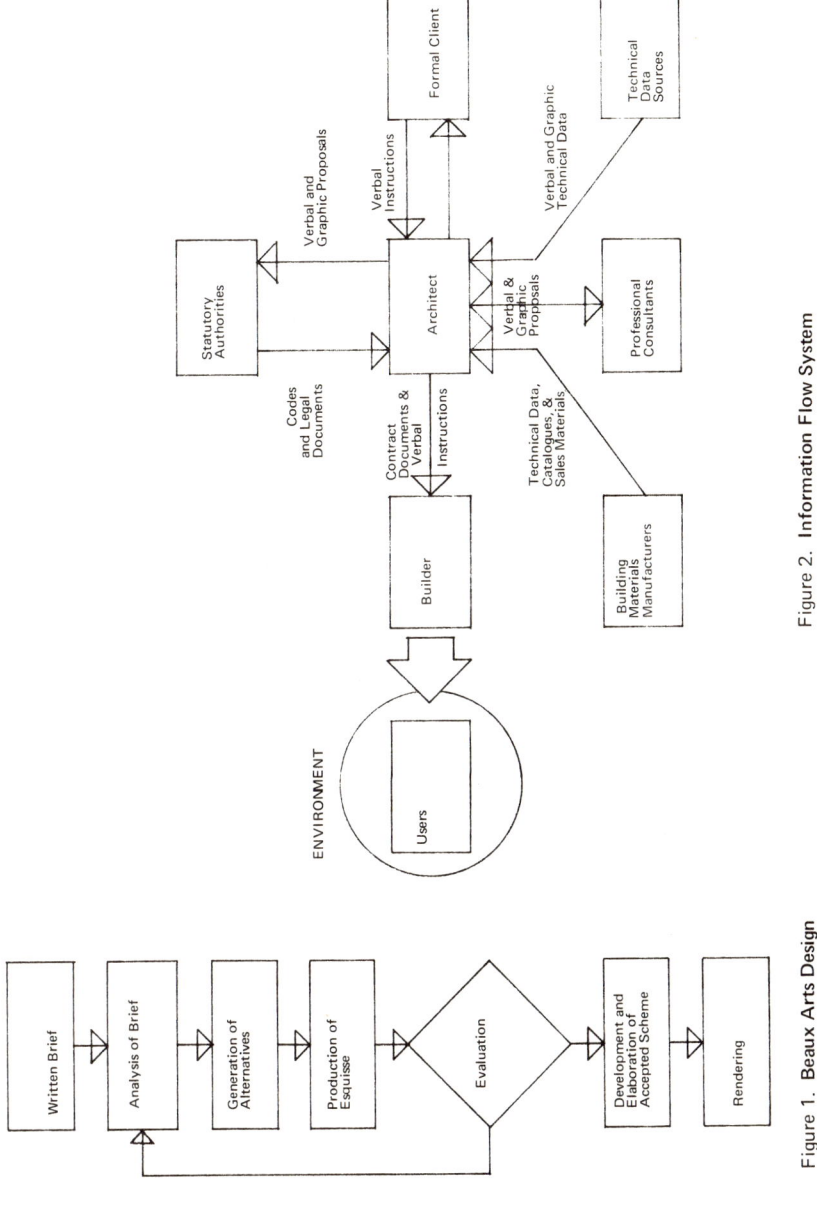

Figure 1. **Beaux Arts Design Process**

Information Storage and Transmission

The effectiveness of communications channels which do exist is being questioned, too. Most of the media described in the communications model, i.e. sketches, working-drawings, specifications written briefs, trade catalogues, etc., either evolved with the professional organizational structure of the last century or even earlier. With much better understanding of information-processing principles and more sophisticated hardware available currently much effort is being devoted to finding ways of improving the efficiency of data channels and, when possible, automating data manipulation. Automated document production and information retrieval systems are now coming into increasing use,[21] and as in most fields today, there is a large and growing range of computer programs available for solving various common standard problems. (Whether it is economical to use them is another question, though.) In addition, sweeping reorganizations of the whole flow system are implied by the development of comprehensive data cross-referencing systems such as CBC; much of the work of traditional architects is simply locating data stored in one format and translating it into another.[22] As these developments progress, the traditional professional structure of architecture will appear increasingly unworkable simply on the grounds of information-processing efficiency alone.

Constraints of the Human Mind

One element that will not be altered much by the restructuring of the information-flow system however, is the human designer. Fortunately, some of the most exciting investigations of design methods have been directed to amplifying his capacity or the capacity of design teams. Some of this work has focussed on the concept of "creativity", drawing either on the broad, speculative and domewhat anecdotal approaches of authors such as Koestler and Ghiselin, the advertising and commercially-directed techniques developed by Alex Osborn, William Gordon, etc. and the investigations of experimental psychologists.[23]

A further significant trend in this direction is the interest in developing problem-structuring techniques. Extraordinary interest has been generated by the idea of hierarchical partitioning algorithms, as in the HIDECS and CLUSTR computer programs, and is still maintained despite the many unresolved problems which this approach has been shown to entail. An attractive aspect about this development is its relationship to the cognitive theories of information-processing oriented psychologists such as Jerome Bruner and George Miller.[24] Another approach, which derives from the development of morphological analysis in the forties, uses a variety-generation selection model to discover the range of possible solutions to a design problem.[25]

To investigators of design systems in general, it has become increasingly apparent that the design process may be divided rather neatly into things that computers do well and things that people do well. Based on this effective bifurcation the idea of an interactive man-machine design system becomes attractive. Some very ambitious interactive systems, such as URBAN 5, have already been developed and tested. All kinds of hardware for interaction via graphic display now exists. Unfortunately, they still tend to be very expensive and to have inadequate software support.[26] With the increasing availability of time-sharing terminals, many architectural routines are being designed for at least some small degree of interaction.

Conclusion

Much more is going on than can be included in this very condensed summary of recent methodological development, however, this review does give some idea of the wide range of sources being tapped, and most important, illustrates the complex ways in which methodology and organizational structures are intertwined. The two evolve in constant interaction. If we seek to alter one, we must be aware of implications for the other. I and many of my contemporaries cannot accept either the structure or the methods of the architectural profession as it presently exists. At its worst, it is socially irrelevant and intellectually bankrupt; at its best, it is good-intentioned but frustrated by lack of appropriate opportunities and techniques. Neither methodological research nor redefinition of designers' roles can alone provide an acceptable new basis for design, but by working with both simultaneously, we move in the right direction.

References

1. T. C. Bannister, *The Architect at Mid-Century,* Volume I *Evolution and Achievement,* Reinhold, New York, 1954.
   J. F. Harberson, *The Study of Architectural Design,* Pencil Points Press, New York, 1927.

2. Harberson, Ibid.

3. D. G. Thornley, *Design Method in Architectural Education,* in Jones *Conference on Design Method,* Pergamon, London, 1963.

4. A representative selection is contained in the following references:
   J. R. M. Alger and C. V. Hays, *Creative Synthesis in Design,* Prentice-Hall, Englewood Cliffs, 1964.
   L. B. Archer, *Systematic Method for Designers,* Council of Industrial Design, London, 1965.
   M. Asimow, *Introduction to Design,* Prentice-Hall, Englewood Cliffs, 1964.
   T. W. Chu, *Some Enquiries into Design Method,* in *Architecture in Australia,* June, 1967.
   C. M. Eastman, *Explorations of the Cognitive Processes in Design,* Carnegie-Mellon University, Pittsburgh, 1968.
   W. E. Eder, *Definitions and Methodologies* in Gregory, *The Design Method,* Butterworth, London, 1966.
   S. A. Gregory, *A More Detailed View of Design* in Gregory, *The Design Method,* Butterworth, London 1966.
   G. Herbert, *A Diagrammatic Summary of the Architectural Design Process,* in *The British Journal of Aesthetics,* April, 1966.
   J. C. Jones, *Design Methods Compared, 1: Strategies in Design* 212.
   T. A. Markus, *The Role of Building Performance Measurement and Appraisal in Design Method,* in *Architects Journal,* December 20, 1967.
   R.I.B.A., *Plan of Work for Design Team Operation,* in *Handbook of Architectural Practice and Management,* Instalment 2, July 1964.

5. A. Feldt, *Operational Gaming in Planning and Architecture,* paper delivered at the A. I. A. Architect-Researchers Conference, 1967.
   Architectural Forum, December 1968, *The New Gamesmanship.*

6. Urban Analysis Corporation, *The Housing Analyzer,* (descriptive paper).

7. A. M. Carr-Saunders and P. A. Wilson *The Professions,* 1933.
   B. Kaye, *The Development of The Architectural Profession in Britain,* Allen and Unwin, London, 1960.

8. Quoted in Kaye, Ibid, p. 80.

9. Quoted in Kaye, Ibid, p. 129.

10. Kaye, Ibid, Chapter Nine.

11. The Ecole des Beaux Arts was established in Paris in 1807, but it was not until much later that formal architectural education was established in English-speaking countries. Among the earliest established chairs in architecture within universities were those of Kings and University Colleges, London (1837), Melbourne University (1861), and M. I. T. (1865).

12 R. Boyd, *The Puzzle of Architecture,* Melbourne University Press, Melbourne, 1965.

13 Lecture to Yale Students, 1968. Reprinted in *Novum Organum,* a student publication of the Yale School of Art and Architecture, December 3, 1968.

14 For a striking parallel in the development of new paradigms in scientific research, see: T. S. Kuhn, *The Structure of Scientific Revolutions,* University of Chicago Press, Chicago, 1962.

15 One of the few philosophers to write in this area (in relation to design) has been Janet Daley, who has published some biting attacks on behavioural research in architecture: *Psychological Research in Architecture; The Myth of Quantifiability* in *Architects' Journal,* 21 August, 1968; *A Philosophical Critique of Behaviourism in Architectural Design,* Paper given at the Portsmouth Symposium on Design Method, 1967.

16 It would be futile to attempt to catalogue this work here, but the recent proliferation of research and publication in this area is striking evidence.

17 P. Goeters, *The Patrician Hangup,* in Perspecta 12, 1969.

18 S. Chermayeff and A. Tzonis, *Advanced Studies in Urban Environments: Toward an Urban Model,* Yale University School of Art and Architecture, 1967.
C. Alexander, S. Ishikawa, and M. Silverstein, *A Pattern Language Which Generates Multi-Service Centers,* Center for Environmental Structure, Berkeley, 1968.

0 T. Markus, Ibid.
H. Sanoff, *Techniques of Evaluation for Designers,* Design Research Laboratory, School of Design, North Carolina State Universiyt, 1968.
C. W. Osgood, G. J. Suci, and P. H. Tannenbaum, *The Measurement of Meaning,* University of Illinois Press, Urbana, 1967.

1 G. N. Harper, Chapter on specifications in Harper, *Computer Applications in Architecture and Engineering,* McGraw-Hill, 1968.
A Ray-Jones, *Computer Development in West Sussex,* in *Architects' Journal,* 21 and 28 February, 1968.
J. Barnett, *Computer-Aided Design and Automated Working Srawings,* in *Architectural Record,* October, 1965.
See also papers presented at the Yale Computer Graphics Conference, 1968 (to be published later this year).

2 B. Bindsley, *C. B. C. Progress,* in *Architects' Journal,* January 31 and February 7, 1968.

3 A. Koestler, *The Act of Creation.*
A. F. Osborn, *Applied Imagination,* Scribner, 1963.
W. J. J. Gordon, *Synectics,* Harper, 1961.
B. Ghiselin, *The Creative Process,* Mentor, New York, 1955.
G. H. Broadbent, *Creativity,* in Gregory, *The Design Method,* Butterworth, 1966.
G. T. Moore and L. M. Gay , *Creative Problem Solving in Architecture,* Department of Architecture, University of California, Berkeley, 1967.

4 C. Alexander, *Notes on the Synthesis of Form,* Harvard, 1964.
M. A. Milne, *CLUSTR: A Structure Finding Algorithm,* paper presented at the DMG Conference, Cambridge, Mass., 1968.

J. S. Bruner, J. J. Goodnow, and G. A. Austin, *A Study of Thinking,* Science Editions, New York, 1962.
G. A. Miller, *The Magical Number Seven, Plus or Minus Two,* in *Psychological Review,* 63, pp. 81-79.

25 K. W. Norris, *The Morphological Approach to Engineering Design,* in Jones, *Conference on Design Method,* Pergamon, London, 1963.
S. A. Gregory, *Morphological Analysis: Some Simple Explorations,* paper given at the 1967 Portsmouth Symposium on Design Methods.
I now have a program, MORPOL (running the DCS at Yale Computer Center), which is a variety-reduction and classification algorithm for application to morphological analysis.

26 Transcript of the Proceedings, *Yale Computer Graphics Conference,* 1968.

Fred I. Steele
Department of Administrative Sciences
Yale University

PROBLEM SOLVING IN THE SPATIAL ENVIRONMENT[1]

There seems to be little question that the physical-spatial environment is an important factor in determining how a person feels, what he can do, and how he interacts with other people. In spite of a steadily growing concern for this relationship, there is relatively little research today to guide psychologists and designers toward an understanding of the impact of the environment on man's behavior and the impact of behavior on the physical environment.

The latter process is the main concern of this paper—specifically, the underused potential that people have for influencing their environments ("problem-solving" about what the surroundings ought to be like). The main theme is that many of the features of our environment perceived as fixed are, in fact changeable but unrecognized as such. Some of the reasons for this are explored, as well as some of the costs of a nonproactive stance toward the environment. Finally, some suggestions are presented both for research in this area and for methods for expanding spatial awareness in order to increase people's understanding of themselves, the environment, and its impact on them and others.

Physical and Psychological Features of Space

One of the most important contributors to this expanding concern with the quality of the physical environment is the cultural anthropologist, E. T. Hall. His *Silent Language* entails, in part, men's use of physical space and objects as a means of communicating with one another, while his *Hidden Dimension* describes the impact and the use of space by a variety of cultures.

In *The Hidden Dimension,* Hall discusses a pair of very interesting concepts: the first, *fixed-feature space,* includes physical environmental features which are relatively non-changeable and non-moveable, at least in the short run, such as permanent walls, monuments, large boulders, streets, and so on. The second, *semifixed-feature space,* includes elements of the environment which are, in fact, moveable, shiftable or changeable in some way, such as tables, chairs, pictures, bureaus, office desks, rugs, and so on. Hall then suggests some of the ways in which each of these kinds of space influences behavior, and the differences between the two. Both of these categories are determined by the objective properties of the physical environment. However, looking at fixedness in subjective or psychological terms, a third type of space emerges which may be of greatest interest. I am referring to *pseudo-fixed feature space*[3] (PFF): those physical features in man's environment that are relatively simply changed but which are perceived or treated as if they were fixed even though their configuration or position may be inappropriate for what is being done. I have observed that in a large number of group and organizational settings there are a great many variable features of space which, in fact are treated as if they were fixed and are never changed once they have arrived at a certain position or use.

Pseudo-Fixed Features

There are many examples of PFF space which could be cited—the list is almost infinite—but let us just consider just a few. We have all observed the tendency of a group to spatially distribute itself in accord with the arrangement of chairs and tables in their meeting room. This is often the case in the early phases of the T-group (self-analytic learning group). Later when they begin to be self-conscious about their behavior and choices in general, they often engage in enthusiastic experiments that involve changing the physical setting. They become aware of having taken the setting as a given through the early stage when in fact it was a variable.

Similarly, I have met in several organizational board rooms which contained tables whose size could be varied depending upon the size of the group. I have, however, never actually seen a group change the table, even where it would have been appropriate. In one such case, five of us conferred around a four section table which must have been 20 feet long. Instead of using one section and facilitating our relationship, we sat strung out along two sides and the end as if to test for formality and stiffness.

Another example is the placement of lighting in a particular room: how many of us have placed a lamp at a particular position and left it there for a very long period even when a real desire had grown to use the different parts of the room or different areas? Yet we continued to read or write in that particular spot. A final example is quite striking in its import for the design process and for the flexible use of space. In the new Ford Foundation Building in New York City where privacy curtains were provided to screen the glass walls, it is rumored that a norm developed that the curtains be left undrawn to indicate that the occupant had nothing to hide. A design feature which was originally created to give people choice about how they would like to be experiencing their own space—as open or as closed and private—became fixed, and the glass walls and the open drapes have now become what I am calling PFF—pseudo fixed features rather than the intended variable features of the space. Observation indicates that although this norm was a trend in the early life of the building, it is not always followed now.

As a behavioral scientist, I find the pseudo-fixed feature space more interesting than either fixed feature or variable feature space for a number of reasons. First, it is an instance where the individual's behavior and assumptions about behavior are, in fact, influencing what the space is like, rather than the space simply influencing behavior. Second, assumptions about PFF space are a determinant of the quality of the individual's environment (and this may be as important as the nature of the particular feature itself). That is, there is an interaction between the nature of a place and people's beliefs about how something ought to be or has to be, with both of these determining what can be done in a setting. Third, a great leverage for change exists if we can understand more about this particular phenomenon. A change in attitudes or behavior around spatial problem solving, especially the variable feature of space aspect, can lead to greater "Environmental Competence" that is, to greater utilization of resources, better fit between activities and the spatial environment, and a richer learning process as people find out more about what can be done in their environment. Improving users' Environmental Competence would make flexible design elements more important and more exciting. Since designers would believe that people would and could use things in multiple ways, they would put their hearts and more energy into the design of flexible, variable things rather than lethargically doing it on the hope or wish that people would see them that way.

Problem Solving and the Environment

In general, the process we are talking about here is the tendency of individuals and groups to take a non-problem solving stance toward their own physical environment. A problem solving stance would be to ask questions such as, "Is the physical setting right for what we want to do now? What could be changed to make it more so? Is this easily done with what we have or what we can get with relatively little effort?" When talking about spatial problem-solving, several steps are involved: (a) Asking what we are trying to do here. (b) Asking what spatial arrangements would be adequate or useful for doing what we want to do here. (c) Asking what the spatial arrangement is right now—how well does it fit with what we need. (d) Asking what we can do to change it to make it more appropriate, if it is inappropriate. What alternatives do we have and how do they fit with the amount of energy we would have to put into this? (e) choosing and acting on our choice.

In my observations of groups and organizations, as well as individuals, I note that these questions are asked infrequently and when they are raised, they often get shunted off into other issues before they can be answered. The observations also raise the issue of why people tend not to problem-solve about space—an issue that will be dealt with shortly.

Potential Benefits from Spatial Problem-Solving

There are many different ways of improving action in the world and thereby deriving benefits if problem solving about space were increased. One obvious area is the urban slum where considerable improvements could be made simply by helping the residents to become more aware of the ways they can influence their own environment and improve it. A second outcome of increased spatial problem-solving would be to influence better design of space and facilities. As people became more in touch with how the environment affects them and how they affect it, they would provide better feedback to designers and planners, resulting in a continually spiraling process of improving quality. A third potential benefit is the often discussed increased work producitvity. Many physical settings could be easily altered to improve the kind of interaction or the climate or mood for work so that people would be more productive in their day to day work. A fourth potential benefit for people is the possibility to learn more about themselves and about the world—that individual growth would be promoted by problem solving about space, as people learned about how they considered alternatives, how they looked at things, what they wanted in the way of physical spaces and in terms of outcomes from their activities. Just asking the question "What am I wanting to do here," would in itself often be a growth-producing mechanism. The final benefit, particularly important as our society gets more complex and less personal, is the ability to individually influence our immediate environmental surroundings. Without an increased inclination toward influencing spaces immediately surrounding us, our cities and suburbs may soon become psychologically uninhabitable by anyone other than people who are fortunate enough to be able to afford great luxuries in the way of space and quality.

Blocks to Spatial Problem-Solving

The next questions to be explored here are: Why does problem-solving about space not occur more regularly? What forces people to define space in PFF terms: And what should one work on to increase environmental problem-solving behavior? We obviously need a great deal of research in order to understand and to answer these questions. I think, however, that it is possible at this time to make some general statements based on my observations and those of other behavioral scientists.

As implied in the preceding paragraph, forces toward PFF space seem to fall into two categories: characteristics of the individual and characteristics of the social setting or climate.

Individual Characteristics

(1) Some people may have a constellation of personality "traits"—relatively stable attitudes which combine to produce a basically non-problem solving stance toward the world. They tend to be non-proactive—to be passively influenced by the world rather than to attempt to influence it. Rotter[4] indicates that there are many generalized tendencies to perceive reinforcement or effects as linked to one's own behavior or as controlled by external forces. Some experiences of psychological success—that is, influencing something in an observable way that builds their sense of competence would be essential to opening up new possibilities for the type of person who assumes external control in his world.[5]

(2) A second block to problem solving results from a lack of awareness of one's experiences in the first place. Many people in fact know little about what they feel or experience in different settings; they tend to develop schemata which focus them not on their experience and feelings in the present but on how they will describe them to someone in the future. Schachtel[6] has called this future orientation which blocks

experience in the present "childhood amnesia" and is nicely illustrated by the tourist with his ubiquitous camera. Many tourists spend their time in an exciting place capturing it on film so they can have the joy of showing the artifacts to others in the future. They are in much closer touch with their cameras than with the unknown places they are visiting. Spending time in a foreign land shopping for trinkets is another example of non-use of present experiences, although the shopping itself may be valued as a way to see a place.

Closely related to an unawareness of internal experience is the process of behavior control through habits. Many PFF assumptions could be described as habitual uses of spaces in certain ways. Once a stable pattern or acceptable use is reached, internal sensing mechanisms are "switched off", in essence, and the new possibilities that change might produce are not examined.

To the extent that a person is unaware of how he is being influenced by the limitations or mood of a particular setting, he is very unlikely to be motivated to think about how the setting might be changed to produce better experiences for him (and, in fact, "better experiences" would not have much meaning to him, since he perceives his own process in the world with few visible highs and lows.

Several people such as Perls, Hefferline and Goodman, and Schutz[7] have developed experience-based training methods designed specifically to re-connect people with their own experiencing of both the internal and external world. These methods will likely lead to an increase in spatial problem solving and a decrease in PFF assumptions.

(3) Even when we are aware of our own experiences, we tend to have difficulty in connecting this experience to spatial causes—we have a blindness to this like a fish has to water. It is with good reason that Hall titled his excellent book on spatial effects the "hidden" dimension. We are imperceptive of our physical surroundings and how they are influencing us—very little of our schools' training is aimed in this direction and almost nothing in work training is specifically oriented toward this process. The recent explorations of consciousness—expanding experiences (through drugs, happenings, light shows, sound, and the like) seem to me to be the start of a trend away from blindness to the effects of the environment.

(4) A fourth individual factor in the lack of problem solving may be a simple lack of concrete knowledge about the technology of changing the variable features of a space—if someone does not know how something can be changed, he is unlikely to consider it as an lanternative in terms of what he is doing. If the person feels unknowledgable, he usually divorces himself from the process and leaves it to the "experts" (who often know a good deal less than he does about what he really needs. Many how-to-do-it magazines, such as "Decorating Made Easy" are attempts to dissipate this one-down feeling.

(5) A fifth factor, which may vary with time, is that many people often have a low awareness or lack of clarity about what they want or what they are trying to do. This makes it difficult to specify what an appropriate setting should be like. Getting at this particular block would entail practice at making choices, increasing attention to one's strivings and goals, and systematically elaborating what needs to be done in order to get to the desired ends.[8] In a related vein, people often have some sense of what they might like in terms of the space to achieve them, yet are unable to articulate them in a way that can lead to choice and action. The languages of both personal goals and of space are much too slim at the present time. Heuristics such as "form follows function" doesn't help much if you can not describe the function.

(6) Another factor involves motivation; some people do not want a better solution in terms of their surroundings—they are not motivated to having a useful, pleasing, or exciting environment. By this I suggest that a person may keep his environment in a less useful shape (in terms of what is in it or its state of repair or cleanliness) than it could be because he feels (usually unconsciously) that is more congruent with his own sense of worth. A person who feels that he is of little value and not much good and also feels that he deserves to be in crummy surroundings, may in fact then rule out alternatives which would make his surroundings better. In essence, he may be trying to reinforce and play out his self concept physically in the surroundings.

It seems that this is a particularly strong component of why slum areas continue to run down or stay the way they are (even after renewal)—there is a sense of rightness or fitness in the resident's minds that they be living in a place that is deteriorating, given their own sense of themselves as people who are deteriorating, and of little value. Given this, we should not change the physical surroundings and build antiseptic high rise storage boxes as first steps toward social change, but we should help the residents become more aware of their self concept and their present self worth, increase that self worth, and use the physical environment as a tool to be influenced by them in the way of helping build a sense of competence. The practice value of the renewal process, in this light, is too important to be given to anyone but those who need it most—the residents themselves.

(7) Another motivational factor has to do with a person's feeling of need to influence the setting in which he operates. In many instances the impact of the physical environment is perceived as a marginal concern; given the importance of what is being done at the particular time, (it is not worth the dissipating energy to think about changing the setting). However, this could be handled in true problem solving about space and would be a logical outcome of the process. Where someone asks the question, "is the setting right?" it would take only a few seconds to say that for what we are doing now there are wide limits that would be all right and we'll live with it as it is. The point is that often this marginality notion of "it's not worth the energy" gets carried back one step and blocks the question of the setting's appropriateness. This stems from the assumption that they already know hte answer in every case and that the space will always be a marginal concern. This may or may not be ture: a change in the physical space in some cases would have a big impact on what is happening—add to the quantity and quality of what is being produced or to the kinds of experiences people are having, etc.—but the question does not get asked because of the assumed marginality. This is one of those areas where it is difficult to tell what a thing would have been like once you have assumed away the problem and not considered it to begin with.

Goffman[9] discusses an interesting related phenomenon: group norms which may prohibit exposure of concern over spatial factors on the assumption that a person's role in the proceedings calls for him to look fully involved in the main task. It often does not matter whether the space is hindering the main task—the person should still express his interest in the content of the main task, not in the surroundings.

(8) Another cause of PFF assumptions is obvious: there may be a low level of general problem solving ability on the part of the individual (or a group). This is often a personality trait like the discussion in (1) above. Rokeach[10] has found evidence that there are differences among individuals in their ability to problem solve and that these differences are related to factors more specific than intelligence, such as tendencies to become locked into particular sets of assumptions which limit the quality of solutions. If the general problem-solving ability were low, then it is likely that it would be low for problem solving about space as well.

This could be improved through training sessions devoted paritcularly to the process of problem solving in general,[11] using physical features as examples of this particular area. Spatial problems are good training for this process since they are concrete and visible; people can, once they get in tune with it, experience the difference a change makes fairly easily and quickly.[12]

(9) Finally, under individual factors, there is what I call the fear of failure (messing something up). There is a tendency to feel that if we make a change in the physical environment, it will be irreversible, which can result in a strong push toward physical changes being "perfect" the first time. We have had little contact with the notion of space as an ever-changing, evolving experimental place. This gap makes us wary of actually stating to change the environment because we might reduce its value or necessitate its replacement. This may also relate to the Protestant ethic notion of the saving and maintenance of resources which in the extreme means preserving something that could be functional and saving it for someone that never uses it because it can never be used—it is always being saved. I have observed this in some churches physical facilities—they are underused, apparently out of the notion that they ought to be preserved for eternity. Research would probably discover ties between fear of failure in making spatial changes and early family experiences and fantasies about parental expectations of perfection.

Social Factors

There are a second set of factors which I would relate to the social setting which is present along with the physical environment.

(1) One clear factor is the group climate of a group or organization. Group norms may operate to inhibit initiation of changes or any kind of concrete suggestions for doing something as a group. My observations indicate that these norms are widespread. If anyone who raises something is mocked or put down, no one raises the issue of an inappropriate physical setting for fear that he may be branded as trying to achieve a hidden end or grab the leadership. Similarly, norms about the degree of open confrontation and discussion of process issues will affect spatial problem solving (and promote a tendency to create PFF assumptions). A member of a staff group once told me that he felt that the rigid seating pattern that they followed at every meeting helped keep interaction rigid and formal and got in the way of change[13]; yet he also said that he would never raise the seating issue in the group—it would be a *faux pas* on his part, since it would focus attention on the taboo subject of power relations in the group.

There may also be relatively clear rules in the systems about who can and cannot change physical arrangements—often with those who have the power being the ones who have least information about the need for change.

(2) A second social factor may be disagreement over goals which can result in maintaining the physical setting because its nature is less uncertain than some alternative. If there is not agreement on what is trying to be done in a group setting, there probably is even less agreement about what the appropriate physical setting would be; therefore, the inertia of the status quo turns out to be the biggest factor in its maintenance.

(3) The third social factor in PFF assumptions is the notion of territoriality.[14] We often do not think of changing the environment if it is on someone else's turf. This includes raising the question of the appropriateness of a particular seating arrangement, light system, etc. I think this factor generally holds true for people in our society, be it in the home, office setting, or in public spaces. If we have a sense of outsidership and non-ownership, we feel that we have no right to try to influence the space, and ought to keep quiet even though the space is having a strong influence over us. One way out of this particular block is fairly obvious; that is, the development of a group climate or an inter-personal relationship where raising such questions is not embarassing, but is considered a part of being alive and connected with the total experience.

(4) Group or social system norms may also operate very specifically to rule out potential uses of a space. If an organization's culture defines certain areas as "off limits" to lower (or upper) level members, the way in which these areas can be used and the activities that can take place there is limited. A norm that the "good" executive always has his desk clear (thereby showing that he is up to the moment on his work) can severely limit what he can do in his "public" office and shift his freer work to his home. This separation of uses in time can be costly, for example when his best ideas occur at home where his associates are not available for immediate discussion.

(5) Finally, the amount of problem-solving about space that people who are together engage in is influenced by their degree of organization. In many instances a social setting is a collectivity, not a group, in the sense that people do not think of themselves as a part of a group that has a purpose, common values, or caring about one another. People in an airport waiting room are not a group in this sense. The lower the degree of "groupness", the less likely they are to be mechanisms of decision making, conflict resolution, and therefore the less likely they are able to problem solve about good use of resources, including physical features. More general societal norms will be the major influence on use of space here. In creating flexible components for a space, designers should ask themselves whether the people who will typically use it will have a high enough degree of organization as a group to really use the flexibility.

Inconspicuous Costs

The last notion discussed and the general categories suggest that variations in the measurement of value as another reason why people do not problem-solve about space. This is the concept about inconspicuous costs of a given process; that for many kinds of processes, the cost of not doing something is much less conspicuous and much less obvious and often much less measureable than the cost of doing it, in terms of energy, time involved, inputs or whatever. At an individual and small group level this is often the case. In many instances, the energy required for a small group to decide where it ought to meet or where it should move psysical features is more obvious and more immediate than the costs in terms of productivity or creativity if it is not done. One staff group, upon analyzing its spatial behavior, noted that they always stopped generating ideas when their blackboard was filled. They speculated that it was likely that valuable ideas remained in the corner of someone's mind because of the limited blackboard space. Since taking action has an immediate calculable cost and the cost of the present pattern is doscounted to the point where the gains resulting from changing the space are not considered, action is not taken in favor of avoiding immediate costs.

At an organizational level, this effect can be even more extreme and more important. There are many examples where a large organization creates a physical setting, such as a 60 story office building, with very little notion of what will happen in that building and little understanding of the functional requirements. Many organizations, when they are planning their new space, will make decisions based on economy of design and construction and never think about the extent to which they are creating significant long run costs, both financial and human. In the typical instance, the costs considered are design, construction and maintenance, not the long run cost to the organization of an inappropriate physical setting or an inflexible one which limits useful interaction and cannot be changed. In this sense, assumptions about PFF Space can occur before the space even exists; i.e. when it is being planned (or not planned). Often this happens as soon as first drawings are created—the framework these create is often assumed as a given that is worked around but not questioned. I have suggested elsewhere that this process needs to be more carefully examined as an integral part of systematic organizational development.[15]

Conclusions

We have been exploring the concept of Pseudo-Fixed-Feature Space—those elements of the environment that are variable but are used as if they were fixed. The major focus here has been on the forces which lead to PFF assumptions about the environment—the individual and social factors which make problem-solving about space fall far short of its potential and block the development of Environmental Competence (the ability to manipulate and use the potential resources in one's surroundings).

That PFF exists seems to me to be obviously validated by both observation and experience (my own and others). But we have little to anchor to in the way of data. We need major research efforts aimed at examining the conditions under which PFF assumptions tend to occur and when they do not, allowing us to evaluate the relative strengths of the various factors discussed above. Other related questions would include ways in which people conceptualize environmental alternatives, physical cues that are more and less likely to signal variability or flexibility, and the psychological roots of the inconspicuous costs phenomenon and its permanence or changeability.

More generally, a major research thrust is needed toward understanding man in relation to his physical/social environment and how the physical and social aspects interact to provide facilitative or stultifying conditions for human performance and growth. We still know relatively little about the impact of different kinds of environmental elements on different social conditions, with different kinds of goals or expectations from past experiences. To research this complex set of variables also requires that we develop adequate and inventive means of measuring both properties of the environment and responses of individuals and groups to these properties. The development of good measures would be one vehicle for defining the conceptual boundaries of problems and allow us to gather manageable data rather than diffuse impressions that can never be pulled into any meaningful patterns. It also seems apparent that the area of spatial behavior and problem-solving is one that can most fruitfully be approached by a combination of controlled laboratory and field research. Defining variables or processes experimentally in the laboratory can have highest payoff when validated by people's behavior in the settings most important to them—their own homes, offices, public spaces, etc., and it seems especially likely for environmental studies that the laboratory can provide only part of the data—it is in fact a specialized environment with certain properties and meanings for the subjects.

Directions for Growth

From the other end, why, other than adding to knowledge for its own sake, is it worth learning more about PFF space and its causes? Why bother to train people to take a more problem-solving stance toward their own environments? One basic assumption here has been that the setting can influence how people feel and behave. Given this, there seems to be useful growth-producing outcomes for several segments or levels of our society if we understand PFF better.

For people in the home, greater awareness of the variability of their environment could lead to a richer life, a better use of resources, more involvement in the process of living in the world, better interaction within the family, and some nicely concrete experiences that would help in building a sense of personal competence. It is also likely that standards of taste and quality would improve organically, rather than being artificially induced by advertising—people would begin to understand for themselves why an object or layout was better or worse for their purposes.

In the work setting, reduction of PFF assumptions would enhance the quality of interaction, allow greater variety in work experience, and again provide a greater sense of self-control over one's physical life space. The trend toward more self-direction for employees in their task environment growing in this country needs to be present in the physical setting as well in order for the values of the organization to be experienced as consistent. For both individuals and groups, spatial problem-solving would also have the serendipitous benefit of improving their general problem-solving skills and also raising the more basic issues of goals, direction, individual needs, and the like—questions that all too often are kept in the background responding only to immediate day-to-day pressures of the job.

For public spaces, an increase in spatial awareness could be one step toward building a feedback process between planners and users. It seldom happens that the creators (or remodelers, or controllers) of a public space such as a railway station ever receive (or ask for) data about the experiences of its users. This is a process that must be encouraged if our facilities are going to be anything but minimal and subminimal quality.

Finally, in high density urban areas, the notion of improving spatial problem-solving has special relevance. PFF assumptions are one reason for the under-utilization of the resources that exist in slum areas. If we want to help in the creation of living areas that have meaning and value to the residents (and the experience of creating high-rise renewal wastelands that have less vitality than the slums they replaced suggests that we must), then it is essential that the residents themselves be connected with influencing their environment, both because of the better solutions their participation can generate and because of the sense of competence it can foster. This process is likely to be a failure (as it has been) however, if people in the slums (or in the monotonous suburbs, for that matter) are simply told, "All right, participate in decisions about your new space." It is essential to recognize that spatial problem-solving skills are likely to be low or non-existent, the PFF assumptions will be strong, and that training and attitude change efforts are a necessary first step toward a growth-producing process of influence in one's own environment.

References

1 I would like to thank Benjamin Schneider, Edward Lawler, David Sellers, and J. Richard Hackman for their helpful comments on this paper.

2 Lynch, Kevin (1960), *The Image of the City,* Cambridge: MIT Press (Joint Center for Urban Studies), (paperback).
Sommer, Robert (1966), "Man's Proximate Environment," *Journal of Social Issues,* Vol. 22, no.4.
Steele, Fred (1968a), "The Impact of the Physical Setting on the Social Climate at Two Comparable Laboratory Sessions," *Human Relations Training News,* Vol. 12, no.4.
Steele, Fred (1968b), "Organization Development and Sticks and Stones," unpublished manuscript.

3 This type of space can be thought of as "Pseudo-Fixed" from the habitor's point of view (they treat it as fixed when it is not), or as "Pseudo-Variable" from the standpoint of the designer or creator (they see it as changeable, but it is not used that way). I have developed the notions here from the point of view of the user, since he is the "problem-solver" with whom I am concerned here. Hence the use of PFF (the alliteration is also better). That these are not completely separate is obvious, as illustrated by a discussion I recently had with an architect, when he said, "I'm designing a unit that can be altered or shifted for different moods or needs—it is a richer environment." My view here is different: that it will not be a "richer" environment until people use the variations and bring the potential richness out of it.

4 Jacobs, Jane (1961), *The Death and Life of Great American Cities,* New York: Random House (Vintage paperback).
Rotter, Julian (1966), "Generalized Expectancies for Internal versus External Control of Reinforcement," *Psychological Monographs,* Vol. 80, no. 1, whole no. 609.

5 White, Robert (1959), "Motivation Reconsidered: The Concept of Competence," *Psychological Review,* Vol. 66.

6 Schachtel, Ernest (1959), "On Memory and Childhood Amnesia," in *Metamorphosis,* New York: Basic Books.

7 Perls, Frederick, Hefferline, Ralph, and Goodman, Paul (1951), *Gestalt Therapy,* New York: Julian Press (also Delta paperback).

8 Miller, George, Galanter, Eugene, and Pribram, Karl (1960), *Plans and the Structure of Behavior,* New York: Holt, Rinehart and Winston.

9 Goffman, Erving (1963), *Behavior in Public Places,* Glencoe: The Free Press (paperback).

10 Rokeach, Milton (1960), *The Open and Closed Mind,* New York: Basic Books.

11 Gordon, William J. J. (1961), *Synectics: The Development of Creative Capacity,* New York: Harper and Brothers.

12 This feature of the physical environment leads me to a prediction that cognitive complexity, although an important personality dimension for many processes, would not be closely related to spatial problem solving.

13 Reports were always given starting on the chairman's left, which meant that the same people always spoke first, second, etc., and others almost never spoke because the subjects were exhausted by the time they got to them.

14 Altman, Irwin and Haythorn, William (1967), "The Ecology of Isolated Groups", *Behavioral Science,* Vol. 12, no.3.
Ardrey, Robert (1966), *The Territorial Imperative,* New York: Atheneum.

15 Steele, Fred (1968b), "Organization Development and Sticks and Stones," unpublished manuscript.

16 A friend told me recently that he noticed one evening that his television set had a handle on it, and he carried it from its pedestal in the living room to a back study. He reports that this change in location changed their viewing pattern from continuous background TV in the evenings to selective (and more enjoyable) viewing of programs of interest to them; the set became a communication medium rather than a time filler.

17 Sommer, Robert (1966), "Man's Proximate Environment," *Journal of Social Issues,* Vol. 22, no. 4.

18 Doxiados, C. A. (1968), "Ekistic Synthesis of Structure and Form," *EKISTICS,* Vol. 26, no. 155.

19 Willems, E. P. (1965), "An Ecological Orientation in Psychology," *Merrill-Palmer Quarterly,* Vol. 11. no. 4.

20 Steele, Fred (1968b), "Organization Development and Sticks and Stones," unpublished manuscript.

Richard Wilkinson
Department of Landscape Architecture
North Carolina State University

WORKSHOP ON ENVIRONMENTAL QUALITY

The workshop on environmental quality dealt principally with the problems of generalization and relevance to real world problems. The three papers were of a general and highly personal nature based on the experience of the presenters. Mr. Euston spoke of the problems involved in applying the idealism and assumptions of this assemblage to the organization and articulation of an effective national policy of environmental quality. As a counter force to the super organized lobbies of the transport, energy and production lobbies, we pale in significance. Theirs is the power to define quality in their own interest. His thrust was for a concept of quality that is a measure of the capacity of the design clientele to actualize themselves and become involved in manipulating their own environment.

Mr. Mitchell spoke generally to this point also. His statements came from his background of design methodology. He has studied the historical precedents of the designer-client relationship from the standpoint of the designer's role of a client actualizer. He is beginning to formulate a model of the strategies designers might adopt to energize the client toward a more interactive role in defining his own environmental problems. He points to the terminal and static nature of the designer in defining environmental quality as an evolving viewpoint of a client whose competence grows with experience.

Dr. Steele was more precise in defining the phenomenon of environmental competence. His research has examined peoples' reactions to and capacity to perceive of the potential of pseudo-fixed features (movable elements) of the environment. His experience points to a lack of competence in people to understand the manipulative power they have to structure their surroundings to satisfy temporary or alternately occurring needs. In his studies the pseudo-fixed features were accepted as offering and not understood as maniputable elements.

Essentially this is the same point made by Euston and Mitchell; the competence of individuals within the society to fashion personal and community environments is abdicaged to special interests. This is not done knowledgably but by default. The task facing designers and their profession is solely concerned with this condition. We must learn how to interact with people and be able to discriminate their desires and their competence to manifest these desires. Once this is achieved, the task centers on the problem of activating people to achieve their own environmental ends. This is essentially the task of the present day advocate designers. It goes beyond this, however, into the need for a basic methodology for research and definition of common goals that permit design to become an interactive rather than a dictatorial skill. Not until this is achieved can we hope to mount an effective counterforce to the dominant functional interests of industry and exploitation. The ultimate aim is to compete at the national level, the apex of efficacy. Prior to this we must learn to create for any one man an environment in which he is a self actualizing and an increasingly effective participant.

This is the only rational and effective measure of environmental quality.

Andrew Euston's paper, "Federal Focus on Environmental Quality" was not available when this volume was ready for publication.

Charles M. Eastman
Institute of Physical Planning, Carnegie-Mellon University

PROBLEM SOLVING STRATEGIES IN DESIGN[1]

Problem solving, thinking, and other forms of mental activity can be dichotomized into two aspects. One is the *content* of the information that is being considered. The other is the organization of the *process* that is treating the information. The obvious examples of this dichotomy come from arithmetic. Once the rules of addition are known, any two numbers can be added. Addition is the process involved; the specific numbers being added at any one time are the content. When we learn arithmetic or any other type of mathematics we are learning a process. When we use arithmetic to solve a problem, we apply content to that process.

In design, the content of a problem is the contextual information defining the goals to be achieved, the constraints and considerations necessary for solving the problem, and the materials and components available for constructing a solution. The process of design involves the sequence in which decisions are made, the means in which alternatives are generated, and the methods of evaluation.

A growing number of psychological studies show that the processes by which a human treats information represent general capabilities. Relatively minor variations in process have been found to distinguish the capabilities of expert from inexpert problem solving. The processes for handling information seem to play a significant role in determining the "intelligence" of an individual's behavior. Important examples of such studies are those of Bruner,[2] Wertheimer,[3] and de Groot [4]. These studies have begun to articulate some of the general processes that people use to reason intuitively and to comparatively analyze their performances. With the articulation of these processes, it becomes possible for educators to develop exercises that specifically develop competencies in processes that are known to be important or particularly effective.

The processes articulated by psychological studies have been found to apply to design. The application, adaptation, and extension of processing theories to predict, explain, and model design capabilities has begun, but are still in their early states of development.[5] Yet even at this stage, there may be benefits for future designers. Current practice in design education treats both content and process together in a succession of "case studies". Yet the direct teaching of both, with an emphasis on the fundamental design processes and on encouraging an awareness of the effects of different processes may significantly benefit a student's capabilities.

This paper reviews the premises of information processing models of problem solving and presents in detail one of its applications to design. The purpose of this paper is to outline the background information useful for introducing problem solving processes as an explicit part of design education. Thus it may be considered as a short tutorial on design problem solving.

Premises

The premises used in most research of design processes are those of complex information processing. Design and all other forms of problem solving are viewed as a particular form of decision making Alternatives are evaluated against goals to determine an appropriate response. Two issues make decision making in a problem solving situation unique: (1) the necessity of making sequences of decisions where succeeding decisions use the results of prior ones as input; (2) the lack of means to directly evaluate single decisions in the sequence. Evaluative ability only initially exists for the results of sequences. Because decisions without complete evaluation are simply *operations*, complex information processing uses the terminology of operations rather than decisions. Design problem solving is concerned with indirect methods for evaluating sequences of operation.

A decision or operation can be represented as a node from which emanates a series of branches. Each branch designates an alternative action. A sequence of actions thus forms a *tree* (only upside down). See Figure 1. Each node of a tree represents the summation of a unique set of previous decisions or operations.

If a decision tree is small, then all possible combinations of actions can be evaluated backward from their final results and the best sequence chosen.[6] Most interesting problems in all fields including design however, involve large decision sequences where such evaluation becomes virtually impossible. All that may be done is to explore those parts of the tree that are most likely to lead to a solution. Exploration involves the making of tentative decisions, carrying out operations, backtracking and iterating decisions that have been made previously, until a sequence of operations and decisions are found that produces an alternative that qualifies as a solution.

A major issue in problem solving is the means used to determine the appropriate sequence of operations. Various strategies which are either commonly used or which have been found to be particularly effective have been identified in the problem solving literature and given the name *search strategies*.[7] A related issue has been the means used to evaluate sequences of operations.[8]

Problem solving processes may involve any level of consciousness. While we have been considering them as explicit decisions or operations, they are involved also in almost unconscious forms of behavior, as in the sequence of actions used to ride a bicycle or drive a car. Design probably falls somewhere between conscious and unconscious levels of decision making. An example of problem solving may clarify the above points. Figure 2 presents a word arithmetic problem. On the right are the steps involved in solving the problem. First, the verbal problem is translated into a problem specification in algebra. Arithmetic offers a variety of operations that can be applied to appropriately represented problems. By selecting an appropriate sequence of operations from those shown on the left of the figure, the initially specified information is transformed into new information that meets the criteria of a solution. Notice that there is no obvious means for evaluating any one of the operations. The criteria for a solution considers the total results of all operations. Figure 2 shows only the correct operations that directly lead to a solution. If some of the incorrect operations were included at each node, the result would have been a decision tree.

The finite set of operations shown in Figure 2 would produce an infinite number of problem states. The same operator could be applied many times, with the same or different constants. Thus, no finite tree could really be defined for this problem; an infinite tree is the general situation. Because of this, many students of problem solving do not conceive of decision sequences as trees, but use the concept of a *problem space*. A problem space is defined as the sum of all possible problem states that can be generated by applying a set of operators to a given problem state. Problem solving behavior generates a tree of operations and decisions that partially explores the problem space. Any particular point achieved in the problem space is known as a problem *state*. Thus a problem state corresponds to the node of a decision tree. Significant mathematical progress has sometimes resulted by being able to prove that some desired solution lies outside of the problem space; thus it can never be generated.

The common method used to study the search of a problem space within a wide variety of task contexts is called *protocol analysis*. By carefully analyzing the behavior of a problem solver while he is treating a problem and by developing models that are capable of simulating the information processing aspect of this behavior, much can be learned about the mental processing that is involved. The most thorough presentation of the techniques of protocol analysis is made by Newell.[9]

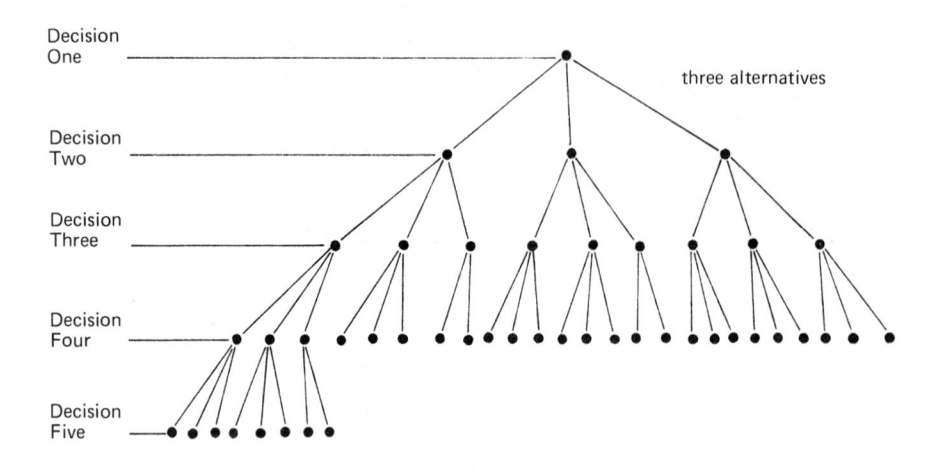

[Figure 1 diagram: a tree showing Decision One through Decision Five with branching alternatives, labeled "three alternatives" at the top level]

Figure 1.

Verbal Description Of A Problem

"John is twenty-four years old. His brother was half his age fourteen years ago. How old is his brother?"

Arithmetic Operators

f = any function
c = any constant

Axioms
A1. $f = f - c + f = c + f$
A2. $f = f - c - f = c - f$
A3. $f = f - c \times f = c \times f$
A4. $f = f - c - f = c - f$
A5. $f + f$ — add operation
A6. $f - f$ — subtract operation
A7. $f \times f$ — multiply operation
A8. $f - f$ — divide operation
A9. any function can be replaced by its equivalent.

Arithmetic Form

$J = 24$, $(B-14)2 = J-14$
A9
   ↓
$(B-14)2 = 24 - 14$
A7
   ↓
$2B - 28 = 24 - 14$
A1
   ↓
$2B - 28 + 28 = 24 - 14 + 28$
A5, A6
   ↓
$2B = 38$
A4
   ↓
$2B \div 2 = 38 \div 2$
A8
   ↓
$B = 19$

Answer:

"His brother is nineteen years old."

Figure 2

Search Strategies in Design

The concepts of problem solving theory presented can also be applied to design. If one were to formally conceptualize the nature of design problems, we could say that they involve *problems where the major task is the selection and arrangement of space and material so as to meet a variety of criteria.* For this paper, this will serve as a working definition of design.

Figure 3 presents a very simple problem that emphasizes the search aspect of design. It concerns the layout of a mechanical room. It is similar to many bathroom, kitchen and other design problems requiring the selection and arrangements of elements in bounded space. It involves one selection decision and a number of arrangement decisions. A set of explicit restrictions identify what relations are required for appropriate functioning. What is the appropriate arrangement that solves this problem?

This problem and ones similar to it, have been given to many student and professional designers in protocol experiments. By analyzing in detail how such problems as this are handled, the aspects they include become better understood. Those aspects of design omitted from this task will be considered later.

In order to be quite precise in our examination of this task, it should be carefully formulated. The initial task consists of

$a ::=$ a space

$b_1, b_2, \ldots .b_m ::=$ a set of elements (e.g. DUs) to locate in that space.
(Some elements may be defined as any member of a set.)

$c_1, c_2, \ldots .c_n ::=$ a set of constraints delimiting acceptable solutions

Three methods that are sometimes used by designers as search strategies will be described in some detail. They are common but do not include all of the search processes used by architects and designers to solve problems.

Generate-and-Test

The simplest (and most naive) search strategy is named in the problem solving literature as generate-and-test. Essentially, generate-and-test involves trying every possible alternative until one if sound that is satisfactory. In the example problem, all locations of machine A would be tried, all locations of machine B, etc. so as to eventually lead to complete generation of all alternative arrangements in an ordered sequence. After a complete arrangement alternative is generated, it is evaluated. If the alternative is not satisfactory, another is generated using the same method. A schematic description of generate-and-test in flowchart form is presented in Figure 4.

When used at all, generate-and-test is used at the beginning of the solution sequence. Its primary effect is to determine if the task is really a problem at all, and if so, how difficult it is. Occasionally it is used as the method of last resort when no other seems to work. It is the only method that is commonly available that is algorithmic—that is, that guarantees to eventually produce a solution. It is the brute force method par excellance.

Generate-and-test includes all the necessary elements of a problem solver. It and the other processes we will discuss incorporate both a generator of alternatives and an evaluation mechanism. The essential feature of generate-and-test is that the generation process operates independently; no information passes to it from the evaluation process. The distinction between generate-and-test and random trial and error is that the generator is structured so as to sequentially produce the total set of alternatives. Usually, the generator produces a fixed sequence of problem alternatives; the nth problem state is always the same. Generate-and-test has been shown to closely model the problem solving strategies of some designers.[10]

Experiment Five

Below is the plan of a mechanical room. Inside of it must be located six machines. A template of each machine is located on the bottom of the page. Notice that there are two D machines. You may use either one of them; both are not required. You are to arrange the templates in the room so as to satisfy the following restrictions:

R1. Machines A, B, and C give off sparks. Thus they should not be seen from the door.
R2. Machines D and F have controll lights on their face panel. They should be clearly visible from the door.
R3. Machines C and E are connected together and must be directly adjacent.
R4. Machines B and D also connect and must be directly adjacent.
R5. Machines B and F are connected together with a high voltage line. The length of this line should not be more than one grid unit in length.
R6. All machines must be accessible from the door via a path at least one unit wide.

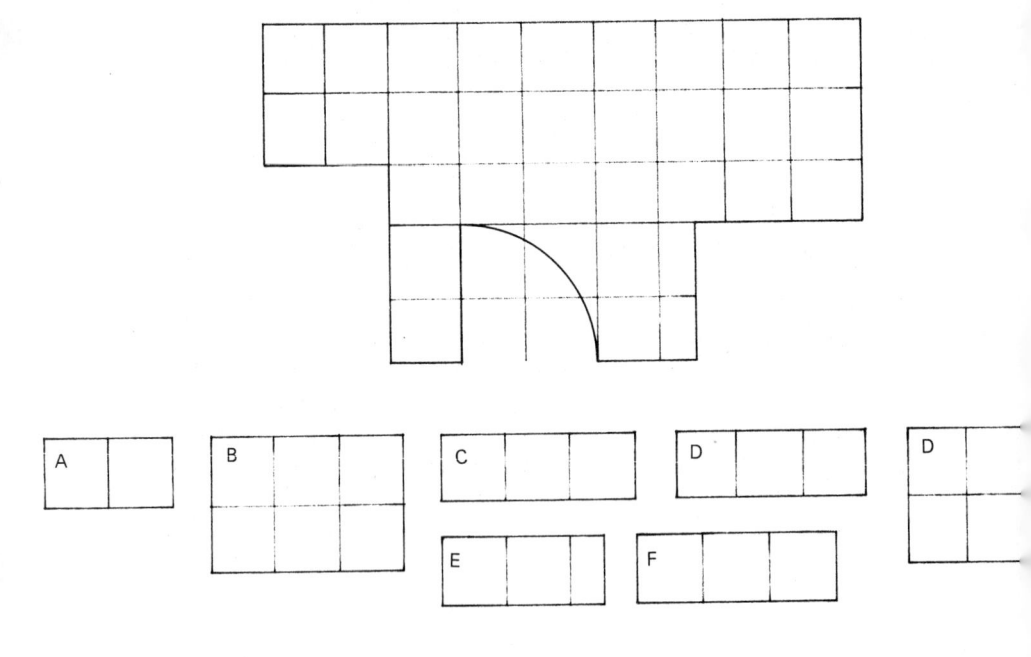

Figure 3

Means-Ends Analysis

One of the simpler problem solving processes that uses information gained from the evaluation process to guide the generation of alternatives is means-ends analysis. It does so by utilizing information that relates the testing of criteria (eg., "ends" to be achieved) with operations for achieving those goals (eg., "means").

Means-ends analysis comes into play when an evaluation fails. In order to efficiently find a means to resolve the failure, information must be available concerning the relationship between "means" and "ends". The simplest form of such information is a statement of the construction "If the condition is X, do Y". Such a statement is known as a *production* and is the basic form of statement used in Markov algorithms which have been used to model a variety of biological behavior.[11,12] Productions take the form

        if the current problem state has   apply Operator
        as a predicate condition X   Y and/or DU Z

The predicates and actions may be simple or complex.

Production systems are usually organized as a table. Because in problem solving they are used to connect means and ends, Simon and Newell call such relational information a *table of connections*.[13] Essentially, a table of connections is a list of heuristics, or rules of thumb. Heuristics are used in decision making to focus search on specific parts of the search tree. They may either direct attention to specific nodes of a tree, or delete various branches from the relevant search space. They identify relationships perceived in experience between problem situations and the actions that seem effective in responding to them.

A table of connections that provides some productions for dealing with the machine room problem are shown in Figure 5. Each describes the response to take for a particular design situation. The table holds three different productions for one condition. The last of the three happens to be required to efficiently solve the problem. Many people seem not to have this heuristic readily available and thus have difficulty in solving the problem. The third production is a special case of the second one; different forms of productions may produce the same design results. More than one production for a given state may be necessary to adequately model human problem solving behavior. A rule must be provided for selecting among them at each problem state.

Means-ends analysis has two variants. In the first variant, means-ends analysis is applied to an already complete alternative generated by some other strategy. Static evaluation of the alternative according to the restrictions it satisfies gives a crude measure of how difficult the problem is likely to be. If only a few restrictions are not satisfied, then the problem is likely to be solvable by the first variant. In this strategy, the constraints are sequentially applied to the problem. When a failure is encountered, the failing constraint is matched against the table of connections to identify appropriate actions. Those actions are implemented. The constraints are then reapplied. The process cycles until no productions can be matched against constraint or until all constraints are satisfied. See Figure 6a.

The table of connections need not be used only for error correction. It can also be used to identify appropriate operations for generating the initial alternative. In the second version of means-ends analysis, each constraint is sequentially applied to the table of connections in order to identify the initial transformations to be applied to generate the first complete problem state. See Figure 6b. By applying the information held in the table of connections at the outset, this version provides a greater likelihood that a satisfactory location of a DU will be determined initially. This version is applied in our protocols only to tasks which seem too difficult for the first version, possibly because it requires more processing effort.

Both versions of means-ends analysis use the table of connections to guide the problem state generator to produce states that are likely to satisfy the constraints. The effectiveness of mean-ends analysis

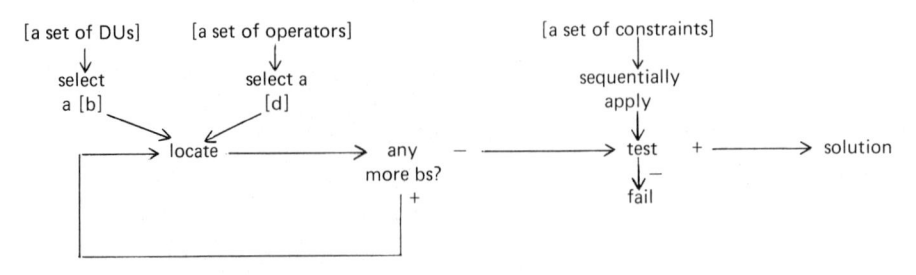

```
[a set of DUs]      [a set of operators]              [a set of constraints]
     ↓                     ↓                                   ↓
   select              select a                          sequentially
   a [b]                 [d]                                apply
        ↘              ↙                                      ↓
         → locate ─────────────→ any ─────────────→ test ──+──→ solution
         │                      more bs?             ↓
         │                         │+                │ −
         └─────────────────────────┘               fail
```

Figure 4. Generate-And-Test

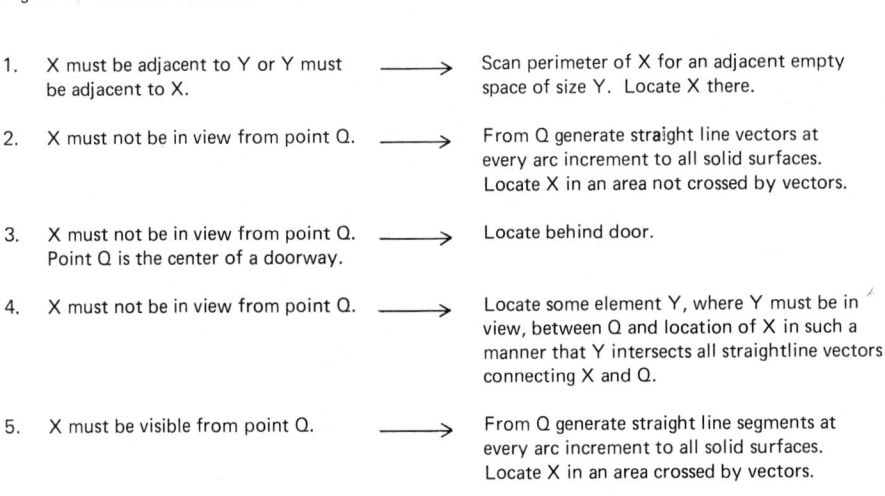

1. X must be adjacent to Y or Y must   ⟶   Scan perimeter of X for an adjacent empty
   be adjacent to X.   space of size Y. Locate X there.

2. X must not be in view from point Q. ⟶   From Q generate straight line vectors at
                                             every arc increment to all solid surfaces.
                                             Locate X in an area not crossed by vectors.

3. X must not be in view from point Q. ⟶   Locate behind door.
   Point Q is the center of a doorway.

4. X must not be in view from point Q. ⟶   Locate some element Y, where Y must be in
                                             view, between Q and location of X in such a
                                             manner that Y intersects all straightline vectors
                                             connecting X and Q.

5. X must be visible from point Q.   ⟶   From Q generate straight line segments at
                                             every arc increment to all solid surfaces.
                                             Locate X in an area crossed by vectors.

Figure 5. Table Of Connections For Experiment Five

depends upon the quality of the table of connections. The table provides a first level of indirect processing that acts to control or guide the state generating process. Further levels of processing could be used to elaborate the system's ability to distinguish appropriate actions. For example, one could conceive of a method by which the table of connections was dynamically generated as a result of information about the current problem state. At each level of processing that mediates between evaluation and generation processes, new information about the relationship of information must be introduced to the system.

Means-ends analysis generates problem states in essentially an unpredictable manner. No guarantee exists that it will generate all possible solution states or will not get into an endless loop. Thus no guarantee exists that it will ever generate a desired solution. Because of this unpredictability, means-ends analysis is not an *algorithm* i.e., a precisely specified process that can be proven to terminate and that provides a result upon termination, but an important example of a *heuristic*, i.e., a process that is effective in gaining results in some cases but cannot be proven to terminate.

Means-ends analysis seems to be a prevalent means for solving the type of design problem that we have been discussing. Without carefully controlled experiments, the number of levels of indirect processing that mediate between the generation and evaluation processes of a designer cannot be precisely determined. For at least some types of problem solving tasks, the level of indirectness provided by only one table of conncetions has been sufficient to model about ninty percent of all processing.[14]

Planning

One further strategy was occasionally found in the protocols of the machine arrangement problem. In studies we have made of more complex problems, this strategy is common. Like the previously described processes, it has several variants. One variant is applicable to the machine arrangement problem in the following way.

An analysis of the restrictions shows that machines D and F must both be located near machine B according to restrictions R4 and R5. Machine C must also be located near E according to R3. We could show these relationships by diagramming them as in Figure 7. The implication provided is that machines B-D-F represent a "subset" of the problem; another "subset" is C-E. Any successful arrangement would probably reflect these subsets as clusters within the total arrangement. Thus, a subset might be initially located in a single operation. When a likely space is found for the total subset then the individual machines comprising the subset may be analyzed and located within the area identified. If the assumptions concerning these actions are valid, then we would expect the trial arrangement generated this way to be close to the solution.

The problem solving literature calls such a process "planning". Planning has the following qualities.

1. It involves analysis of the problem structure in order to find those elements that are most related to other elements. "Related" may mean spatially if distance relations predominate, or relations may be in terms of control if communication is the central issue. Other specific forms of interaction are also possible. But the type of relations may not be mixed.
2. Planning involves search of an abstract problem space to provide a guide for problem solving. Because only one kind of relation is used to structure the problem in planning, many details are ignored. An abstract "language" for representing the problem in the planning phase is common. A common form of planning language for physical designers are graphs known as "bubble diagrams" or schematics.
3. The search strategies for solving the limited relationships of the problem considered in planning are the same as those for other aspects of problem solving. That is, they include generate-and-test and heuristic search processes. Planning strategies may themselves utilize planning.

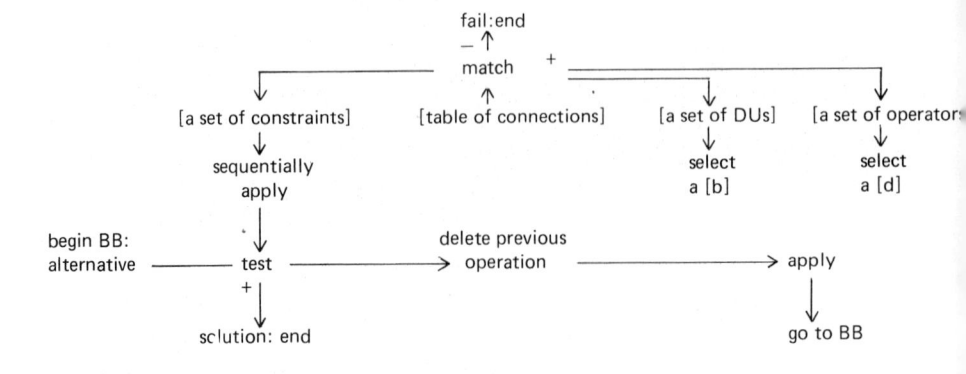

Figure 6a. **Means-Ends Analysis (version one)**

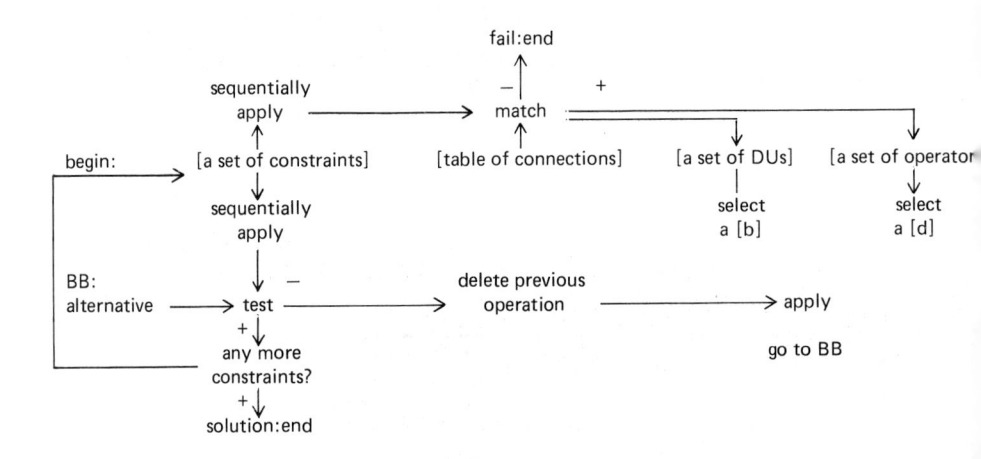

Figure 6b. **Means-Ends Analysis (version two)**

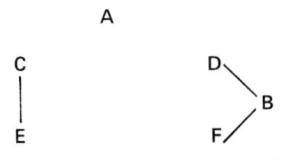

Figure 7.

4. Once the planned aspect of the problem has been resolved, its resolution of the limited relationships is used as a guide to generate the fully specified problem.

A more detailed description of planning is presented in Figure 8. It shows two stages. The first stage resolves the relationships of a problem that are being planned and achieves this by means-ends analysis. The second stage maps the planning solution into the complete problem space, after which another problem solving process takes over.

Planning also is of two types. The first mode consists of using known relationships between DUs which interact more densely with each other than with those outside the set. This mode of planning requires that the primitive DUs and their restrictive relations be initially known to the designer. If we view design as a hierarchy with all primitive elements that make up the design at the bottom and the single design unit that is being designed at the top, then the first mode of planning aggregates DUs in a "bottom-up" sequence.

Planning need not be limited to a single set of relations. The aggregations provided by analyzing one type of relationship can be combined with others. The simplest kind of combining assumes that the relations between different kinds of DU interactions are linear. By using weighting functions for each type of interaction and adding them together, comprehensive definitions of interaction are defined. Aggregation based on these interactions provides a means for defining a comprehensive system-subsystem hierarchy. Still, weighting functions that combine different types of interaction can only be identified in terms of some specific form of meta-interaction. Little work has been done that identifies the bases for such comprehensive system analysis.

Seldom are the basic components of a design available at the outset of the problem. What is known at the outset is the complete unit that is at the top of our imaginary hierarchy. The second mode of planning does not aggregate units, but decomposes them in a "top-down" manner. Top-down planning is not found in problems like the machine room problem but is used primarily to develop assumptions as to the components of a large scale project.

To gain a better understanding of top-down planning let us hypothetically assume that "bottom-up" planning has been applied to housing problems for a period of years. Thorough analysis of the relations required to carry out household and family activities defined a hierarchical set of DUs. One path through such a hierarchy would provide DUs like: lot, house, bedroom wing, master bedroom, closet, door, hinge. After each analysis the DUs were stored that were aggregated at each level of design. At some later time, a new housing problem arose, but this time the specific relationships between DUs were not available. DUs could not be analytically aggregated. Could it be assumed that the aggregations used in earlier problems would be relevant here? I would suggest that an affirmative answer is appropriate. In this case, the weighting of sets according to the frequency of their past generation offers hypotheses concerning what the current sets should be. But no means would be available to prove that the sets produced from this induction were valid or optimal.

This example suggests the situation that faces designers in top-down planning. Top-down planning is common in most large design projects. It relies on the designer's past experience, but when it is applied in reality, experience is not limited to fully analyzed aggregations of activities. Thus the examples from which subsets were learned cannot be validated. Alexander[15] has rightly criticized the unquestioned use of top-down decomposition and has proposed a heuristic procedure for generating alternative decompositions. The significant point to be made here is that procedures are available to form "bottom-up" aggregations of primitive DUs. It is possible to derive complete and "objective" set aggregations for any well-defined system. The requirements for such aggregation are the determination of: (1) all primitive elements that will be used to construct the total solution; (2) all restrictions or operational goals that the solution must satisfy; (3) a representation or language for manipulating DUs and restrictions, both for problem solution and planning.

Planning Space

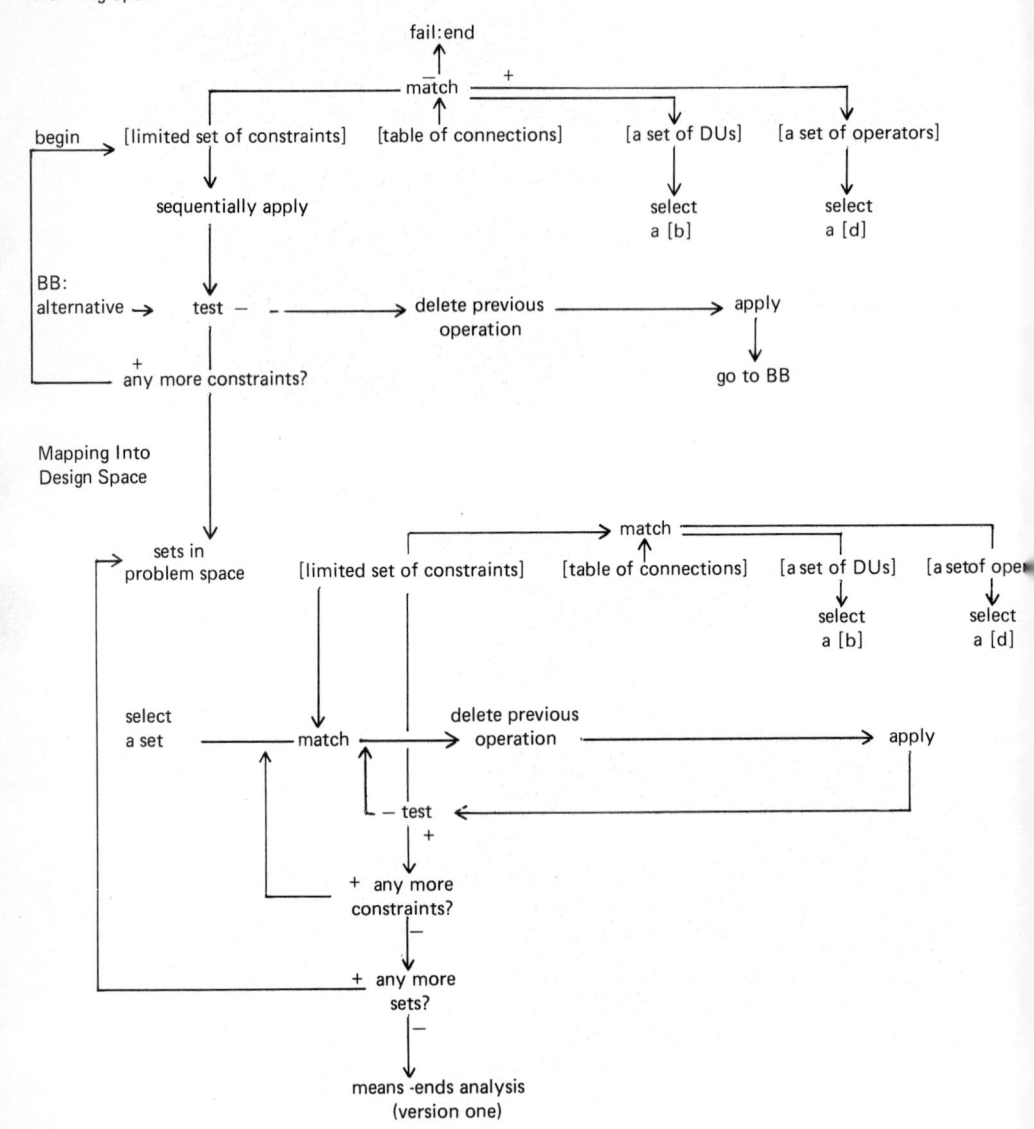

```
                              fail:end
                                ↑
                          ─ m̄atch ════════ + ══════════════════╗
                                ↑                               ║
begin    [limited set of constraints]  [table of connections]  [a set of DUs]   [a set of operators]
  │              │                                                   │                │
  │              ↓                                                   ↓                ↓
  │      sequentially apply                                       select           select
  │              │                                                a [b]            a [d]
BB:             ↓
alternative →  test  ─ ─ ─ ─ ─ ─→  delete previous  ─────────→  apply
  │             │                     operation                    │
  │          +  │                                                  ↓
  └─── any more constraints?                                   go to BB
```

Mapping Into
Design Space

```
  ┌──→  sets in                        ─────────→ match ═══════════════════╗
  │    problem space    [limited set of constraints]  [table of connections]  [a set of DUs]  [a set of oper
  │         │                   │                                                  │              │
  │                             ↓                                                  ↓              ↓
 select                                                                         select         select
 a set  ─────────────── match ─────→ delete previous ─────────────────────→ apply   a [b]          a [d]
  │                       ↑           operation                               │
  │                       │                                                   │
  │                    ─ test  ←──────────────────────────────────────────────┘
  │                       │
  │                    +  │
  │         + any more
  │          constraints?
  │            │ −
  │            ↓
  └────── + any more
            sets?
              │ −
              ↓
         means-ends analysis
           (version one)
```

Figure 8.

Planning, whether bottom-up or top-down, provides another method by which the combinational possibilities of a problem are delimited. It focuses on particular types of arrangements that are likely to provide a solution by factoring out for preliminary analysis those relations of a problem that are critical. The value and efficiency of planning depends upon a priori identification of thise relationships that most constrain the total problem.

Computer Implementation

In order to indicate how an automated designer might operate on a simple problem, Figure 9 presents the sequence of operations and solution states created when the table of connections shown in Figure 5 and means-ends analysis are used to solve the example design problem. The second version of means-ends analysis is used. The strategy works as follows:

The first restriction is taken along with the first machine. It is associated with Production 2 in the table of connections. The designated operations are carried out. (We assume a left-to-right scan for "hidden" space.) The A Machine is located and evaluated. Then the B Machine is identified from the restrictions. An initial location is tried, but evaluation of it fails. The next relevant production in the Table of Connections is tried, but is not applicable. The third one is tried. It is applicable and its evaluation passes. (The rules are not quite complete here. The previous operation should not be removed when Production Number Four is applied.) The strategy proceeds in this manner until it comes to the last restriction. After a fairly intelligent beginning, the program fails because it does not have enough information. Although the system could begin work on Restriction R6 if only one more heuristic is included in the Table of Connections, several are actually needed to allow efficient solution of this problem.

The Table of Connections and process this system relies on are general; they may be used to solve a variety of simple space planning problems and are not limited to the example presented here. Ultimately the validity of an automated design process depends upon its effectiveness in solving particular problems and the breadth of the class of problems it can resolve.

Aspects of Design Not Covered in Search

Upon reflection, search of a problem space is not the only significant process involved in design problem solving. The arithmetic problem presented in Figure 2 suggests what is missing. Once the arithmetic form of the word algebra problem is formulated, then the rules of arithmetic provide the operators for searching the problem space. Search strategies similar to those presented here are able to solve this kind of problem quite easily. But how did we get from the verbal form of the problem to the arithmetic form? The translation of an intuitive statement of a problem into a problem solving language that allows its resolution can be called *problem specification*.

In the arithmetic problem shown in Figure 2, the science of mathematics has provided us with a set of operations that have been proven not to alter numeration. Though often not explicated in school, mathematics even provides search strategies for problems such as the one shown. And though not precisely formulated, crude rules are offered for translating word problems into their algebraic form. The following quote from a high school textbook provides an example of such translation rules.

In solving problems about motion affected by wind or water currents, use two unknowns: one to represent speed in still air or water, the other to represent the speed of the current.[16]

With these operations search strategies and translation procedures commonly available and formally agreed upon, the specification of an algebra problem is fairly straight-forward. When the problem specification is agreed upon, we say that the problem is *well defined*. But the same is not true of a design problem.

An Example of the Operation of a Design Problem Solving System:
  —sequence of operations and tests defined by version two of means-ends analysis when applied to experiment number five—

Assumptions:

1. Restrictions are checked in their given order.
2. Machines are considered in alphabetical order.
3. The Table of Connections used are those presented in Figure 5. It is scanned in its given order.

Sequence of Operations:

1. Identify R1, Machine A.
2. Match Table of Connections (TOC) No. 2.
3. Apply TOC No. 2 to Machine A.
4. Test to R1-A: Passes.

5. Identify R1, Machine B.
6. Match TOC No. 2.
7. Apply TOC No. 2 to Machine B.
8. Test to R1-B: Fails at R1-B.
9. Match Toc No. 4. Go to R2, Machine D.
10. Apply D.
11. Test to R1-B: Passes.
12. Identify R1, Machine C.
13. Match TOC No. 3.
14. Apply TOC No. 3 to Machine C.
15. Test to R1-all: Passes.

16. Test to R2-D: Passes.

17. Identify R2, Machine F.
18. Match TOC No.5.
19. Apply TOC No. 5 to Machine F.
20. Test to R2-all: Passes.

21. Identify R3, Machine E.
22. Match TOC No.1.
23. Apply TOC No. 1 to Machine E.
24. Test to R3-all: Passes.

25. Identify R4, Machine B.
26. Match TOC No. 1.
27. Apply TOC No. 1 to Machine B.
28. Test to R4-all: Fails at RI-B.
29. Match TOC No. 1.
30. Apply TOC No. 1 to Machine D.
31. Test to R4-all: Passes.

32. Test to R5-all: Passes.

33. Identify R6, Machine A.
34. No match of TOC. Program fails.

Problem States:

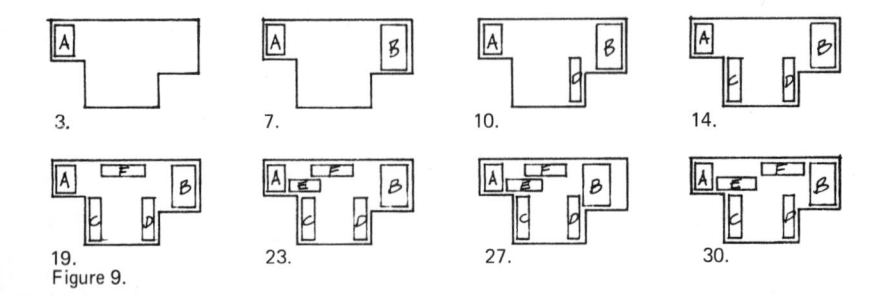

3.

7.

10.

14.

19.

23.

27.

30.

Figure 9.

No formal language with precise operations is available. The goals to be achieved and restrictions are open to interpretation. Design problems are usually *ill-defined.* The machine room problem shown in Figure 3 varies from being a realistic design problem because the representation of the problem and the restrictions and goals were explicitly presented.

Some initial explorations of the problem specification processes used by human designers[17] or that could be used by a machine to design[18] have been presented in the literature. Basically, the processes proposed consist of information retrieval systems that store experiential information, learned problem solving languages—such as those of mathematics and computer programming, orthographic projection and maps—and the specific goals and constraints applicable to different problem situations. Retrievals are made by processing the given problem information. The retrieved information is then mapped into an appropriate problem solving language. Essentially, these efforts attempt to model human long-term memory. While several computer programs have been written that handle the translation of mathematics problems,[19] operational knowledge of the specification process for a task as difficult as architectural design is almost completely lacking.

Another issue not treated here is how the heuristics used in most design processes are originally acquired. It was pointed out that the quality of design is strongly influenced by the availability of appropriate heuristics to guide search. One way that heuristics might be gained is as follows. Let us assume that strategies utilizing little information and no heuristics—like generate-and-test—are used at the beginning of a design problem solver's experience. Each time an operation is followed by the successful evaluation of a constraint that had previously failed, a reinforceable memory records the action and the constraint. Specifically, if each time an evaluation fails, the problem solver records the goal that is not satisfied and then later records the actions taken just prior to the first case where that evaluation passes, those two records constitute a heuristic. The number of times a particular heuristic is encountered determines its priority on a table of connections.

No heuristic generating process for design has yet been explored by computer simulation. The details of such a process, including how each hypothesis should be weighted and how many operations previous to the successful one should be stored, would likely have important effects on the quality of the heuristics derived. Eventually, the mechanical derivation of heuristics may allow more effective automated or manual design.

This paper has presented detailed descriptions of three search strategies often used in design. Certainly other search strategies exist. But the general issues outlined in the study of these three—the need for two processes, the use of evaluation information to guide generation, hierarchical aggregation or decomposition—are relevant to the others. The strategies described here seem to be general processes used by humans in solving a wide variety of problem types. Exploration of the power, limitations and logical implications of these and other problem solving processes may allow human problem solvers to more effectively apply such processes.

Studies such as this suggest that the synthesizing activities of designers are open to scientific investigation. An important outline of the proper concerns of such a science is presented by Herbert Simon in *The Sciences of the Artificial.* Architects have seriously considered their special area of expertise, the synthesizing processes, by which diverse criteria and materials are combined to result in a comprehensively considered physical product. It is hoped that designers will play an active role in the development of the Science of the Artificial.

References

1. This work was supported in part by the Advanced Research Projects Agency of the Office of the Secretary of Defense under Contract No. F44620-67-C-0058.

2. Bruner, J. S. Goodnow, J. J. and Austin, G. A. *A Study of Thinking*, New York: Wiley, 1956.

3. Wertheimer, Max, *Productive Thinking,* Travistock, (enlarged edition), London, 1959.

4. de Groot, Adriaan, *Thought and Choice in Chess,* Mouton, The Hague, 1965.

5. Eastman, Charles, "Explorations in the Cognitive Processes of Design" Carnegie-Mellon University, Department of Computer Science ARPA Report, DDC No. 671-158, 1968.

    Eastman, Charles, "Cognitive Processes and Ill-Defined Problems: A Case Study From Design" Proceedings of the International Joint Conference on Artificial Intelligence, Walker, D. and Norton, L. (eds.) Washington, D. C. 1969b.

    Krauss, Richard and Myer, John, "Design: A Case History", Center for Building Research, M.I.T. 1968.

    Moran, Thomas, "A Model of a Multi-Lingual Designer", G. Moore ed. *Emerging Methods in Environmental Design and Planning,* M.I.T. Press, 1969.

6. The exhaustive search of a decision tree essentially is the method of design proposed by Norris (1963).

7. Minsky, Marvin, "Steps Toward Artificial Intelligence" *Computers and Thought,* J. Feigenbaum and J. Feldman (eds.) McGraw-Hill, N. Y., 1963.

8. Samuel, A. L. "Some studies in machine learning using the game of checkers", *IBM Journal of Research and Development, 3,* July 1959, pp. 221-229.

9. Newell, Allen, "Studies in Problem Solving: Subject Number 3 on the Crypt-Arithmetic Task," Department of Computer Science Working Paper, Carnegie-Mellon University, Pittsburgh, July, 1967.

10. Eastman, Charles "Cognitive Processes and Ill-Defined Problems: A Case Study From Design" Proceedings of the International Joint Conference on Artificial Intelligence, Walker, D. and Norton, L. (eds.) Washington, D. C. 1969b.

11. de Leve, G., *Generalized Markovian Decision Processes of Design,* Mathematisch Centrum, Amsterdam, 1964.

12. Newell, Allen, "Studies in Problem Solving: Subject Number 3 on the Crypt-Arithmetic Task," Department of Computer Science Working Paper, Carnegie-Mellon University, Pittsburgh, July, 1967.

13. Newell, Allen, and Simon, "GBS: A Program That Simulates Human Thought" in *Computer and Thought,* Feigenbaum, J. and Feldman, J. (eds.) McGraw-Hill, New York, 1963.

14. Newell, Allen, "Studies in Problem Solving: Subject Number 3 on the Crypt-Arithmetic Task," Department of Computer Science Working Paper, Carnegie-Mellon University, Pittsburgh, July, 1967.

15 Alexander, Christopher, *Notes on the Synthesis of Form,* Harvard University Press, Cambridge, 1964.

16 Freilich, J. Berman, S., and Johnson, E., *Algebra for Problem Solving,* Book Two, Houghton Mifflin Co., Boston, 1957.

17 Eastman, Charles, "Explorations in the Cognitive Processes of Design" Carnegie-Mellon University, Department of Computer Science ARPA Report, DDC No. 671-158, 1968.

   Eastman, Charles, "On the Analysis of Intuitive Design Processes", G. Moore, ed., *Emerging Methods in Environmental Design and Planning,* M.I.T. Press, 1969a.

   Eastman, Charles, "Cognitive Processes and Ill-Defined Problems: A Case Study From Design" Proceedings of the International Joint Conference on Artificial Intelligence, Walker, D. and Norton, L. (eds.) Washington, D. C. 1969b.

18 Moran, Thomas, "A Model of a Multi-Lingual Designer", G. Moore ed. *Emerging Methods in Environmental Design and Planning,* M.I.T. Press, 1969.

19 Bobrow, Danial, "Natural Language Input for a Computer Problem Solving System" M.I.T. Ph.D. Thesis Project MAC Technical Report MAC-TR-2, 1964.

20 Simon, Herbert A., *The Sciences of the Artificial*, M.I.T. Press, Cambridge, Mass., 1969.

Joseph M. Ballay
College of Design, Architecture and Art
University of Cincinnati

VISUAL INFORMATION PROCESSING IN PROBLEM SOLVING SITUATIONS

There is a fundamental difference between the way in which one teaches a science or technology and the typical teaching methods in the "creative" fields of design, architecture, etc. The sciences, and even the liberal arts, have a fairly well defined body of knowledge that must be transmitted. The design student, on the other hand, is required to "practice" his art until he is good at it. This approach though it may have some humanistic side effects, is very slow and unpredictable. We have little way of knowing what aspect of a problem a student is attending to and what knowledge he is assimilating. It seems that we could go a long way toward organizing a body of knowledge about design if we knew more about what a designer does when he designs.

In order to progress, it is necessary to consider something less than the entire scope of human design activity. In becoming a designer one ability which the student must achieve, is that of conceptualizing and manipulating forms in space. Attempts are made to teach this by having the student mentally manipulate the spatial relations among a few basic shape components (cubes, spheres, complex surfaces) and then gradually represent these geometric images typically as perspective view. The interesting aspect of this exercise is that (except for the component parts) students are drawing objects they have never seen.

Of major importance in such an exercise is the teacher's ability to uncover some of the basic processes and organizations of knowledge which allow an individual to conceptually build an image and to identify some of the deficiencies of those students who can not accomplish this task. These objectives can be more specifically stated: a. An information processing theory of human problem solving has been proposed[1] which utilizes the concept of applying a series of internal processes to succeeding states of knowledge. A state/process explanation seems at first glance to be valid. How well does it hold up under investigation? b. In order to deal with graphic images there must be some way of dissecting this image and storing the data internally. What is the nature of this data structure? c. Our experience has shown that some people are better at this task than others. What individual differences might account for these individual abilities and to what extent are they influential? d. In addition to differences in people, there are differences in problems particularly in terms of difficulty, thus—what constitutes difficulty?

The first step in this investigation was to determine a technique of gathering information for analysis. We needed to have a test or situation which would be sufficiently revealing and controllable to provide reliable results. The general approach used was the "problem solving protocol." A subject was put into a problem solving situation and instructed to " think out loud" as he worked his way toward a solution. More specifically, the task and the task environment were structured as to enable him to describe his steps toward a solution in as much detail as possible. This description was recorded and analyzed to identify the structure of the subject's problem solving process, and thus answer some of the above questions.

The task in this case was to complete a series of "missing-view" problems. In each problem, two of three necessary orthographic views of a geometric solid was given to the student.[2] The subject's task was to provide the missing view based on information included in the other two views. This requires that the subject not only recode graphic information internally, but use coded data to construct a new external graphic image.

In the initial runs, the subject was required to solve the problem without the aid of pencil and paper. The intent here was two-fold: (1) to make the subject rely fully on internal procedures (including internal memory); and (2) to maximize his verbal output by not providing an additional activity (drawing) which might divert him from his verbal report. In later runs, the graphic constraint was dropped because it was believed that useful information would be gained from a graphic protocol which proceeds in step with the verbal protocol. In addition, each subject was given the opportunity to look at an orthographic information sheet.[2] This information is from French & Vierck's engineering graphic book which is a classic explanation of the orientation of views in engineering drawing. This intended to prepare the subjects for the task without giving them the opportunity to rehearse on a sample problem.

In total, three different problems were used. The best results (and those covered by this paper) were achieved using the problem shown in Figure 1.

Information Processing System

Thus far, we have been looking at what the subject does and have not discussed how he does it. As a foundation for doing the latter, we must first outline the components of a general information processing system. If our subject's problem solving behavior corresponds to this foundation, then we may be justified in proposing a visual information processing system. (IPS)

The following definition of an IPS leans heavily on a description of such systems by Newell & Simon.[1]

Generally, an IPS consists of a memory, a processor, effectors, and receptors (Figure 2). The memory is a component which is capable of storing symbol structures. A symbol structure can be thought of as "instances of symbols" connected by a set of relations, that is, structured information.

The processor, the hub of the system, consists of:

a. A set of elementary information processes
b. A working memory—sufficient to hold I/O symbol structures for the
c. An interpreter which determines the set and sequence to be executed based on the symbol structures in working memory.

This combination forms a system which can take in information from the external world and encode it, via a symbolic code, into an internal representation of the external problem. The processor, under direction of the interpreter or control program, is then capable of operating on this and any other information drawn from long-term memory by choosing and applying the appropriate method from a store of possible methods. This results in a modification or enlargement of the internal representation which may require a manipulation of the external environment via effectors. This model, which broadly but accurately describes computing machines, is, I propose, also applicable to human problem solvers.

Implicit in this system are several components which should be made explicit.

1. The internal representation—a very complex representation of the external world or "task environment." It reflects not only the input information but a "problem space" or organization of the information befitting the task as the problem solver sees it, and eventually a restructuring of information which constitutes a solution to the problem.
2. Data structures—symbol structures useful in encoding information for retrieval when needed.
3. Programs—a special kind of symbol structure which organizes a series of information processes into a functional unit.
4. An external environment of sensible readable stimuli.
5. Elementary Information Processes—our processor must also shave a set of process which it can apply. From our knowledge of computing machinery and human processors we need at least the following processes:

Top View

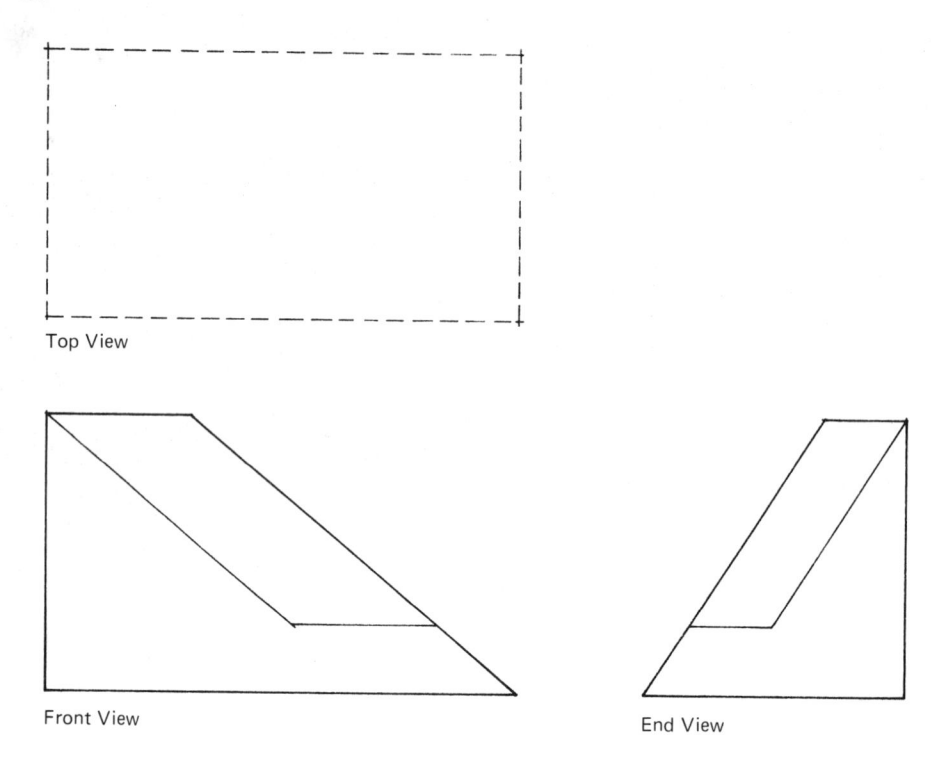

Front View

End View

Figure 1

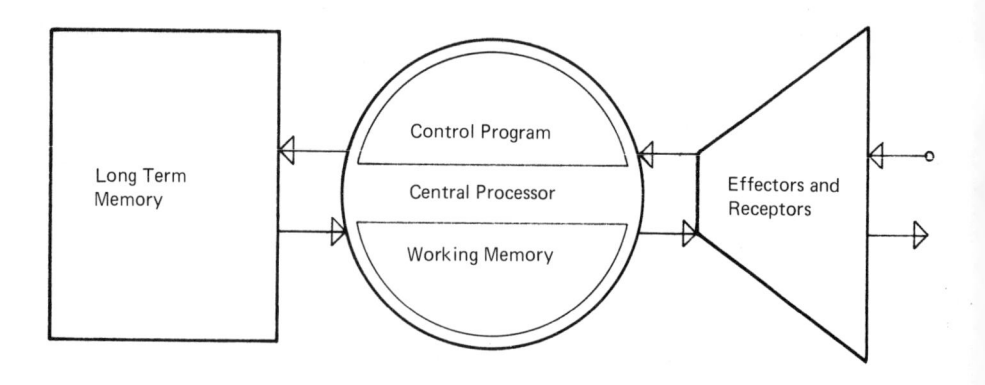

Figure 2. An Information Processing System (I. P. S.)

a. Reading & writing externally—receive information from the external world and work upon it.
b. Discrimination (or identification)—the ability to determine what its behavior should be, dependent on the input symbols.
c. Comparison (matching) of two instances of symbols belonging to the same symbol type
d. Symbol creation—in essence, the ability to restructure previous symbol structures.
e. Storing symbol structures.

Nevertheless, the general question remains as to whether the human problem solver fits the model and more specifically:

1. What changes occur in the internal representation?
2. How do these influence his choice of problem solving method and thus his progress?
3. What data structures seem applicable to the task?
4. Can we identify processes that account for his problem solving ability?
5. What might be his program for ordering of processes?

The protocol analysis as outlined by Newel[3] suggests a method for answering these questions.

A Visual Information Processing Theory

The core of the visual IPS involves the derivation of a program which can account for the behavior of human problem solvers. The chart (Figure 3) indicates the proposed general flow of control in the problem solving IPS. The main axis of the process embodies the fundamental search and test processes. The tests can fail, however, leading into a series of hypothesis generating loops. The inner loop is iterative in nature; the outer loop is also iterative but includes a recursive capability. It allows the IPS to change the level of detail to which it is attending.

Start. Though the process may be initiated at any point and at any level of detail, the probable starting point is at the most general and inclusive level of detail. (It will become clear why this is not critical to the process.)

Attend to Part. Once a point of attention has been established, the system must take note of the important data about that part of the graphic image.

The data which must be symbolically expressed includes such things as the location of attended part, its position relative to nearby parts, and the level of inclusiveness represented. To clarify the last concept, let us assume that there are four fundamental levels of inclusiveness. The highest level (4) is represented by the total shape or its representation in the existing graphic information. Below this in level (3) are the basic subdivisions or simple modifications of the form; these may include major geometric components, or views, with confusing lines removed. Level (2) consists of smaller segments, generally line segments, vertices, angles, etc. The lowest level, (1) is represented by individual points in the graphic information.

Search Data for Match

This procedure is primarily concerned with retrieving information about geometric forms from long term memory, inserting it in working memory, and testing to see if it matches information in the symbol structure derived from the part being attended to. Implicit in this process is the assumption that the retrieved information is pertinent only if it is at the same level of inclusiveness as the attended part.

Does a Match Exist. If our system is able to find a match between stored and read information, we can say that the system has recognized the input information. That is, it has meaning within the general understanding of geometric space. At this point we can say that the IPS has generated the hypothesis that the part of the graphic information which is attended to represents the geometric qualities associated with the piece of stored data that achieved the match.

Build IR Links. Here we refer to the IR of the form which is externally represented by the graphic information. Essentially, points are being scored for the above hypothesis. Let me propose that the proper structure is a tree with loops. Building the structure can consist of adding new branches or links, if a successful hypothesis is on a different level from the previously successful one, or adding strength to existing branches, if hypotheses are on the same level.

All Parts Accounted For. This is fundamentally a test for completion. If the IR symbol structure thus far completed involves each part of the graphic information (at some level) then the IPS has completed its job. This does not demand that all parts be understood on the highest level of inclusiveness. This is implied from the fact that subjects were able to draw perspective and top views when they reported that they were not able to "see" the total form.

IR Complete. If the symbol structure passes the test in the section above, then the IPS reports that it has solved the problem. In this case, the symbol structure contains sufficient information about the referent form that particular attributes, such as top or perspective views, can be generated at will.

Redirect Attention. It may be, however, that the last test indicates that much of the graphic information is unaccounted for. It then becomes necessary to correct the deficiencies. In order to do this efficiently, the IPS must be able to direct the course of its subsequent problem solving activities. Thus, this phase is perhaps the most complex and detailed of the entire program. Among the things which must be accomplished here: determining the present level of inclusiveness, deciding whether to move up, down, or stay at the same level; and specifying possible locations on the graphic information for useful data. This box represents a sub-program whose objective it is to store a number of rules, such as the rules of orthographic drawing, to be invoked in order to efficiently direct the search for information upon which to build new hypotheses. Assuming that the sub-program is successful, control moves back to Attend to Part, and a new cycle begins.

Delete IR Links. This phase is similar in nature to the building node but is the result of a failure at the hypothesis testing node. There must be specific rules for deleting the structure; it is clear from the protocols that an IPS does not necessarily erase everything at each failure.

Any Unsearched Data. If the IPS fails to find a confirming match, two courses of action can be taken. If all branches of the data structure on geometric forms have not been searched, the IPS can return to the search node and continue searching at the present level of inclusiveness. If however, that level of data has been exhausted, control must be turned over to the program in the redirect node for further direction. This path accounts for recursive behavior which moves to lower and lower levels of inclusiveness.

Failure. This node indicates the failure to achieve a symbol structure which accounts for all parts of the graphic information. This generally occurs when the store of rules and techniques in the redirect node is exhausted without success. It corresponds to the subject reporting that he gives up because he has no idea what to do next.

An incorrect IR is really a special case of a complete IR. Although its symbol structure accounts for all parts of the graphic information, certain parts may be accounted for by links that were established by incorrect application of geometric principles, lack of discrimination (bad matching) or failure of the system to adequately confirm its hypotheses. Except in the case of ambiguous figures, errors such as these could be eliminated by extensive feedback and review. For ambiguous figures, any IR is a correct IR, as long as it fits the graphic data.

We now have a basic progarm which simulates visual problem solving behavior. Specifically, it has the capability of (1) searching for and attending to particular data, (2) generating and testing hypotheses about relationships within the data structure, (3) iterating through the generate/testing process, and (4) recursing to different levels of inclusiveness.

In addition, there are several points where more specific systems must be introduced. Fortunately, much of this work has already been done by researchers in the fields of cognitive psychology and IPS generally.

In Search Data for Match we must propose a data structure which is amenable to searching long term memory. Feigenbaum has proposed what he calls EPAM (Elementary Perceiver and Memorizer).[4] It is essentially a discrimination net (or tree structure) which can be modified by learning (Figure 4). As information moves down this tree structure, it is categorized by succeedingly finer distinctive features. If the system is required to make finer distinctions than its existing branches allow, it can, through learning, grow new and finer branches at a deeper level of discrimination. If, on the other hand, only a gross discrimination is required, the tree need not be traversed to its ultimate depth. This is a very flexible kind of data structure and has been applied to other kinds of human behavior, for example, in verbal learning experiments. It should be added that other data structures will work as well such as ordered lists, for instance which are isomorphic with trees.

In Figure 4, information at the deepest level of discrimination has been shown in a graphic rather than verabl form. This illustrates one source of error for the IPS. The "symbol structures" at the terminal nodes include perspective and orthographic views of each of the three form concepts at two different object/observer orientations. If one is concerned with topology rather than proportion, one part of the graphic information could be easily matched at several locations in the tree. This situation corresponds to ambiguous information, where the interpretation of graphic data is not solely dependent on having achieved a match but also depends on the path by which that match was achieved. Therefore, an IPS might pursue a cubic or tri-prismatic interpretation of graphic data depending along which path his earlier matching successes lie.

In order to clarify the Building-Deleting of IR Link phases, a data structure must be specified for internal representation. To correspond to the EPAM net of long term memory we are proposing what Moran has called a POSET (Partially Ordered Set) structure.[5] This can be thought of as a tree where some of the branches converge. The ordering, from top to bottom, is with regard to level of inclusiveness but other structural relationships, such as adjacency, can also be indicated. (Figure 5).

Figure 5 is a POSET for a cube that might be built by an IPS which is concerned with a cube as a framework of edges. Hence, it first branches into a "top", a "bottom", and four vertical edges. The cube is not decomposed in a clean geometric way; rather, the top and bottom parts also include some information about where additional edges may be attached. This kind of decomposition is typical of what human problem solvers report. That is, a form seems to be decomposed by informal "parts" rather than by strict geometric components. These parts then retain a maximum amount of pertinent topological information. It should be noted that Figure 5 is not the only decomposition of a cube which is possible. If the subject were "set" to consider the form as made of paper planes or a block of wood, his decomposition of the form would probably quite different.

The completely interconnected POSET is a very complex structure, rich in redundancy; the minimum structured data needed to generate a view of a cube is far less than shown. All that is required is that each part be accounted for by at least one link. The evidence is that subjects were able to "construct" or generate views of fairly complex forms, even when they reported that they did not "see the whole thing."

With the exception of the terminal nodes, the basic problem solving flow chart is concerned with the direction of attention and choice of operators. It should be possible, therefore, to collapse these nodes (and some other considerations) into a sub-program called the Control Program. This is what Newell and Simon refer to as the interpreter (Figure 6).[1] The Control Program outlined in Figure 6 does not simulate the total behavior of human problem solvers, but it does help to define its functions. First, it provides a place for the IPS to start and specifies conditions under which the IPS may stop, both successfully (IR Complete) and unsuccessfully (Failure). Second, it directs the ongoing progress of the IPS after each modification of the IR symbol structure, it determines whether a change is needed in the level of inclusiveness and what that change should be. Third, it directs the retrieval from long term memory, of operators which may modify the input data structure in a useful way. One important function which the Control Program does not cover is direction of search for data, both internal and external, to update the working memory. Presumably, this function implies, for this task, a knowledge of orthographic drawing rules which would indicate where useful information might be located in the external data (drawings and sketches).

Figure 4. Discrimination Net Structure (EPAM)

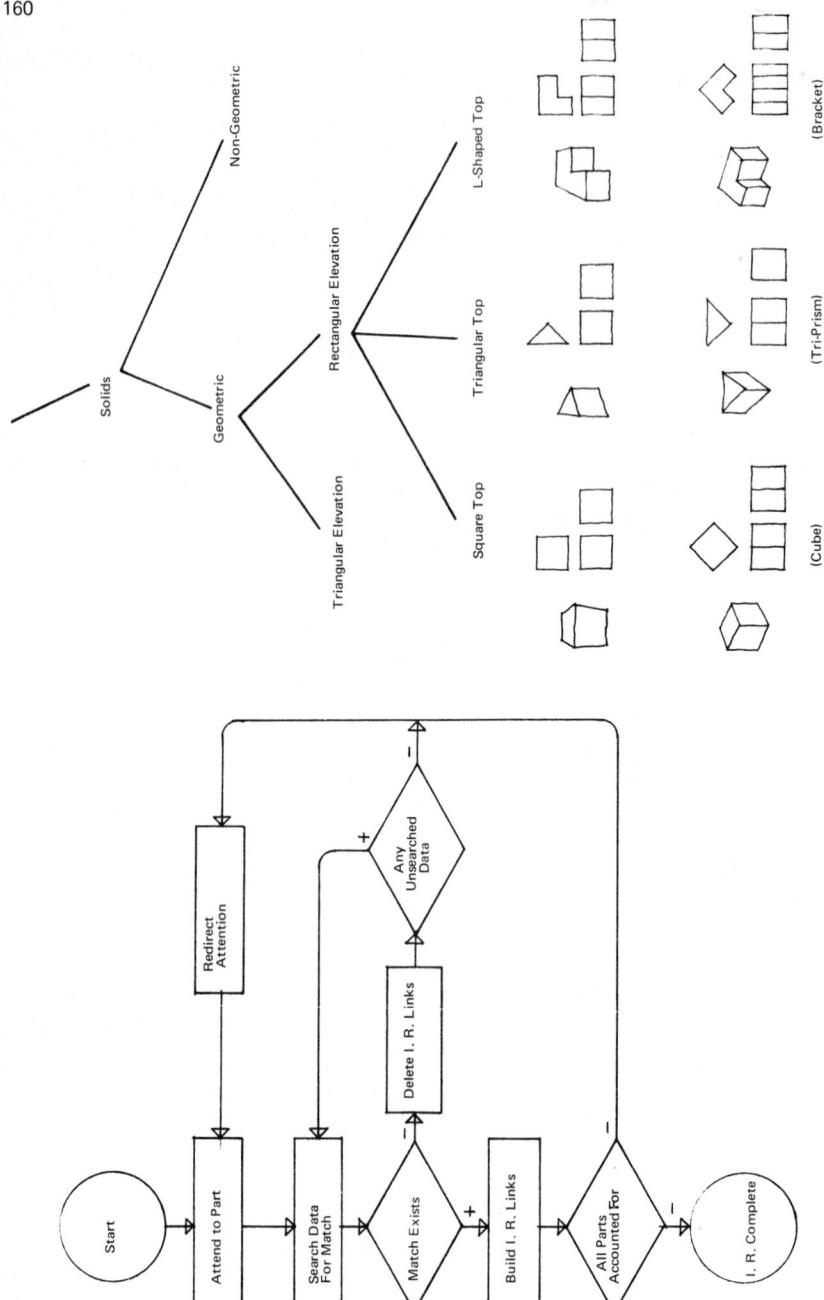

Figure 3. Basic Problem Solving Flow Chart

161

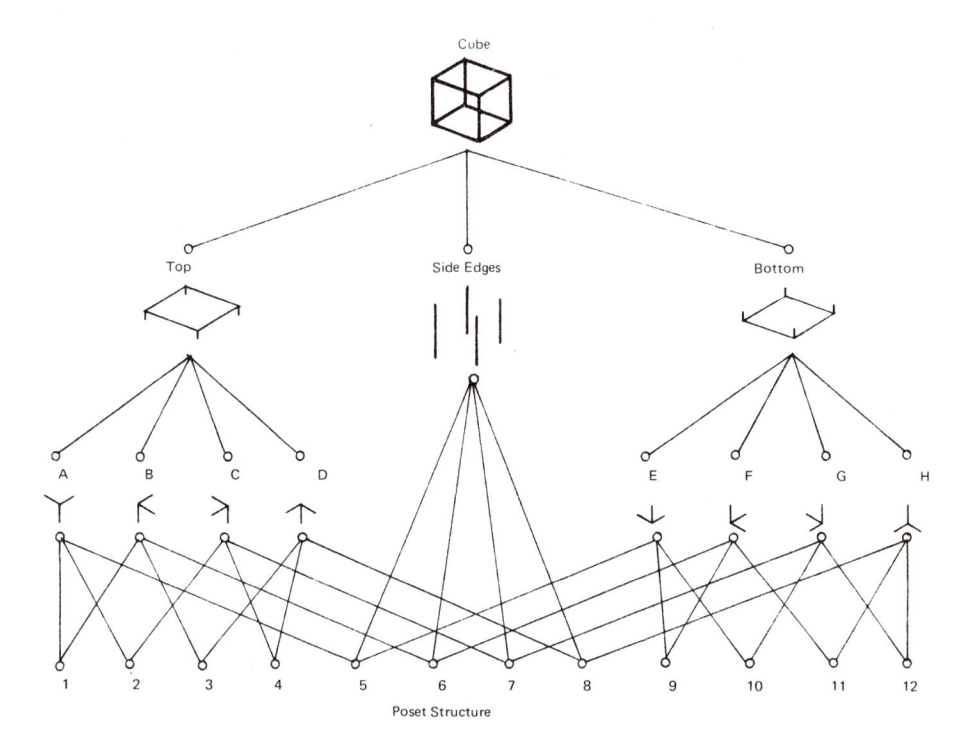

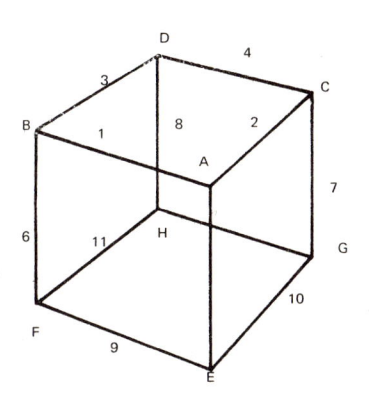

Figure 5. **Reference Cube**

This proposal for a visual IPS had led into several diverse branches. It should be helpful to the understanding of the total concept to reconnect these ideas. In the general view of a visual IPS (Figure 7) an attempt has been made to reconcile those concepts and processes which have grown out of the problem solving protocols with the general concept of an IPS as presented in Figure 2. The central processor contains the control program and other elementary processes (search, matching, discrimination, symbol, structure manipulation) and working memory. The working memory includes the internal representation which is being continually updated and stored in a similar structure in long term memory.

Long term memory also includes, presumably in an EPAM-like structure, information about geometric forms and problem solving heuristics and potentially useful data-manipulating operators. All of this is accessible, via the Control Program, to the working memory. Finally, the system has access, via its receptors, to the external data which includes both the original problem sheet and any additional data which the system may store externally through the use of its effectors (sketches).

This proposal has a parallel in the field of maching pattern recognition. Guzman has designed a program called "See"[6] which analyzes a field that has been divided in a manner that is equivalent to a perspective drawing of a stack of blocks. The analysis leads to a decomposition of the scene into a set of "blocks" by grouping or linking regions which constitute one block. This is not accomplished through complex mathematical geometry, but by the application of a few sub-processes which are similar to those herein attributed to human problem solvers.

There are two fundamental kinds of information which are processed: information about the regions and information about vertices. The information about a given region includes ordered lists of neighboring regions and vertices belonging to the region. Vertex information includes x and y coordinates plus ordered lists of adjoining vertices and adjacent regions. In addition, there is a store of general knowledge about the topological implications of typical region/vertex situations. From this and the ordered lists of region/ vertex information, it is possible to establish "links" between regions which imply that they are part of the same block. Thus we may say that the internal representation of the scene begins as a list of regions which is unstructured with regard to its three-dimensional meaning, and that solving the problem consists of establishing this structure by finding evidence for links between related regions.

Future Directions

In conclusion it should be noted that individual differences, the nature of difficulty, and external representations have not been adequately investigated. The proposed visual IPS includes three specific elements which seem to be particularly subject to individual differences. First, the contents of long term memory depend on an individual's background and training. Although differences in ranges of forms distinguishable by the discrimination net exist, our experience suggests that the quality of general knowledge about problem solving may be more important. Expert visual problem solvers seem to have a large repertoire of manipulative operators at their command and a set of problem solving and search heuristics which enable them to choose an effective operator and apply it to the proper subset of symbols.

Second, differences in the control program may be important. Problem solving protocols have revealed instances where the subject failed to find a solution because he apparently was unable to effectively redirect his attention.

A third factor which will undoubtedly affect performance is psychological set. Subjects in these protocols were told that they were dealing with representations of geometric solids. If they were told nothing, or that the drawing represented planes floating in space, then their performance would certainly have been different.

The Nature of Difficulty involves a different question. For a given subject under consistent instructions, what makes one problem more difficult than another? It has been suggested that this is a function of the complexity of the shape and familiarity with the form. This is certainly true, but familiarity represents a dif-

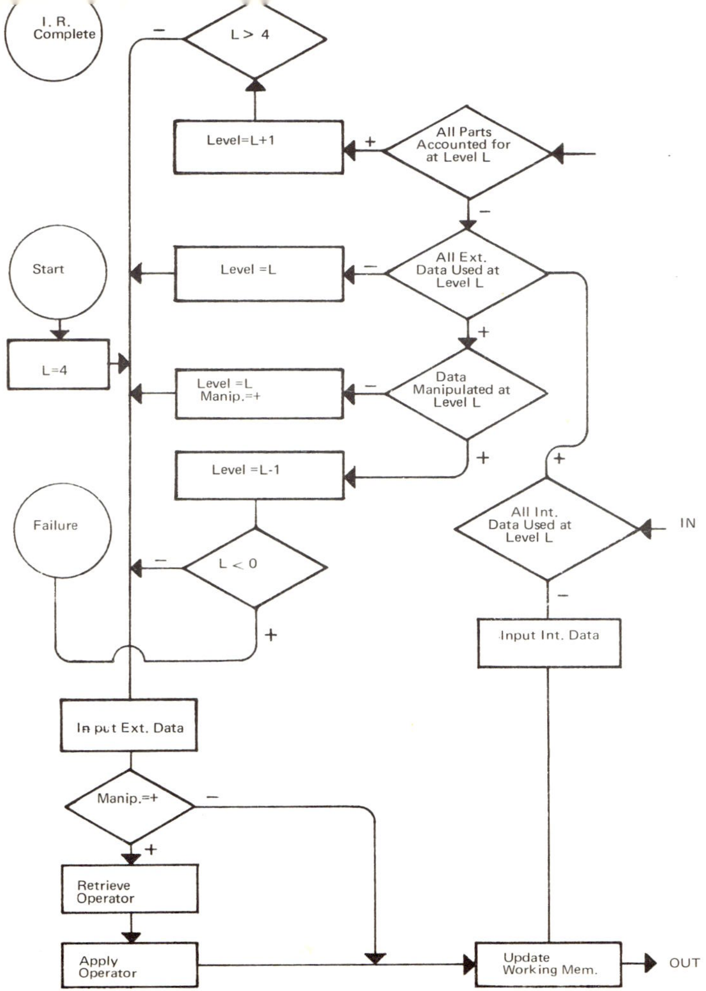

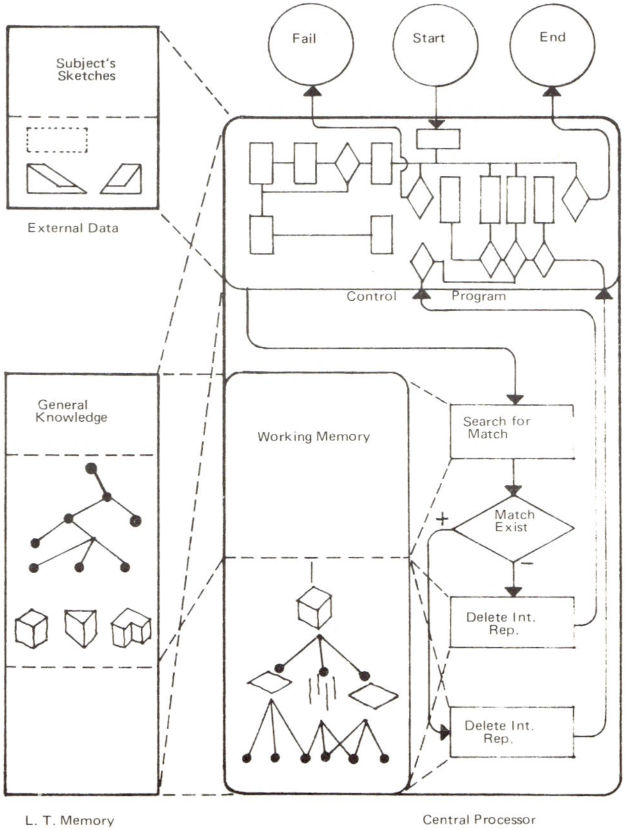

Figure 7. General View of Visual I. P. S.

Figure 6. Control Program

ference in the problem solver, not the problem. A familiar form may be one for which the subject has large well-rehearsed set of operators. The best place to look for differences in problem difficulty is not in the referent form, but in the information about that form. Redundancy will clearly aid and ambiguity will clearly hinder problem solving.

We have been primarily concerned with determining the process by which a person builds an internal representation of a form based on external information. To do this we have gathered data by what may be a reversal of this process. We have been asking subjects to produce external information, graphic and verbal, which we assume reflects the state of their internal representation. We can gain a great deal of confidence in the IPS theory if we can show that the building of external images, from an internal representation of coded symbols, follows a similar IPS program. This may be the next interesting question: How do we draw?

References

1   Newell, A. and H. Simon, *An Information Processing Theory of Human Problem Solving,* Working Draft. Carnegie-Mellon University, Pittsburgh.

2   French, T. and C. Vierck, *Graphic Science,* McGraw-Hill, New York (1958).

3   Newell, A., "On the Analysis of Human Problem Solving Protocols", Department of Computer Science Paper, Carnegie-Mellon University, Pittsburgh (1966).

4   Feigenbaum, E., "The Simulation of Verbal Learning Behavior" in Feigenbaum, E. and J. Feldman, *Computers and Thought,* Mcgraw-Hill, New York (1963).

5   Moran, T., "A Grammar for Visual Imagery", Working Paper, Carnegie-Mellon University, Pittsburgh.

6   Guzman, A., "Decomposition of a Visual Scene into Three-Dimensional Bodies", AFIPS, Vol. 33, Thompson Book Company, Washington (1968).

Isao Oishi
Harvard University, Cambridge

TOWARD THE COMPUTERIZATION OF ARCHITECTURE:
A THEORETICAL FRAMEWORK AND A PARTIAL COMPUTER PROGRAM

The paper comprises three major sections. First there will be a brief discussion of a theoretical framework for the design process, embodied in Figure 1. This will be followed by a hypothetical version of how the process can be simulated. Finally, this paper will touch upon a partial computer program of the process and its results. It should be noted that the framework and the intent of the proposed simulation are stressed —the partial program is but a first anticipation of a comprehensive program.

Figure 1 has four major sections. On the far left is the "Activity Systems" block, a broad category which only asserts that all physical facilities are designed to satisfy some activity.

The three squares which follow the "Translation" bubble comprise the second major section, the "Selection and Arrangement." The elements within each square are identical; they are differentiated by function only. The first square is involved in the "Selection of Patterns," the second in the "Selection of Units," and logically, the third square is concerned with the "Arrangement of Units in Patterns." Different elements within the squares are called to perform the different functions.

Above the "Selection and Arrangement" is the third section, the "Weighing and Trade-offs from Constraints" function. Every problem is limited by scarce resources, constraints, and conflicting demands. Hence, priorities and trade-off functions must guide the design process and resolve the conflicts. Any conflicts that are found following the selection are sent to this section.

Finally, section 4 involves the translation of the design into specifications and working drawings.

One of the primary concepts integrating Figure 1 is the iteration from the broad pattern to the smallest infilled detail. Simulation of the design process can be expected to initially select the broadest pattern and the broadest units for arrangement, then continually iterate, selecting ever smaller patterns and units with each pass. These smaller units will continually infill the larger patterns and units until all levels are exhausted and a computer design results. This iteration is represented by the arrow returning from the diamond at the end of the three squares, back to the front. Every section of Figure 1, then, must operate at the same level of generality at any given instant. Conflicts and trade-offs can be expected to occur in interfaces within and between levels of aggregation, or generality.

The Activity Systems

In addition to the models and flow diagrams, three concepts must be noted:

(1) Analysis of activity systems will produce the input to a computer program. There will be, as indicated in Figure 1, generalized data for any particular category of client as well as need for real time information to cover uncertainty and variables for a customized product. This is contrary to some procedures which offer highly aggregated, non-defined "relationship values" between rooms as input.

(2) Differentiating between and interfacing levels of aggregation is essential to any organization of infor-

166

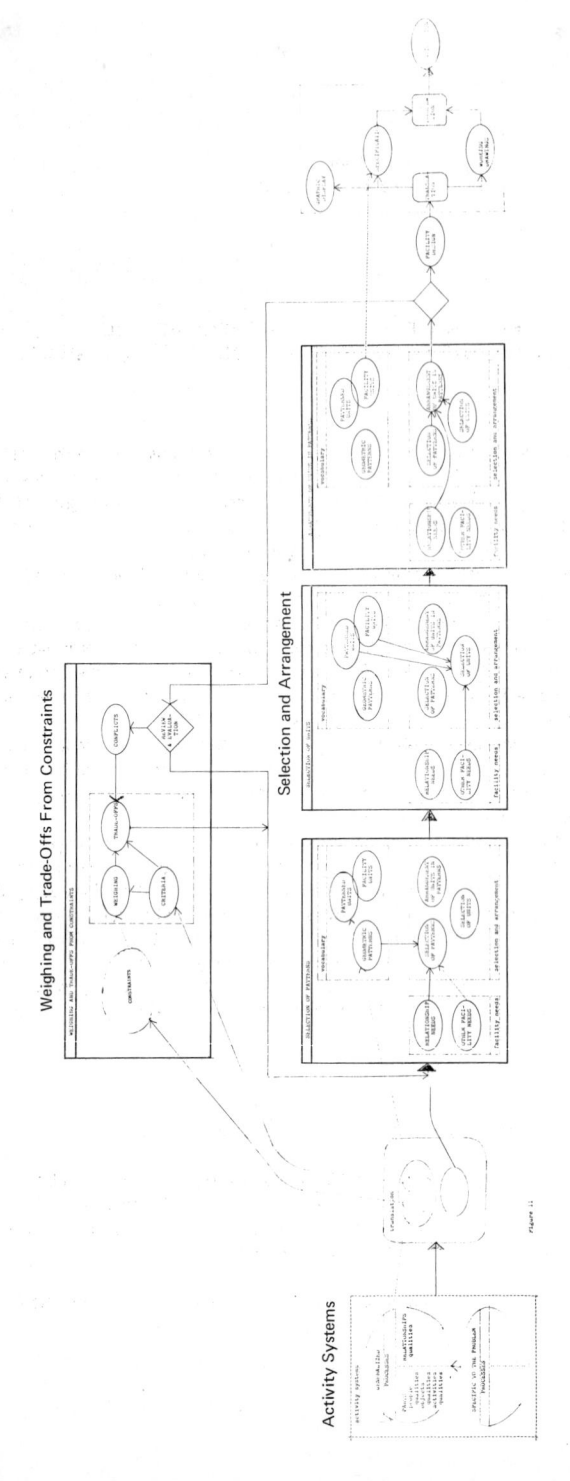

Figure 1. Theoretical Framework for the Design Process

formation and to successful design simulation. Each level is a constraint of those below it; each lower level will modify and refine the product of the upper levels. The upper levels will likely define the more generalized parameters while the lower levels will not only determine the precise selection and arrangement of small units, but will likely be the key to customization for particular clients.

Once the activities for any general category are properly defined and modeled at their respective levels, the translation into "Facility Needs" (performance requirements) can occur much more smoothly and consistently. Again, equality of the level of aggregation for all sections at any one pass is important. Otherwise, increased conflicts will develop in the process and inferior design will result.

(3) Although objectivity is always difficult, increased efforts must be made to differentiate between the activities empirically recorded, the value systems inferred or extracted from the activities, and the designer's own value system. The understanding of how each interacts becomes increasingly important since the designer's priorities and trade-offs are, in essence, a reflection of his value of the simulation of the design process.

Selection and Arrangement

The three elements in each box are the "Facility Needs", the "Vocabulary," and the "Selection and Arrangement" process. The Facility Needs are translated from the activity systems, level for level. At the broadest level, they are better recognized as performance requirements. At lower levels, they are recognized as "specifications." At every level, they must be processed into a form capable of suggesting specific patterns and units. Facility needs are separated into those between patterns or units (Relationship needs) and those unique to one pattern or unit (other facility needs). (1) Cumulative studies over time will find generalized and real time facility needs. It is stressed that a study resulting in facility needs, performance requirements, or whatever, is disappointing indeed if it fails to interface well with other levels.

The three subsets of the vacabulary component range from the broad to the detailed in terms of their specificity. The "Facility Units" are the least common denominators—chairs, appliances, lamps, etc.—which can be quantitatively defined or rated in terms of dimensions, qualities, etc. The "Patterned Units" are comprised of the smaller facility units. They cover the range from small conversational areas to various living room arrangements, house arrangements, dormitory arrangements, campus arrangements, etc. The larger the unit, the fewer the constants and parameters. Patterned units should represent more or less modifiable blocks or modules of data which can be called and manipulated en masse. The "Geometric Patterns" are simple in concept but when used in conjunction with the process of iterating from the broad to the detailed levels of specificity and aggregating at the different levels, they provide a powerful tool towards guiding computer simulation. These geometric patterns are linear (shopping center), circular, (courtyard plan), cylindrical (3-story mall), spherical (great space building) and combinatorial (Salk Institute plan). Some additional characteristics of the geometric patterns need mentioning:

(1) a pattern at any one scale may be composed of patterns at lower scales
(2) they apply at all scales
(3) they are all-inclusive
(4) at the higher scales these patterns are frameworks only and are extensively modified and refined by those patterns and units selected at lower levels of aggregation.

Selection and Arrangement: in the "Selection of Pattern" component **a pattern is selected from** highly aggregated facility needs. subset, in particular, relationship needs. Next, in the "Selection of Units" component, the units are selected principally from "other facility needs." At the higher levels, patterned units representing highly variable clusters of subunits are selected. At the lower levels, specific furniture, materials, etc. are selected. Finally, the units are arranged in patterns in the third square. One notices that more relationship needs are necessary.

Evaluation

Next, the computer asks if all levels are exhausted. If not, it proceeds up to the third section.

Weighing and Trade-offs from Constraints

At this stage the program will check for conflicts identified above. If they exist, the computer makes a decision and proceeds through the "Selection and Arrangement" process.

The simple diagram in Figure 3 represents a truly complex phenomenon the concepts of which are fundamental, integral components of other disciplines, such as engineering and economics. Everything in this particular section occurs at every phase of decision-making. One can divide the problem into the complicated procedures of deriving the proper facility needs, priorities, etc—the entire programming phase before the translation of its results into architectural form. The results of the programming phase is a product itself, and as such, has implicit within it the process of weighing and making trade-offs within constraint One makes trade-offs from values which themselves were the result of trade-offs of less aggregated value systems. This does not mean that operationally, one cannot take the derived values on faith; it is only that researchers must make the basis of all values explicit and keep track of their flows and transformations at each phase of the design process. This is not an easy task. The effects of the constraints must not only be evaluated spatially, but on cost-benefit terms as well. Trade-offs and costing implies that the elements must be measured on a common metric. Apples and oranges, for example, must be converted from ordinal measures to cardinal measures for comparability.

Specifications and Working Drawings

Many of the specifications can be expected to come directly from the facility units in the vocabulary of this section on "Selection and Arrangement."

Before proceeding, let us review the model. The computer will read in certain activity relationship needs and pack a pattern; read in more matrices and select a large cluster, or patterned units. Next, it will read in additional matrices from the relationship needs and arrange the units within the pattern. It will then go back and check for conflicts, resolve these if they exist, and return to the start for the selection of a smaller pattern. This will iterate until all levels are exhausted. Though not discussed here, future programs must have the capability to determine the effects of alternative decisions, or trade-offs, before a choice is made.

The Hypothetical Simulation

A comprehensive program does not exist, and therefore one cannot spell out the exact procedures of a run. Nevertheless, before discussing the partial program, let us run through one hypothetical example.

First, let us assume a problem involving the design of a center on an elliptical site. We take in information on the number and types of prosepctive occupants, the quality and quantity of relationships, between the occupants, particularly service and pedestrian traffic relationships, the parking requirements, the number of entry and service points, and the shape of the site. Upon analysis, the computer recommends a linear pattern and it reads in the traffic flows, the surrounding street patterns, and suggests a grid across the site which is not exactly co-axial to the site. A broken line is then selected. This line, of course, could be long or short, straight or curved.

Next assume that the shopping center can be divided into three clusters of shops. This is determined by research into the activities within and between these shops and the resultant facility needs. We find that these clusters are relatively self-contained and distance and are located around a major commercial establishment. (For instance, a Sears Roebuck buttressed by a number of specialty shops, barber shops, magazine stores, etc., or a grocery store with peripheral and traffic-sensitive stores). Let us label the clusters "A", "B", and "C". This knowledge was taken into consideration when selecting the pattern.

Now assume that the computer can distribute these along the line so that those most closely related to each other, at the broadest levels, are placed near each other. For instance, in Figure 2, "A" is more closely related to "B" than "C".

Next let us assume that there are relationships between "A" and "B" and "C" which can be expressed in terms of shops, circulation space, service docks, and other space consuming functions. These elements can be located between the cluster interfaces. At this stage, let these elements be points representing the first approximation of room centers. Symbolically, we draw the relationships as overlapping circles. The related or common elements are placed in the area of overlap, or intersection. Now, this overlap can be physically and non-physically expressed. Let us assume that at this level we can express them physically so that the expressions represent actual approximate locations of these elements in physical space. Next, the elements which are within each cluster must be distributed. These elements can also be distributed according to a pattern. In this case, let us investigate two alternatives for this distribution, a linear and a circular pattern.

Obviously a linear development results in a more extensive axis of development than does the circular pattern. This may or may not be poxxible. Further the linear pattern does not allow for as great an intensity of physical interaction as does a circular pattern. The circular pattern was elected for all three clusters for two following reasons: Let us assume that (1) there is not enough space for the linear sub-pattern along the main line and a perpendicular appendage would disturb the traffic flow; and (2) that the volume of pedestrian traffic, the need for resting and playing areas, and the advantage of visual identity and surveillance favored the circular development. Hence the computer distributes these elements in the non-intersecting areas of the circles. (See Figure 2).

Now we have two choices: we can assume that these subelements are subclusters of stores, or we can find that at this stage it is proper to define them as the first approximation of the centers of individual stores. Let us assume that the centers of stores are distributed in a circular pattern. Next, let us draw rectangles around each center. These rectangles will represent approximate space allocations for each store (Figure 3). Now, theoretically, lower level matrices cannot only distribute sub-clusters of stores and distribute individual stores, but also sub-clusters of activities within the stores and so on until it distributes the walls, windows, and furniture.

However, let us use only matrices which will arrange these rectangles into tighter fits. Overlapping rectangles are moved if there is no further facility relationship between them, walls are checked for adjacency and moved were appropriate, entry ways are connected to the linear mall, etc. This results in the real-world example (Figure 4). The program would then distribute the trees, the mall, the bricks, the grass, the light fixtures, etc. Now let us step down in scale and come to the partial program and to residential design. It can be appreciated that one important problem in the preceding sequence is the efficient distribution of the points, lines, or areas according to the matrices that are read in. Purely random methods are too unstructured for complex problem solving; what is needed is a program that will provide a satisfactory distribution, for every level, every time.

The Technical Theory

A subprogram which could translate the various matrices into some spatial significance on a plan was seen as the first problem. If a subprogram could analyse a relationship and accordingly locate points, lines, or spaces on a two-dimensional plane, then the problem of locating patterns would be a natural next step.

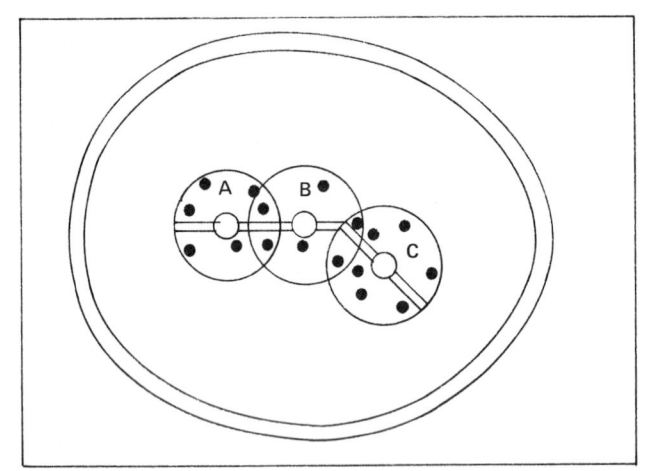

Figure 2

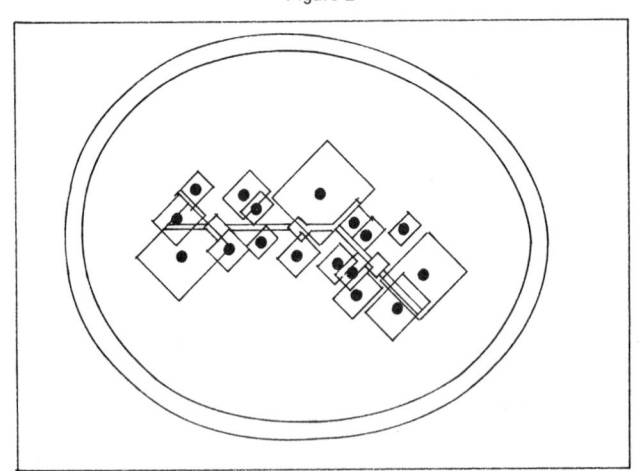

Figure 3

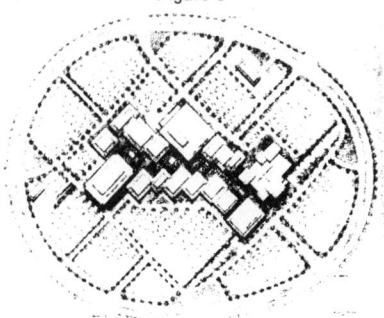

Figure 4

Following the locating patterns would be the Weighing and Trade-off from Constraints section, the Vocabulary, and the selected matrices could be slowly added in the form of other subprograms or additional inputs. Ideally, this location subroutine would locate the elements as a reasonable translation of Activity Needs on each and every run. This implies that for every element affected by a set of Activity Needs, there is an area or areas of solution where it can be located. If this area could be found and an element placed within it, the making of a Location Program is at hand.

A set of Activity Needs is required for testing and developing the Location Program. The horizontal distance Relationship Need was chosen. Since it was broad and other subrelationships would modify it considerably, the area of sulution would be relatively large.

The location Program would take the horizontal distance range and define it in terms of a minimum distance and a maximum distance. For instance, the minimum distance center to center, of a 6' x 8' kitchen to a 8' x 10' dining room may be seven feet; the maximum, twenty-nine feet. The shaded area within the concentric circle is the area of the solution. However, one finds for this family that the living room needs to be within ten and fifteen feet of the dining room and thirteen and thirty feet of the kitchen. Let us assume that the dining room and the kitchen are located satisfactorily. Then we have Figure 5 with concentric circles of ten and fifteen feet from "D" and thirteen and thirty feet from "K". The intersection, or shaded area, is the area of the sulution, hence, the living room must be in this area. Given the center points "K" and "D", and "L", one can draw more concentric circles to find, say, the playroom, and so on.

Given this extremely sequential or linear process, what can handle a number of elements simultaneously? Again, the sequence from the higher levels of aggregation to the lowest solves the problem when one locates a cluster of subunits, one is inherently locating simultaneous relationships as a unit. There is another great advantage of working with clusters. A program may iterate through only a few passes, yet solve for a large number of passes. For instance, one can locate ten clusters, then ten subclusters, then ten sub-subclusters. Hence, in three great runs, one can theoretically locate 10 x 10 x 10 = 1,000 elements.

The procedure for selecting a point in the solution area has recently been abandoned for a much shorter, improved subroutine capable of locating an element with reference to a large number of other points. However, the old procedure found points by selecting an angle and a radius. In Figure 6 point "A" or "B" can be selected as a reference point for the angle. Select "A". From "A", the program randomly selects an angle @ between angle OAO'. Next, it selects a radius between AM and AN. The endpoint of the radius identifies the new point "C". This becomes more complex when more concentric circles are introduced. Time limited the effort to only the selection of the third point. Additions will slowly be added to the new subroutine to handle variable solution area ranges and to further limit the random generator.

The Computer Program

The Location Program became a subroutine. Values were read in from the main program to the subroutine and new locations of points were returned. The first run located the first three clusters, just as "A", "B", "C" in the hypothetical example. Then the common points between the clusters were located. Since it is only a three-point program, no more than two other reference points could be read in to find a third. The general sequence, then was to find three points; then given two located points, find a third; given two more (or the same) points, find another third; given two more points, find still another third until all points are located.

Sometimes one can select a third point and check it for minimum or maximum distances from other points. If the conditions were not satisfied, a new point could be selected. Still, some relationships will remain unsatisfactory due to the inability of the program to locate a point with reference to more than two points at one time. This also increases the importance of sequence of selection; the problem of which space is to follow which.

The Results

The output resulted in points distributed on a grid of 6:10, Figure 7. This was due to the letter spacing of the computer type. However, with each run, the output was manually transferred to a 1:1 grid, each unit representing two feet. Next, room walls were drawn around each point and adjusted until a more reasonable layout occurred—a layout that is feasible given a few more matrices and/or subroutines. (Figure 8, 9).

To close, the important notes are the extraction of data at every level from activity systems as computer input, the iteration from the broadest or highest level of aggregation down to the lowest, and the implications of scarce resources at each phase of a problem. These are essential guides to a flow diagram which will outline all the subroutines, options, and capabilities that a first, relatively sophisticated, technical program to spatially manipulate must contain.

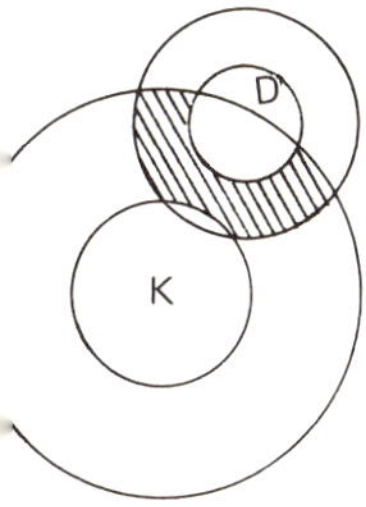

173

Figure 5

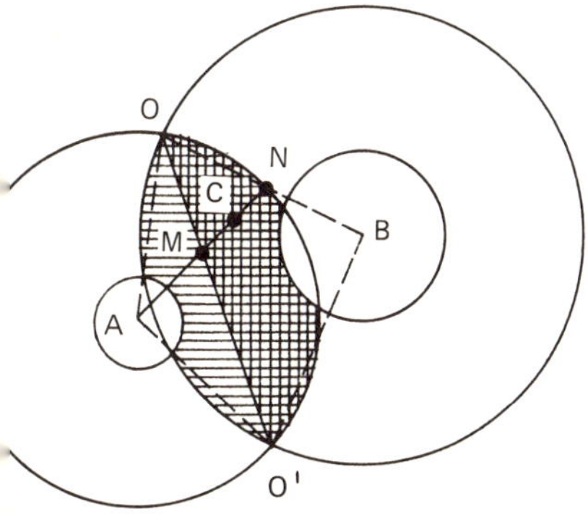

Figure 6

Figure 7

174

Figure 8.

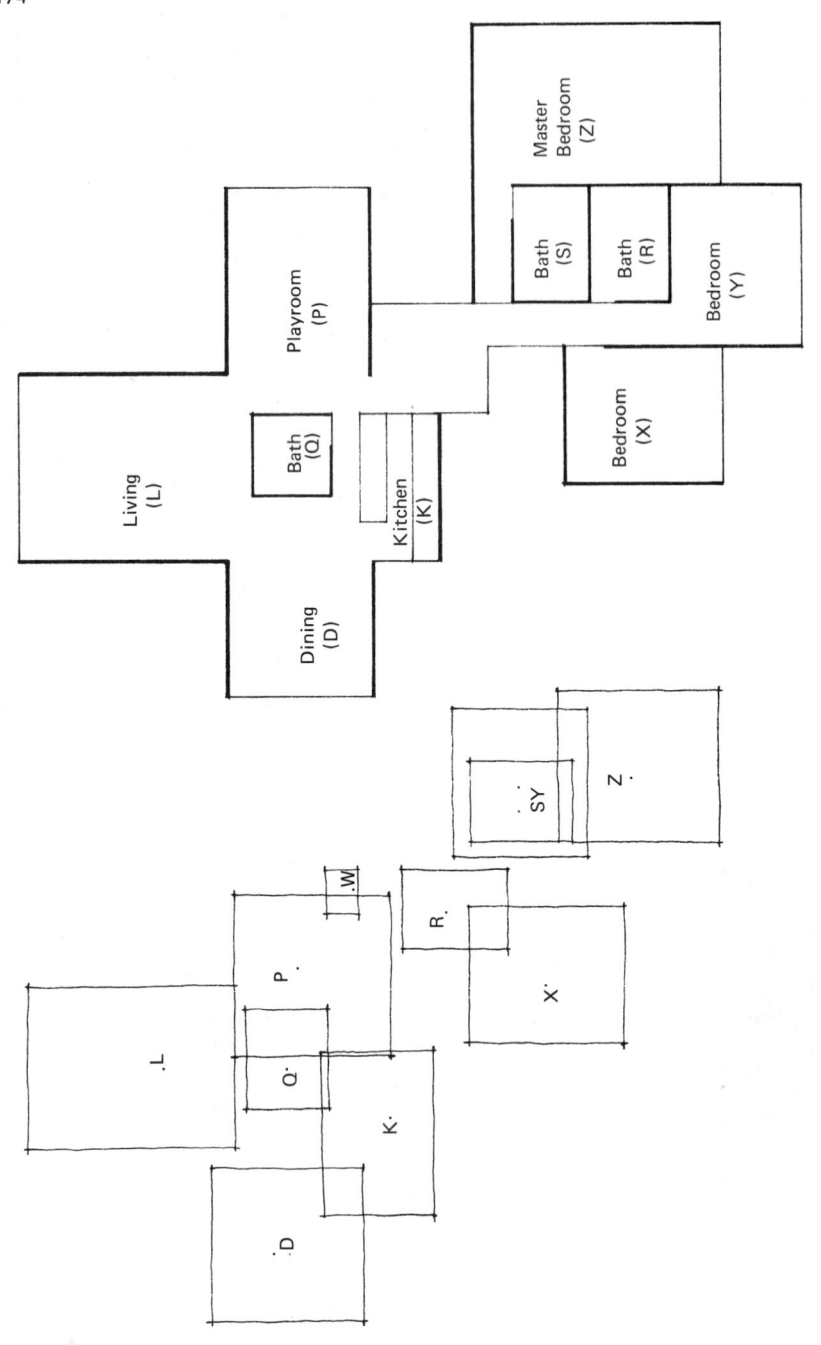

Figure 9.

John Grason
Department of Electrical Engineering
Carnegie-Mellon University

FUNDAMENTAL DESCRIPTION OF A FLOOR PLAN DESIGN PROGRAM

This work is intended to be a brief description of the salient features of a computer program, currently being completed, which takes a unique approach to solving certain space planning problems. Much of the theoretical background for this program and the details of its components was covered in a paper presented by this author to the DMG Conference in June of 1968 (1), and so these aspects will not be reviewed in detail here.

Basically, however, the objects that the program designs are two dimensional floor plans for a special class of buildings, i.e. buildings which are rectangular in shape, and which are composed of rooms which are themselves rectangles. The input to the program is a set of design requirements of two types:
 (1) *Location requirements,* which specify the adjacency of rooms, one with another or with certain outside walls of the building.
 (2) *Size requirements,* which specify the allowable range of physical dimensions for each room.
As its output, the program is capable of producing all floor plans (up to some specified limit N) which satisfy the set of input requirements. Arguments are presented by this author (1) and by Eastman (2) which indicate that the class of buildings which the program treats is a useful and quite general one, and that the two types of requirements which are treated constitute an appropriate basic set.

The Representation

The most unique feature of this program is the manner in which it diagramatically represents a partially designed floor plan in the computer. The representation used is that of a "colored," directed, linear graph, which is the dual of the floor plan, itself treated as a linear graph. (See Figure 1)

The basic motivation for using the dual graph representation is that it allows a relatively independent treatment of the adjacency and size requirements. Adjacency between rooms is indicated by constructing edges between the nodes of the dual graph which correspond to rooms, while room sizes are indicated by assigning weights to these same edges, which correspond to the wall segments separating adjacent rooms. In this way the two requirement types can be satisfied more directly and independently in the dual graph representation than in the literal diagrammatic representations usually used in floor plan design.

For the purposes of this description, the importance of the representation is the manner in which it affects the structure of the computer program. The complete description of a representation is the manner in which it affects the structure of the computer program. The complete description of a representation includes three components:
 (1) A set of *objects* suitable for describing the current state of the solution of the problem.
 (2) A set of *operators* for transforming one solution state into another.
 (3) A set of *prefixes*, or tests, for determining whether a given solution state has certain properties (e.g., satisfies certain design requirements).

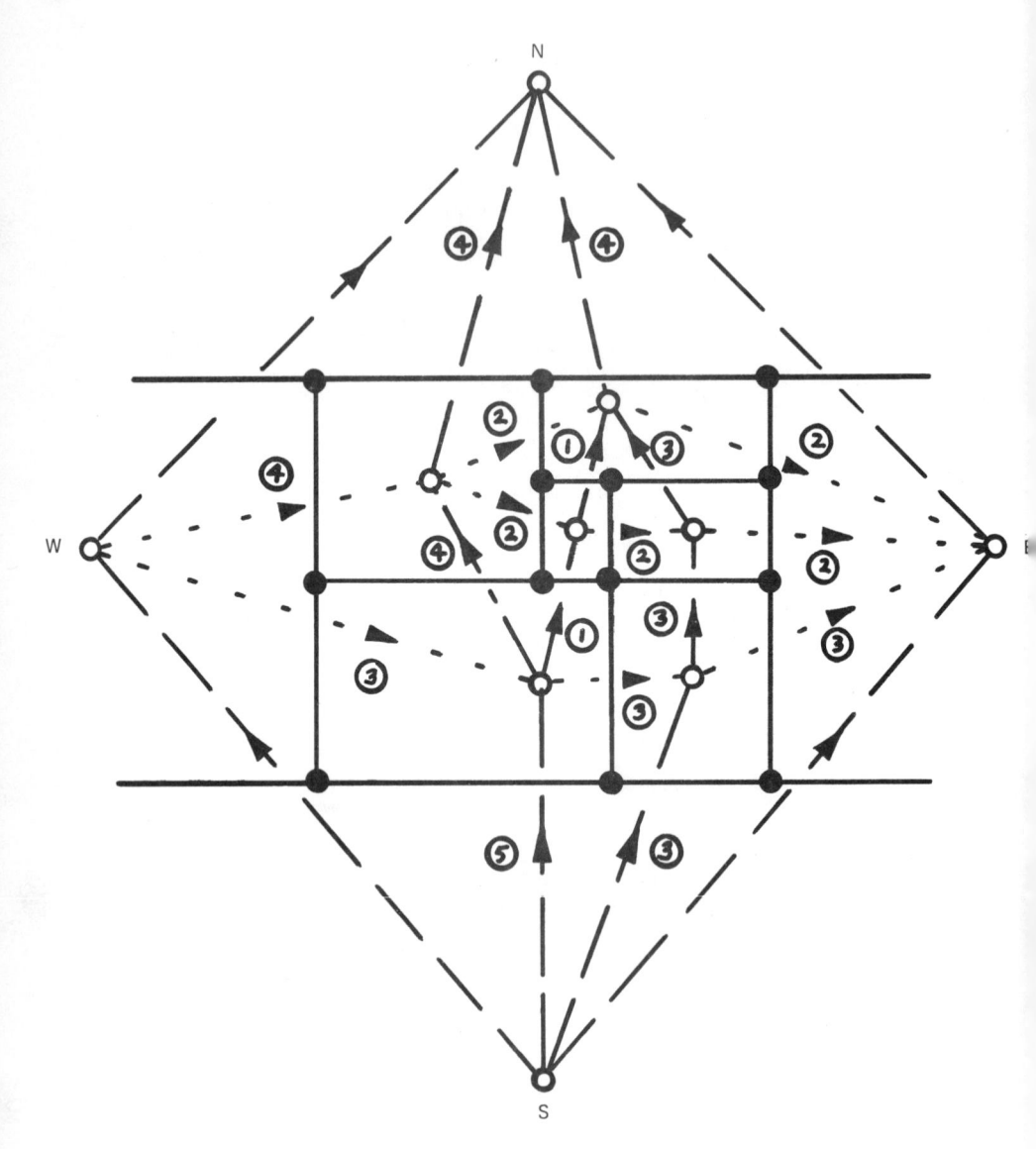

Figure 1. A Floor Plan Graph with its Dual Graph

The Objects

The basic objects of the dual graph representation are the *nodes, edges,* and *regions* which comprise the dual graph. However, to keep the complexity of the dual graph organized, a hierarchical structuring of its components has been devised. This structuring involves the introduction of three further objects into the representation:
(1) A *region structure* (abbreviated RS) consists of a set of regions, each of which is bounded by a circuit of edges which shares at least one edge with a circuit bounding some other region of the same set. (See Figure 2A)
(2) A *tree structure* (abbreviated TS) is a non-closed chain of edges in which any given edge position can alternatively be occupied by a region structure. (See Figure 2C)
(3) A *pivot pair set* (abbreviated PP) is a set of tree structures, all of which share a common pair of endpoints. Furthermore, the definition of a region is now generalized so that any edge position in the circuit bounding a region may alternatively be occupied by a pivot pair set. (See Figure 2B)

A graphical hierarchy of composite structures "belonging to one another" is set up using these three new structures along with edges, nodes, and regions. This hierarchical structuring is extremely useful in that it organizes the planarity information in a graph in a manner well suited to the design operations done in the program.

In terms of these objects, a given *state* of the solution of a floor plan design problem consists of a partially completed dual graph, which contains nodes corresponding to the rooms to be designed and some set of weighted and unweighted edges connecting those nodes. The dual graph is considered to be partially completed if not all room adjacencies and room sizes needed to completely specify a floor plan have been included. (See Figure 3)

The Operators

The operators which transform one solution state into another are those which add to the graph or subtract from it appropriately colored and directed edges, and those which add weights to edges or delete them. In the initial stages of a design solution, these operators are applied in response to design requirements. However, a given set of design requirements rarely completely specifies a dual graph, so in later stages of a design solution these operators must be applied to transform a partially specified dual graph into a completely specified one.

The Prefixes

The ultimate goal in the completion of a dual graph in the design process is that it correspond to a physically realizable floor plan. Through the use of formalizations, the author has developed a set of well formedness rules for determining whether a given partially completed dual graph satisfies a set of necessary but not sufficient conditions for physical realizability. These rules are the prefixes of the representation. Another use may be made of the prefix notion as well. That is, it may be used to test the dual graph *a posteriori* for the satisfaction of some alteration or combination of requirements from the basic set, such as the negation of an adjacency requirement. It should be noted that, although the prefix method of satisfying design requirements is much less efficient than the operator method, it is the only method used by some alternate floor plan design techniques, namely those which use some sort of "generate and test" scheme.

Structure of the Program

The program being described approaches the solution to a floor plan design problem in two main steps:
(1) The solution of the location requirements.

178

(B) A Typical Pivot Pair Set

(A) A Typical Region Structure

(C) A Typical Tree Structure

Figure 2.

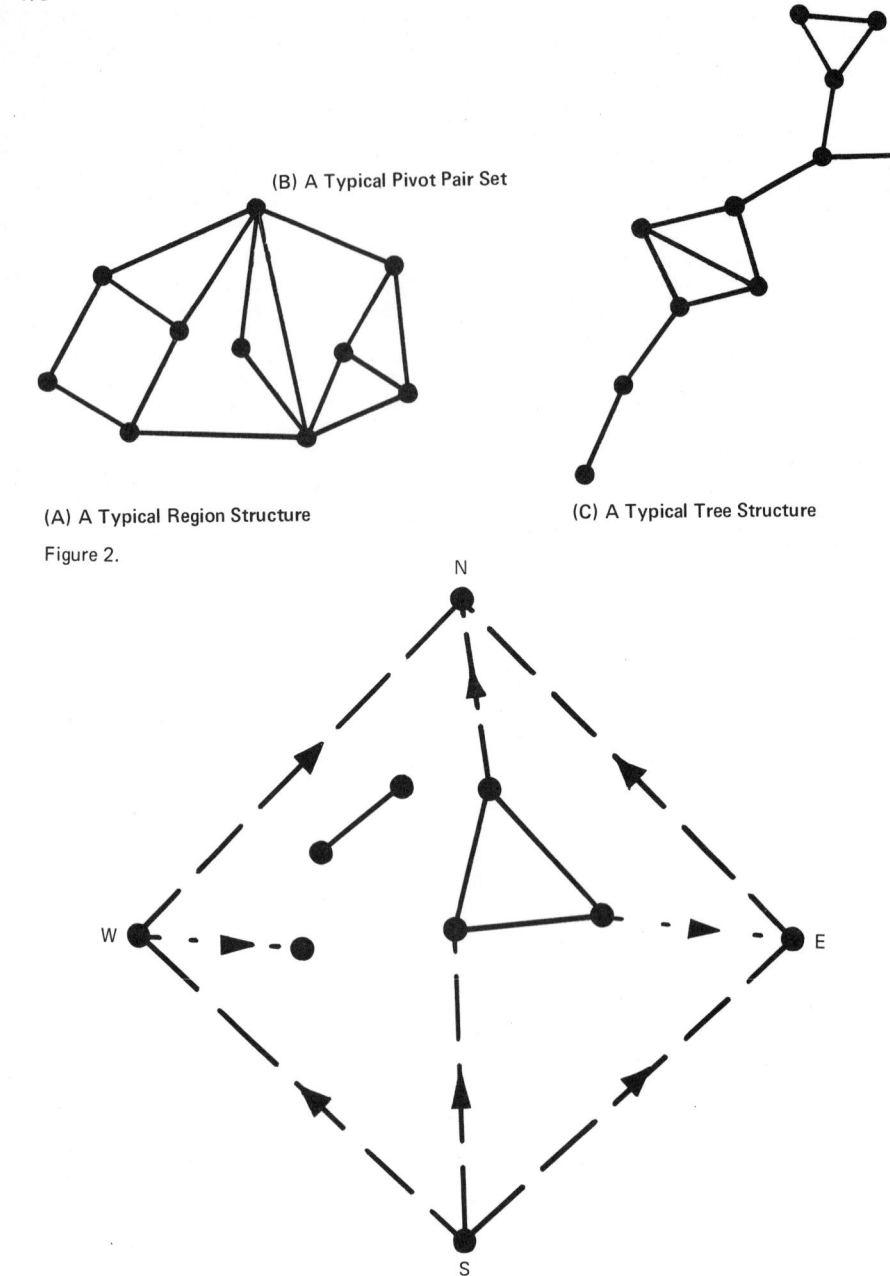

Figure 3. A Partially Completed Dual Graph

(2) The simultaneous solution of the size requirements and completion of the topology of the floor plan.

The Location Requirements

This portion of the design porcess is carried out by an operator which is capable of introducing a single edge into a partially completed dual graph in response to an adjacency requirement. It is assumed that the graph to be added to is described in terms of the hierarchical structuring mentioned earlier in this paper. For each location requirement, this operator is applied in conjunction with several prefixes in the following way:
(1) A test is made to see if the addition of the proposed edge will result in a planar graph, one of the necessary conditions for physical realizability. The hierarchical structuring of the dual graph greatly facilitates this test. In fact, the amount of computation required to determine planarity increases *less than linearly* with the size of the graph, while most standard planarity tests increase *exponentially* in difficulty with the size of the graph.
(2) If step (1) is passed, the location requirement operator adds the required edge to the dual graph and updates the hierarchical structure. This is the most important single step in the design process. This operator is capable of making simple deductions about the color and direction of uncolored and undirected edges.
(3) Prefixes are applied to the resultant dual graph to insure that its nodes and regions are well formed (see Reference (1)).

If any of the prefix tests result in failure, appropriate action must be taken. For this early version of the program, that action is simply termination.

The Size Requirements

The output of the location requirement part of the program is a partially completed dual graph, with unweighted edges. The second phase of the program has the double task of filling in those edges of the dual graph which have not been specified by location requirements, in order to completely specify the topology of the floor plan, and of adding weights to all edges of the dual graph in response to the size requirements. There are several different ways that this can be done within the framework of the representation. An initial method that is being attempted for this program is to carry these two tasks out in parallel. This actually requires the coordination of three different search activities:
(1) A given partially completed dual graph, since it is a planar graph, can usually be drawn in several different planar configurations, each containing a different set of regions (although each has the same total number of regions). Each such configuration can be called a different realization of the dual graph. The hierarchical structuring of the dual graph which is accomplished by the program allows one to tell how many different realizations a given graph will allow, and provides an algorithm for generating them one by one.
(2) Given a particular realization of a dual graph, not only must colors and directions be decided upon for all uncolored, undirected edges, but a sufficient number of missing edges must be filled in to create a completed dual graph, corresponding to a completely specified floor plan. All this must be done in accordance with the rules of well formedness so that the resultant floor plan is physically realizable.
(3) Assignments of weights must be made to the edges of the dual graph in accordance with the room size requirements. Since each room may have several allowable sizes, a search space is involved here too.

The coordination of these three search activities allows the best opportunity for the use of heuristic techniques in this program. To provide a basis of operation, an exhaustive search procedure is first being tested, as follows:

(1) A given realization of the dual graph is selected by the generation algorithm.
(2) Given this realization, a suitable incomplete region of the dual graph is selected. Then, following simple selection techniques, rooms are assigned sizes and edges are introduced and assigned colors and directions in an alternating fashion until the region is completed.
(3) A neighboring incomplete region of the graph is selected and process (2) is repeated, until the dual graph has been completed.
(4) For the addition of each new room size or edge, an operator is used which appropriately updates the data structure of the graph, making deductions about edge colors, directions, and weights where possible.
(5) After each application of this operator, a set of prefixes is applied to insure the physical realizability of the dual graph. If a failure occurs, the search cycles back to the closest previous decision point.

The net effect of this search method on the floor plan design is that of solving a jig-saw puzzle, working around the edge from the outside in. This method allows the application of such search pruning techniques as testing for total dimension and total area limitations for a building.

Program Extensions

It should have been noticed that this program constitutes a unique but nevertheless bare framework for the solution of space planning problems. As such, it is intended that it not only be built upon to handle a more complete range of requirement types, but also that some of its basic elements be experimented with, such as the search techniques used in the solution of the room size requirements.

References

1   Grason, John, "A dual linear graph representation for space filling location problems of the floor plan type," in *Emerging Methods in Environmental Design and Planning,* Moore, G. (ed.) M. I. T. Press, Cambridge, 1969 (in press).

2   Eastman, Charles, "Representations for Space Planning," internal working paper, Department of Computer Science, Carnegie-Mellon University, March, 1969.

Charles Eastman
Institute of Physical Planning
Carnegie-Mellon University

WORKSHOP ON PROBLEM SOLVING METHODS IN DESIGN

Two issues constantly seemed to be in the air during the workshop and repeatedly initiated questions and occasional debate. I think both have a similar basis. The first was the contradiction between scientific study and architectural practice. Traditionally, an architect's training includes a conscious attempt to teach him to loosen the constraints of the problems expressed to him. That is, he is encouraged to not accept the assumptions of the client, not to accept the standard way of doing things. Rather, he is taught to question assumptions; he should broaden his solution space and make the range of possibilities as large as possible so that he has "room" to innovate. Even the words used to describe a problem situation are brought into question. For instance, he may question the meaning of "living room" or "office" so that he may explore alternative designs that mix traditionally disjoint functions.

But this way of approaching the external expression of a problem is antithetical for the carrying out of research activities—for at least two reasons. First, in research we are trying to record knowledge with words or other symbols. To accurately convey an idea, the meanings of words must be agreed upon and to be precise. This use of vocabulary is quite different from the reliance on poetic words full of private meaning that are the resource of so much intuitive inspiration. Also, in order to study something, a specific phenomenon or set of phenomena must be intensely viewed. Some issues must be identified for study and others omitted; in other words, bounds must be put on the problem. Again, this is the opposite of what the designer normally tries to do. Where the scientist must deal with a part of an issue, the designer is normally concerned with its totality. Both aspects of research are the opposite of the normal tendencies of designers. Small detailed analyses of an aspect of design seemed to bother some people because they were specific, detailed, and did not attempt to be all inclusive.

The other related issue was the semantic one concerning what is the definition of a problem. In all of the presentations, the kind of problems that were treated were problems that were well-defined. The goals to be achieved were known. In the task situations presented the problem always was to find the appropriate means to achieve the specified end. Several people at the workshop voiced the opinion that this type of problem was not important in design. They felt that the problem needing attention was the identification and definition of the problem to be solved. That is another aspect of problem-solving. Recent studies presented elsewhere have attempted to treat both problem solving strategies for identifying problem goals and those for determining the best means to achieve those goals.[1] But in negating the search aspect of problem solving I think the implicit idea being expressed was that some people think design is primarily the deciding of goals concerning what should be done. This assumption implies that we designers can rely on technology to provide the means for implementing the goals.

Of course, this is only another way of saying that environmental design theory often emphasizes the big concept, "If the big issue can be proposed the other aspects of the problem can be put into place." This view is realistic in an area like building technology where the structural constraints are almost all resolved by available technology and thus is valid for most design situations. But problem solving studies made by designers also have had the propensity for emphasizing the big idea—the overall schematic outline for automatic design. In the last few years, we have seen several such systems proposed.

I question the value of such exercises. In no case has a design system been completed, because of the technical problems that are quickly run into. No one has been willing to focus on these technical problems that are quickly run into. Specifically, some of the technical problems are: (1) representations for handling spatial, visual and material qualities; (2) specification of a general command language for describing design actions; (3) a language for specifying design goals and constraints, and (4) the development of efficient search strategies for solving design problems. If the development of operational tools for augmenting design problem solving are to come into being, these technical problems must be resolved. And because of their special requirements, designers must be involved in their development, if they are to be of any value to him. This focussing on details will require a new attitude among at least a few designers. No one else is likely to solve a designer's problems in the way he wants them solved. I hope that designers will become more active in dealing with the details. It is a necessity if design research is to make progress.

---

[1]See Eastman, Charles, "Cognitive processing and ill-defined problems: a case study from design", Proceedings International Joint Conference on Artificial Intelligence, Washington, D. C. May, 1969.

William Michelson
Department of Sociology
University of Toronto

ANALYTIC SAMPLING FOR DESIGN INFORMATION:
A SURVEY OF HOUSING EXPERIENCE

The work reported here was conducted under a research contract given by the Central Mortgage and Housing Corporation (Canada) to the Centre for Urban and Community Studies of the University of Toronto. I am indebted to the principal investigator, A. J. Diamond, Architect, for his interest in the sociological substudy. Much of the responsibility for the study methodology goes to the twenty-two students enrolled in 1967-68 in my graduate seminar in urban sociology; the analysis presented here, however, is my sole responsibility. I am grateful to Paul Reed, Janet Lytle, Carole Cox, and William Babchuck for their assistance at various stages of the project.

Few today will quibble with the assertion that special markets for housing reflect special social needs in housing. Almost everybody recognizes the need for some "behavioral input" to design. The social component is on the verge of becoming a conventional part of design programming.
But while few would demean the importance of social considerations in the design program, the source of the content remains a more serious question. Social scientists have long chided designers for their use of personal, class-biased, and unsubstantiated insights as a chief source of social facts. On the other hand, the results of potentially exhaustive social science research projects are meaningless to designers without starting almost from scratch to explain the contextual limitation of the research. Despite agreement on "motherhood", it is rare to find social data available to "plug into" any specific, ongoing design program.
The study described in this paper is an experimental attempt to provide relevant original sociological materials for an ongoing design study.

Background

In the summer of 1967, Canada's Central Mortgage and Housing Corporation contracted to the Centre for Urban and Community Studies of the University of Toronto for a project to plan for lower cost, higher density low-rise multiple dwellings than are currently on the market. These dwellings were to be built without requiring retooling of the housing industry and were to provide the aspects of "environment" that people commonly choose when they seek single family units. The target population for these units was the middle-income Canadian family (earning $6,500 to $9,000) who was too rich for public housing but yet too poor to receive government mortgage assistance on new single family homes, which were thus placed beyond the reach of most of them. Although a design study planned to incorporate the findings of related disciplines in the scheme.
The problem of sociological content was present at the outset. Although sociological data were considered desirable for the study, it was an open question as to what kind would be useful and how they would be collected. In response to this problem, the practicing in a one year graduate seminar in urban sociology was committed to this question. The class agreed to design and implement a study to fulfill, within the limits of time and finances, the expressed needs of the architect for social data, which in addition, would be of intrinsic interest to sociologists.

It was obvious that we were not able to provide all desirable sociological information to the architect; indeed, given the amount of time and manpower available, the focus had to be narrow. Put in perspective, the study was intended as one which would answer a few of the architect's most pressing questions for his potential clientele, but which would have to be supplemented by more standard sources of relevant sociological knowledge.

The Study

In response to the architect's request for information about aspects of environment that serve positive purposes in the lives of residents, we first created a frame of reference. We decided to study the effect on this kind of family of various physical components of environment which could be incorporated into the design. Components of environment were viewed as either facilitating or hindering (i. e. not determining) social contact, routine activities, and the like. Our first question was, "All else being equal, what differences can we find in the lives of middle class families which are accountable to clearcut differences in environment?

However, to answer the architect's needs, a way was needed to evaluate the answers to this question. How important to family satisfaction are these relationships and activities found in specific settings? What aspects of current environment are related to satisfaction? What do people hold to be ideal—and how is this related to current surroundings and activity?

To answer these questions required that we study families who were sociologically homogenous but who lived in settings which contained the necessary physical variation. Thus, our frame of reference centered on the relationship of physical setting, social interaction and activity, and conceptions of satisfaction among a homogeneous stratum of middle class families currently subjected to varying home environments.

The subject matter of the study required intensive interviewing, yet the analytic framework required substantial numbers of respondents. Therefore, we needed a sampling technique which maximized analytic power while at the same time minimized the total number of people sampled.

Conventional sampling procedures would not accomplish this objective. Random sampling of the population would have produced a set of families whose environments were highly skewed and would not have minimized social differences among families.

Social Area Analysis, commonly used to select subjects with specific social characteristics (Shevky and Bell, 1955) takes census tract aggregates as its smallest unit and is insensitive to differences of physical environment within the tract.

Stratified sampling is commonly used to compare phenomena with respect to subgroups which are not found in equal numbers in the population. In the present study, a stratified sample with respect to social characteristics would have been inappropriate, but the same approach could be applied to the components of physical environment. Taking steps to assure relative social homogeneity, we decided to choose a sample of families on the basis of physical characteristics of their home environment.

We searched out relatively small clusters of housing which differed as follows:
1. *housing type (single family, town house, maisonette)*. Inherent in these variations of low-rise housing are the presence and absence of party walls and the presence and absence of direct access to the outside.
2. *open space (private vs shared*. Some units (i.e. all single family homes and some maisonettes) had well defined private open space; the others did not.
3. *access to community facilities (close vs distant)*. Some housing units (of all types) had stores, schools and the like immediately adjacent; the others were sufficiently removed to require transportation to reach them.
4. *tenure (owned vs rented)*. Some of all housing types were rented. A number of single family homes and town houses were "owned."[1]

The selection of housing clusters was such that these several components of environment, while at times related to one another, were not synonymous with one another—an important difference in assessing relative effect. Units were selected for study so there would be sufficient numbers for comparisons with respect to all these factors, while staying within the maximum possible sample size of 230. Within each stratum, houses were selected by random means.

At the same time, however, criteria were established to homogenize the sample. All housing units were in the mid-suburban band around Toronto, and monthly rents, where applicable, were in the $180-220 range. Only families with children living at home were studied, and all housing units had either three or four bedrooms. Homes were chosen for study whose interior space and bedroom count paralleled those of the apartments and whose monthly cost, if sold today, would be comparable. Known "ethnic" areas were avoided. The resulting families interviewed strongly approximated the architect's target population.

Within each household, the wife was selected for interviewing--on the basis that she was most subject to environmental influences.

The total number of completed interviews was 173 (75 per cent). The sample interviewed represented the physical strata as intended. About 50 percent of the respondents lived in town houses, with about 25 per cent each in single family houses and maisonettes. About 60 per cent were located distant from community facilities; about 40 per cent were close. About 70 per cent rented; about 30 per cent owned. About 35 per cent had private open space; about 65 per cent did not.

Each interview consisted of several parts: 1) factual material about the family, 2) residential history, 3) satisfaction with environment and plans for the future, 4) nature and extent of interpersonal relations, 5) activity patterns, and 6) perceptions of the ideal. Interviews averaged 45-60 minutes.

The interview period was the last two weeks in February. It included the coldest week of the year. Since we believed that use of environment is affected by weather conditions, most of the interviews were repeated in the last week of June. The differences were substantial at times and suggest a greater impact on interaction and activity of immediate environment in the winter than in the summer.[2] Nonetheless, with the exception of only one or two stable, objective factors pursued only in the summer interviews, the data reported here is uniformly based on the February interviews.

Given the complexity of the frame of reference and the multiplicity of variables considered, a total list of our expected results would be overly long. In brief, we expected that housing types with direct access to the outdoors would be associated with greater satisfaction and use of outdoor environment than homes without direct access; town house dwellers would find their situation better for sociability but worse for overall satisfaction than residents of single family homes due to party walls. We expected that neighbor contact would be higher among those with private open space, as opposed to those without private open space. We expected that people close to community facilities would both value and use them more than people who lived more distant from them. Finally, we anticipated that owners would make more formal and informal contacts with others in their neighborhood than would renters, and would put down roots in the process.

Findings

The results are of three types. First, what differences in social activity vary systematically with aspects of the home environment? Second, what is the relation of these physical components to residential satisfaction and to conceptions of ideal environment? Third, in what way does the relation of environment and satisfaction reflect an activity pattern consonant with that environment? I shall outline the results with respect to these questions one by one.

1. Environment and Behavior

The interrelations found between each component of environment studied and relevant behavior will be summarized and then discussed.

a. Housing Type

First, residents of single family and town houses predominantly meet their neighbors outside while people in maisonettes meet them inside. This appears largely a function of direct access, or lack of it, to the outside. Given the winter setting, it is perhaps no wonder that the two categories meeting neighbors outside are the same ones who resort most frequently to the telephone. It would appear that having common indoor space is at least con-

	housing type			
social activity	single family	town house	maisonette	chance occurrence of difference (from X^2 calculations)
1. where meet neighbors	outside	outside	inside	.02
2. use of telephone	highest	high	low	.001
3. hours spent sewing or knitting	high	high	low	.01
4. effect another home would have on undertaking desired activities	little difference	some difference	some difference	.01

ducive to contact, if all else is favorable for it.[3] Nonetheless, it is quite another question as to whether this environmentally enhanced contact is regarded as favorable contact, even in the winter; evidence to be introduced below would indicate the contrary.

A further evidence of relative isolation among those living without inside space shared with neighbors is the amount of time devoted to sewing and knitting. Those in single family homes and town houses spend more time in this fashion than do those in maisonettes.

While *contact* with other people seems associated with access to the outside, activity appears related to another component of housing—the existence of party walls. When asked, "What difference would living in another type of home make to doing more of what you want?", about three quarters of single home residents answered "no difference." In contrast only about 50 per cent of the residents of the multiple dwellings thought that there would be no difference; 50 per cent thought that there would be one.

I interpret this to mean that those living in multiple dwellings have had a chance to experience desires for activity which have been frustrated by their real or potential impingement on proximate neighbors. Residents of single family homes, on the other hand, have had less of this type of frustration as well as the experience of closer physical contact with neighbors. This is particularly supported by the past experience of the people sampled. Residents of single family homes have lived in significantly fewer different types of housing than have the others, with town house dwellers the most "well-rounded;" fully 72 per cent of the latter have lived in three or more different housing types, compared to only 32 per cent of the former.

Furthermore, research elsewhere indicates that residents of multiple dwellings find self-imposed need for self restraint in activity even more odious that actual nuisences emanating from neighbors (Raven, 1967).

Table 1 gives some additional perspective on the differential impact of housing types. Respondents were asked why they chose their present home. Several reasons were typically given, and these were categorized as to whether they referred to the home itself or to the surrounding neighborhood. People were also asked if they planned to move again within the next five years. Although this question can be interpreted literally (and should be in cases), it is also of great use as a barometer of residential satisfaction. These items indicate that the internal aspects of environment are more pressing (and distressing) to residents of multiple dwellings than are external aspects, while the reverse is true of residents of single family homes. To be precise, 44 per cent of those choosing a multiple dwelling for a reason having to do with the unit itself now intend to move, while only 33 per cent of those choosing one for neighborhood reasons are so inclined. This compares to a 17 per cent future mobility rate among home dwellers who cited reasons having to do with the home and a larger 29 per cent rate among those citing factors in the neighborhood.

Although these differences by housing type are logical and the sample was designed so as to provide similar respondents in each physical category, it must still be questioned whether the findings discussed might not find their explanation in whatever minor demographic or socio-economic differences remain to differentiate respondents from one another. For this reason, similar tables were run substituting some of these latter variables for the physical variables. Are these differences in activity tied more closely to age, family size, occupation, education, and the like?

In fact, only one of the differences is related to one of these personal attributes. The more skilled the occupation of the husband in the family, the more his wife feels a new home would affect the commencement of new activities.[4] Nonetheless, the relationship of proposed new activities to housing type is far stronger than it is to husband's occupation, thus making it highly unlikely that occupation is a crucial intervening variable in this sample.[5]

   b. Access to Community Facilities

One mark of this particular sample is their lack of participation is specific leisure time activities. Nonetheless, there are substantial differences in activity according to access to community facilities such as schools, stores, etc. According to the nature and intensity of the activity, categories were devised to divide active participants from minimal or, in some cases, non-participants. Uniformly, the percentage of participants in the activities listed among people with easy access to community facilities was 20 points higher than among people distant from them.

This does not mean that people living right next to suburban centers are active solely by virtue of that fact. Indeed, some of these activities are neither found in or aided by the particular facilities found there. What it may indicate more accurately is that people with more generally active life styles seek out convenient locations— even within suburban areas. They may not be as active as downtown cliff dwellsrs, but they are more so than their fellow suburbanites. Given that there are forces in the housing market that send young families to the suburbs, there are still life style differences among them that require differential physical surroundings.

In advancing reasons for having chosen their current home, people were much more likely to specify factors having to do with the housing unit than the surrounding neighborhood. Nonetheless, significantly more people[6] living close to community facilities cited neighborhood reasons than did those distant from them.

Table 2 hints more closely at the relationship involved between life style and access to community facilities. We would expect that people living close to facilities who are active and people living distant from them who are inactive would both be more satisfied with their environment than inactive people living close and active people at a distance. With respect to two discretionary activities in which there is some degree of participation, this relationship holds. Those who shopped for clothes or went to restaurants at least once a month were considered active; those going less frequently were deemed inactive. The active-close and inactive-distant categories were considered congruent; the others, incongruent. As Table 2 points out, people in congruent categories are less likely to want to move in the future than are those in incongruent categories. The same relationships hold within all categories of access. Although these phenomena must obviously be approached with caution by such indirect measures, it would appear that the match between life style and environment is one that entails affective consequences.

In addition to satisfying a particular life style, location near community facilities is without question of some instrumental value. In suggesting reasons why they might move in the future, only 6 per cent of those close to facilities cited "neighborhood" reasons for moving, while 21 per cent of those distant from them did so. Furthermore, we expected that this would be particularly relevant for young mothers who might otherwise be housebound. In families with children only 1-6 years of age, the same 6 per cent of respondents close to facilities referred to the neighborhood as a reason for moving; however, among those more distant, 47 per cent of the young mothers noted the neighborhood in this connection.

| | access | | chance occurrence of |
social activity	close	distant	difference (from X^2 calculations)
1. arts and crafts	some	little	.001
2. pleasure reading	much	some	.02
3. educational courses	some	little	.01
4. attend plays and concerts	majority	majority	.01
5. associations and meetings	some	few	.05
6. general use of neighborhood	extensive	moderate	.05
7. frequency visit relatives inside neighborhood vs. relatives outside	almost equal	outside much more	.01
8. frequency visit friends vs relatives outside neighborhood	friends more than relatives	equal	.001

In addition to the information on participation in selected activities, visiting patterns also shed light on the respondents. First, relatively few respondents had relatives living in their neighborhood. Slightly more people living close to facilities had the bulk of their close relatives living inside their neighborhood than did those "distant"; however, these people with easy access to facilities saw these relatives far more frequently than did those without easy access—far greater than would be suspected merely from the numerical availability of relatives. The presence of these facilities may serve as a medium in which close relatives spend time without making inroads in their respective time schedules.

However, when people go outside their own neighborhood for contact, those located close to facilities, the ones with generally more active life styles, spend time more frequently with non-related friends than with kin. "Swinging" is apparently not rooted in kinship. Those distant from facilities, who in addition do not have an immediate location in which to meet their relatives, are evenly divided in the time they spend with kin and non-kin.

None of the differences involving location are explained by whatever demographic or socio-economic differences exist within the sample. In addition, residential background does not vary by location.

c. Open Space

There is a parallel between some of the findings on housing type and on open space. With respect to housing type, the presence of shared walls was related to a feeling among many that another housing type would permit respondents activities in which they desired to participate. The present case differs from the earlier one in that sharing of outside space rather than inside acoustics presents the same difference in the sample. Those with private open space do not, by and large, think another type of home would enable new activities, while those sharing space are more likely to think so.

	open space		chance occurrence of difference (from χ^2 tests)
Social activity	private	shared	
1. desire to undertake new activities	great	near	.001
2. effect that another home would have on undertaking desired activities	little difference	some difference	.05
3. number of neighbors known	many	some	.05

There are several reasons why this could be interpreted as more a function of house type than of open space. First, the question about the effect of another home refers specifically to the home, even though outside space is often considered one aspect of it. Second, the statistical relationship of house type with effect on activity is stronger than that of open space with effect on activity, while at the same time, there is strong but not complete association between current house type and type of open space.

Nevertheless, data from this study ties family activity to private open space and adds to a literature which stresses that people cherish private open space for what they wish to do in it (Wallace, 1952; Kumove, 1966; Michelson, 1966). Although the majority of respondents claimed that they wanted to participate in new activities, the percentage of people without private open space who wanted this was significantly higher than those who had their own plot; in fact, only 5 per cent of this "frustrated" group did not want to add some new activity. It would appear that lack of private open space does inhibit some activity, although its existance is no automatic panacea. Furthermore, this desire is not one which is related to housing type according to the present data.

We anticipated (c.f. Fleming, 1967) that private open space would serve as a catalyst to bring neighbors together more than would shared open space. People do not need an excuse to remain outdoors on their own turf; casual contacts made there are not hurriedly terminated through lack of a socially acceptable excuse to stay put. This expectation was backed by the data, in that people with private open space knew significantly more neighbors than did those who shared open space.

d. Tenure

	tenure		chance occurrence of difference (from χ^2 test)
Social activity	own	rent	
1. plays and concerts	don't attend	do attend	.001
2. frequency visit relatives outside neighborhood vs frequency visit relatives inside	outside much greater than inside	outside greater than inside	.01

In this context, we expected a variety of evidence connecting home ownership to interactions with others in the neighborhood and to activities in that area. The data did not confirm these expectations. Of the two differences in the sample statistically related to tenure, one is substantively irrelevant, and the other, if considered relevant, is converse to expectations; owners have less to do with relatives in their neighborhood than do renters. The small package of findings generally backing expectations with respect to the three previous aspects of environment is lacking for tenure.

Environment, Satisfaction and the Ideal

a. Satisfaction

The previous section outlined findings on environment and behavior. In the present section, I shall temporarily ignore the previous one, asking if there are relationships between current environment and satisfaction and between current environment and people's conception of the ideal. Then in the third section, all these factors will be treated simultaneously.

Two measures of satisfaction—both indirect—were used. The first, as already detailed, is intention to move within the next five years. The second is an index calculated from answers to questions asking whether the following were satisfactory, unsatisfactory or of no importance: a) layout of interior, b) size of rooms, c) front entrance, d) interior noise transmission, e) noise from outside, f) parking, g) outside space, h) outside design and appearance of housing, i) quality of schools, j) distance to shopping, k) distance to family, l) recreation, m) privacy, n) type of neighbors, and o) cost or rent. The index was computed as follows:

$$\frac{\text{number of unsatisfactory items}}{\text{number of unsatisfactory plus satisfactory items}} \times 100$$

It is more properly an "index of dissatisfaction."

As Table 3 points out, the first, intention to move, is related to current dwelling type, tenure, and open space. Satisfaction is distributed pretty well as expected. For families of this type, single family housing has long been the ideal in North America, and the more self-contained the unit which approximates it, the more the satisfaction; the greater satisfaction with the town house than with the maisonette reflects this. Ownership has likewise been an ideal held high and an economic advantage. Possession of private open space, while typically a concomitant of housing type, no doubt reflects this with respect to satisfaction, although it may reflect family activity as well.

The index of dissatisfaction is not related to any of the variables mentioned so far. That it does not work similarly to the first measure can be explained with reference to its content, in that discrete items may carry different weight with respect to the several aspects of current environment, even though their sum varies directly with intentions to move. It is worth noting, however, that the index varies significantly with the length of time people have lived in their current residences. The longer they have stayed, the lower (i.e. more satisfied) is their index score.[7] This, of course, gives a partial vote of confidence to market mechanisms. I shall return later to factor of length of residence.

b. The Ideal

As in all studies I know of (c.f. Lansing, 1966), a strong majority of these respondents stated that they would like to live in a single family home in the future. This choice was made by 85 per cent of the families, with the remainder opting for a duplex, triplex or row house. No-one wanted to live in a multi-family walkup apartment. The reasons advanced most strongly for this choice were privacy and independence, the desire to own or to invest in real estate, the single home as a status symbol, internal space, and private open space.

As in the other studies, too, variations within the sample did not account for differences which upset the strength of this ideal (c.f. Michelson, 1968).

However, to probe more deeply the basis on which people choose their ideal, we asked people to select the best and worst of four photographs illustrating, respectively, a single family house, a town house a maisonette, and an unfamiliar "futuristic" multiple dwelling. Respondents were told that each dwelling was of equal size internally and would cost the same to by or rent, as desired. One question about the pictures was general, "In which would you most like to live?" As expected, consistent with the nonpictorial question, 83 per cent chose the single family home, with most of the remaining choices falling to town houses. The major reason advanced was privacy.

The other basis for judging these pictures was much more specific. While the single family home was given a plurality of positive choices with respect to each, the variation in its strength of supporting is revealing:

> Privacy — 91% chose single family home
> Best for raising children — 90%
> Most easily do the things you want — 87%
> Best outside design — 50%
> Best contact with neighbors — 40%

It is obvious that of these factors, privacy and child-raising are paramount considerations underlying the preference of this type of sample for a single family home. The respondents most commonly cited the lack of party walls in explaining why they considered the single family home more private. For childraising, 31 per cent cited having a private yard, 19 per cent claimed more living space, and 15 per cent valued freedom of interference from landlords. Maisonettes were felt the worst for childraising due to a feeling that there is no place for children to play; they are also felt crowded and confining.

For activities, single homes again led because people did not feel there was a need there to restrain their own noise. A substantial minority also felt they would be more free due to ownership status.

With respect to contact with neighbors, single family and town houses were mentioned almost equally, because, although proximate, people were not forced into contact. In contrast respondents in maisonettes, with access out only to central hallways, believed themselves to be forced into contact with neighbors. Indeed, our data in section 1 above bears them out.

In short, the traditional popularity of the single family home was upheld by the present data. Moreover, these data went on to suggest which of several housing goals these preferences reflect most strongly, as well as what environmental aspects are associated with them.

However, the analytic framework of this paper is to ask what influence current environment has on these conceptions of the ideal. Since there was such wide agreement with respect to privacy, childraising, and home activity— always crucial factors in family choice of single homes—there is insufficient dissent for analysis. There is, however, considerable difference of agreement with respect to outside design and contact with neighbors. Furthermore, since the housing market frequently prevents people from achieving their ideal, it is quite conceivable that people prefering some other type of housing on these additional grounds may well be open to the possibility of living in something other than a difficult to achieve ideal.

Current location and open space were strongly related to preferences for the outside design of something other than single family homes (i.e. usually town houses).. People living close to community facilities and/or without private open space were more likely to prefer the town house than single family.[8] Since these people have already opted for a convenient location and higher densities, doubtless at the expense of other available environmental goals, they may be the first to compromise from their structural ideal.

Some data on past residential experience lends credence to this assertion. Table 4 shows the variation in percentage of people living in single family housing intending to move by the housing type which immediately preceded the current one. According to this criterion, people who have had experience in multiple dwellings are less "satisfied" with single family housing than are those who moved from another single home. They might be more likely to deviate from the public ideal in the future.

The reverse is true among those distant and/or who have had private open space. Sixty-four per cent of those distant, for example, prefer the design of the single family home, compared to only 45 per cent of those close to facilities.

With respect to contact with neighbors, the second aspect of housing choice on which there was a diversity of opinion, current housing type strongly influenced preferences. Those in single family homes thought these dwellings best for contact, while a majority of residents of both town houses and maisonettes thought the town house best.[9] The maisonette was downgraded by all for the reasons advanced earlier.

In addition, we asked about people's desires to rent as opposed to buying in a hypothetical new high density development. Again, there was great difference in the performance of respondents according to location. Of those now close to facilities, 86 per cent would rather rent such a residence, while only 53 per cent of distant respondents would want to rent.[10] Although current location vis a vis facilities is not related to desire to rent or own in such a future development—however not as strongly as current location.[11]

We investigated as well whether demographic or socio-economic variations within this sample varied with choices of the ideal. The level of the husband's education did vary with preference for outside design; the better educated the man, the more likely he was to choose a town house. Education does not, however, appear to be a strong independent factor because its relationship with outside design is no stronger than that of two physical variables, one of which is also related to education (present possession of open space) and one of which is independent of education (present access to facilities).

What education and occupation did explain solely is whether extra features or low cost is a more important goal in housing (within limits). The relation was in the direction anticipated; lower strata emphasized cost more strongly.

Environment, Activity and Satisfaction

I have traced relationships between current environment and activity, then of environment and satisfaction. It is entirely possible for various components of environment to accommodate or make difficult given activities, as in the first section, without general housing satisfaction being involved in any way. It is also possible for environment to have intrinsic forms of satisfaction or dissatisfaction, as in the second section, without having to involve activities. Nonetheless, the question remains as to whether an activity congruent with a given setting makes a difference with respect to satisfaction, as would be expected from the model proposed.

Given the results of the previous two sections, the first positive finding regarding the interrelationship of these three types of factors comes as no surprise. A desire to undertake activities currently dormant, closely related to the absence of private open space, is also significantly related to intention to move.[12] In other words, people who lack private open space are more likely both to want and to desire to move, with the desire for activity intimately related to the desire to move. In short, as measured here, the restriction of activity brought on by lack of private open space matters with respect to residential satisfaction. In this case, the shoe pinches.

Somewhat more positive is the relationship involving education and preferred outside design of housing. Within our sample, the lower a man's education the more hours a week he watched television. Further more, the more he watched television, the more likely he was to prefer the outside design of a single family house. What this adds up to is that education varies directly with preference for some other housing design and inversely with television viewing, which itself is inversely related to an unconventional design preference.

These two packages of interrelationships indicate that leisurely activity is a mediating variable operating between formal characteristics of a person or his setting on the one side and his satisfaction or preferences, on the other. It is clear also that the bluntness of the measures of satisfaction used in this study may have brought to light only a small part of this effect in comparison to the number of differences in activity and interaction shown related to components of environment in the first section of findings.

Postscript

Although the intent of the study design was to maximize variation in physical environment and to minimize it with respect to the demographic and socio-economic characteristics of the people interviewed, at least one potentially relevant factor was not controlled—length of residence. Several writers, for example, have recently stressed than many aspects of behavior at any single time in a new housing development are a consequence of the newness of the place; when lawns aren't yet seeded and friendships haven't matured, life is different than when they are. (c.f. Clark, 1966; Gans, 1967)

As indicated before, length of residence turned out, quite logically to be related to residential satisfaction. As a postscript, there are two aspects concerning length of residence which deserve mention.

First, to what extend does this factor actually underlie one of the environmental variables which accounted for differences in activity? Not at all. It does not account for any of the cited differences more strongly than the environmental variables.

However, it is worth noting that participations in associations and meetings acts as a mediating variable with respect to length of residence and satisfaction. The longer a person has been in his present residence, the more he participates in associations[13] and the less he wants to move,[14] with the extent of his participation inversely related to moving.[15]

Second, length of residence may be an index of different ways of life. It would be tautological to claim that recent movers had lived in more houses lately than non-movers. What we found, however, was a huge difference between these groups in the number of different types of housing which they have experienced in their lifetime. As Table 5 demonstrates, those who have lived only a short time in their current home have lived in significantly more types of housing than have those whose stays are of longer duration. Sixty-four per cent of those who have lived in their present homes less than three years have lived in three or more different housing types, compared to exactly half that percentage (32%) of those who have lived there more than three years. They also make more distant moves than the more stable subgroup.

One possible explanation is that the residents of short duration are "organization men"—transferred frequently from one managerial niche to another—or families undergoing changes in composition. However, length of residence is not related in any way to socio-economic or demographic differences within the sample. Therefore, in closing, we can only speculate about the existence of a segment of the population which, although indistinguishable by standard sociological variables, changes environment more and enjoys it less.

Table 1—Per Cent Planning a Future Move by Current Housing Type and Reasons for Choosing Current Housing

	Current Housing Type	
Reasons for Choice of Current Housing	Single Family	Multiple Dwelling
Home	17% (N=66)	44% (N=178)
Neighborhood	29% (N=21)	35% (N=48)

Table 2—Per Cent Planning a Future Move by Activity and Congruence of Participation

	Participation Level and Location	
Activity	Congruent	Incongruent
1. Eating at restaurants	31% (N=91)	41% (N=61)
2. Shopping for clothing	33% (N=72)	37% (N=82)

Table 3—Per Cent Planning a Future Move by Selected Aspects of Environment

aspect of environment	intend to move	do not intend to move or don't know	%	N	chance of occurrence of difference (from X^2 test)
A. Housing type:					
1. single family	18%	72%	100%	45	.001
2. town house	33%	67%	100%	81	
3. maisonette	60%	40%	100%	40	
B. Rent-Own:					
1. rent	41%	59%	100%	118	.02
2. own	25%	75%	100%	53	
C. Open Space					
1. private	25%	75%	100%	60	.05
2. shared	44%	56%	100%	103	

Table 4—Intention to Move by Immediate Past Housing Type (Single Family Residents Only)

		Previous Residence	
Intention to Move		Single Family	Other
Plan to move within five years		13%	27%
Do not plan to move within five years or don't know		87%	73%
	100%=	23	22

Table 5—Number of Different Housing Types Inhabited in Lifetime by Length of Present Residence

		Length of residence	
Number of housing types		under 3 years	over 3 years
1 or 2		36%	68%
3 or more		64%	32%
	100%=	119	53

$X^2 = 13.78 \quad X^2_{.001}(10.83)$

References

1. Although the town house units preceded the legalization of condominia in Ontario, an arrangement was made between the developer and his customers which amounted to the functional equivalent.

2. See: "Space as a Variable in Sociological Inquiry: Serendipitous Findings on Macro-Environment," paper prepared for presentation to the 1969 meeting of the American Sociological Association.

3. The literature suggests neighbors will have intense interaction if they perceive themselves as in the same boat and have mutual needs requiring assistance.

4. $.05\ x^2\ .02$

5. Some additional differences in residential background by housing type are worth noting, although they would not appear to affect the foregoing arguments. Maisonette, town house, and single family house residents represent a decreasing order of distance of their current home from the immediate past home. Although no more than 30 per cent of people in any of these current house types owned their previous home, twice as many single family and maisonette dwellers owned than did town house dwellers.

6. $.001\ x^2$

7. $F = 5.06\ F_{2,\ 158}\ (.05) = 3.07$

8. It is important in interpreting this result to know that "location" and "open space" are independent of one another in this sample. x^2 test differences in both cases are better than the .05 level with respect to design preferences.

9. $.05\ x^2\ .02$

10. $.02\ x^2\ .01$

11. $.05\ x^2\ .02$

12. $.05\ x^2\ .02$

13. $.05\ x^2\ .02$

14. $.05\ x^2\ .02$

15. $.02\ x^2\ .01$

16. Clark, S. D. 1966, *The Suburban Society*. Toronto: University of Toronto Press.

17. Fanning, D. M. 1967, "Families in Flats." *British Medical Journal,* 18 (november): 382-386.

18. Gans, H. J., 1967, *The Levittowners*. New York: Pantheon.

19. Kumove, D. 1966. "A Preliminary Study of the Socail Implications of High Density Living Conditions." Toronto: Social Planning Council of Metropolitan Toronto.

20. Lansing, J. B. 1966, *Residential Location and Urban Mobility: The Second Wave of Interviews*. Ann Arbor: University of Michigan, Institute for Social Research.

21 Michelson, W. 1966 "An Empirical Analysis of Urban Environmental Preferences." *Journal of the American Institute of Planners,* 32 (November): 355-360.

22 Raven, J. 1967, "Sociological Evidence on Housing (2: The Home Environment)." *The Architectural Review,* 142 (September): 236 ff.

23 Shevky, E. and Bell, W. 1955, *Social Area Analysis.* Stanford, California: Stanford University Press.

24 Wallace, A. F. C. 1952, Housing and Social Structure: A Preliminary Survey with Particular Reference to Multi-Storey, Low Rent Public Housing Project." Philadelphia: Philadelphia Housing Authority.

R. D. Worrall
Peat, Marwick, Livingston & Co., Washington, D. C.

G. L. Peterson
The Technological Institute, Northwestern University

M. J. Redding
Northwestern University

TOWARD A THEORY OF ACCESSIBILITY ACCEPTANCE

One important dimension of the residential environment is the spatial distribution of "neighborhood services"—shops, parks, churches and similar focii of activity. On the one hand, it may be argued that the individual is anxious to have such services convenient and close at hand. On the other, however, he may equally wish to avoid the concomitant irritants of noise or traffic delays which are often associated with undue proximity to the activity. The result, in terms of his locational preference, is a tradeoff between a desire for easy accessibility and a complementary desire for insulation. This paper examines some of the implications of this tradeoff, in terms of the locational preferences of 240 Chicago families for eight selected neighborhood services.

The neighborhood services selected for study were:

1. Local shopping centers
2. Church or place of worship
3. Children's Park
4. Fire Station
5. Emergency Hospital
6. Entrance ramp to metropolitan freeway system
7. Stop or station on metropolitan transit system
8. Close friends' homes

Field data were collected by means of a simple game (see Figure 1). Each respondent was asked first to locate each service optimally *with respect to his home,* assuming that all other elements of the environment remained constant. Location was expressed in terms of average travel-times from the home by the most convenient mode. *The respondent was then asked to indicate whether successive changes in the accessibility (i.e. travel time) to each service in turn would cause the environment to become unacceptable, assuming that the other services remained in their optimum locations.* Note was made of the minimum and maximum values of travel time at which this occurred. The process was repeated for each respondent. The result was a set of (240.6.8) = 11,520 binary choices, covering 240 individuals, six time-increments and eight selected services.

For each service "i", values were computed of the number $f_a(t)^i$ and proportion $p_a(t)^i$ of respondents "accepting" each of the six levels of accessibility "t". These values are listed in Tables 1 & 2. The values of the proportions $p_a(t)^i$, interpreted as conditional probabilities of acceptance, are plotted as functions of the travel time "t" in Figure 2.

Note the approximately log-normal form of each of the functions in Figure 2—i.e., the functions may be expressed algebraically as $p = ke^{a(\log t-c)}$, and the apparent existence of both non-negative veritcal intercepts for zero travel-times, and horizontal assymptote for large travel-times. This suggests that, although there is a relatively wide "acceptable" range for eacn service, there are also upper and lower bounds beyond which the respondent considers his environment unacceptable, and that there is disutility associated both with too much accessibility as well as with too little. Further, there appears to be a portion of the population which is indif-

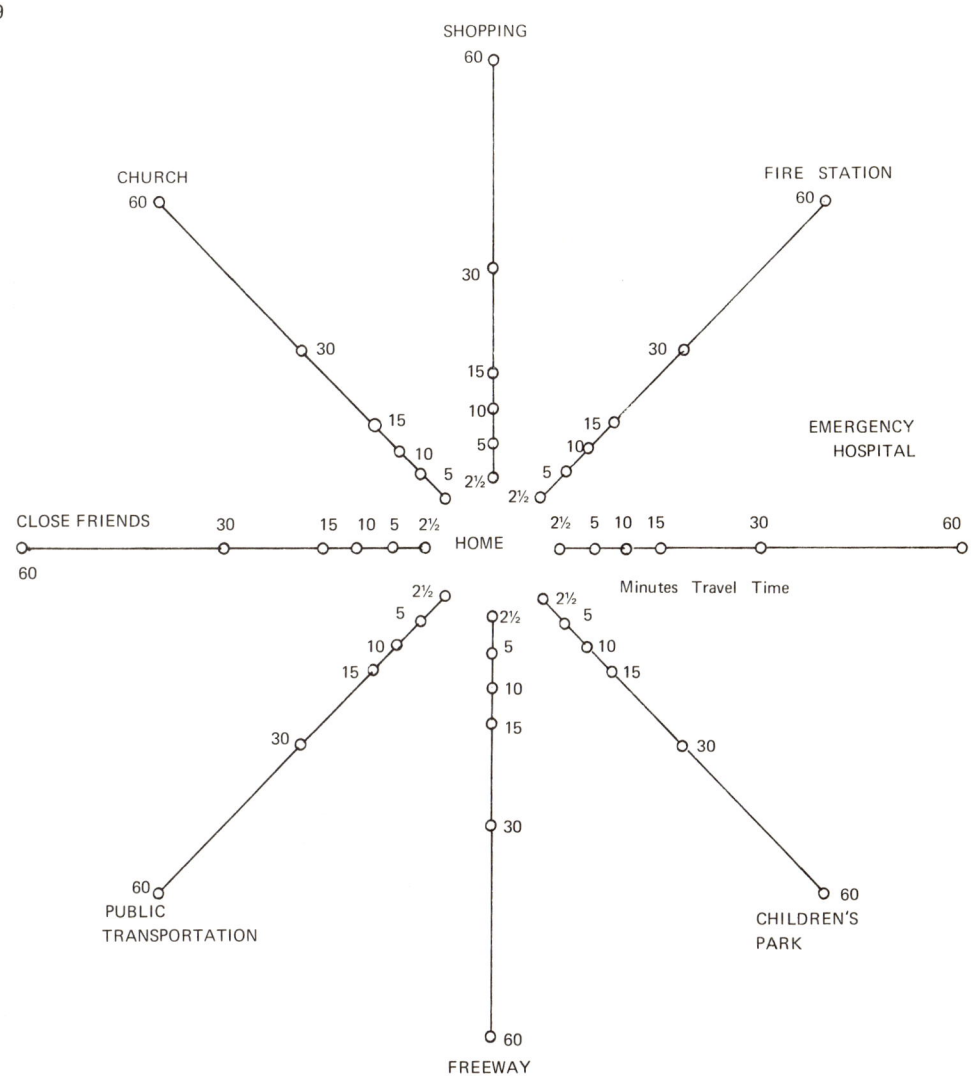

Figure 1.
Accessibility Game

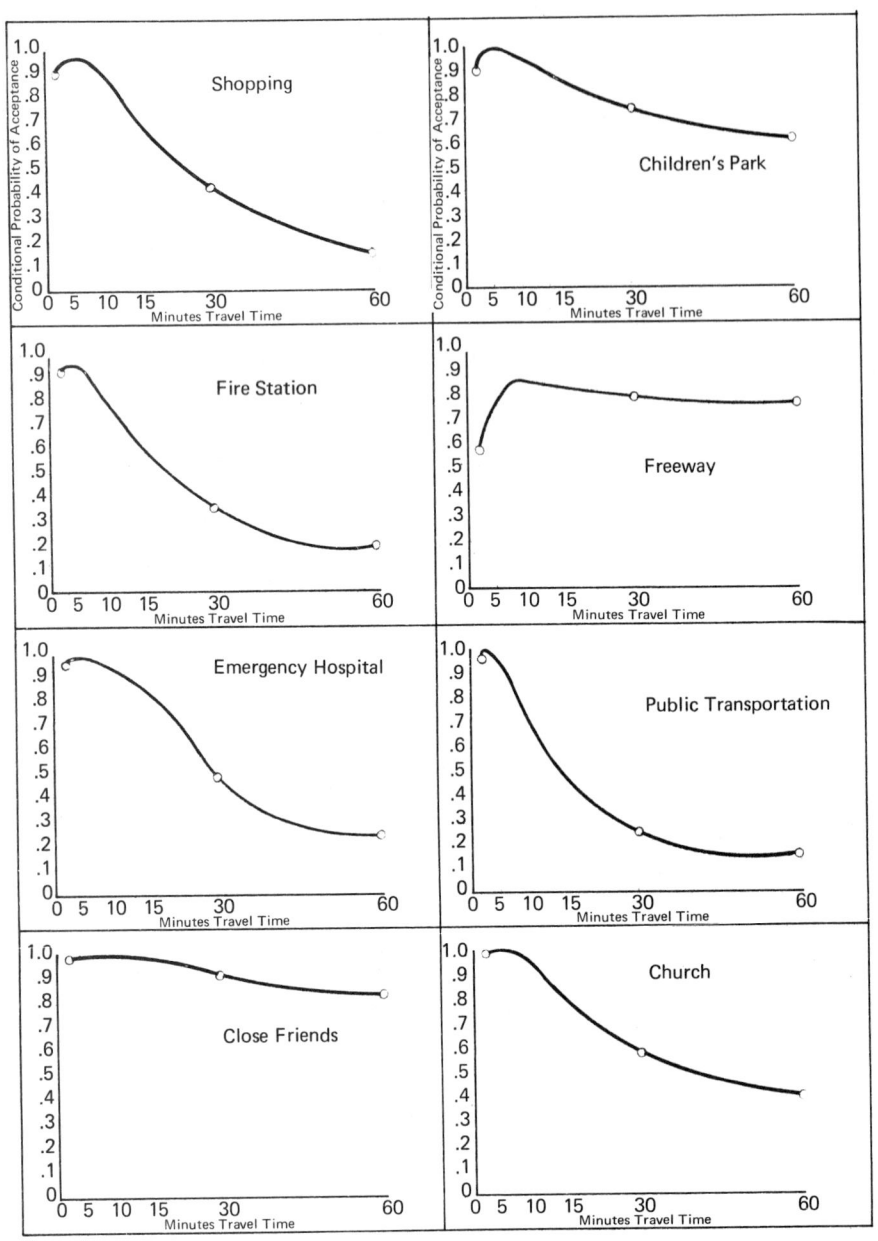

Figure 2

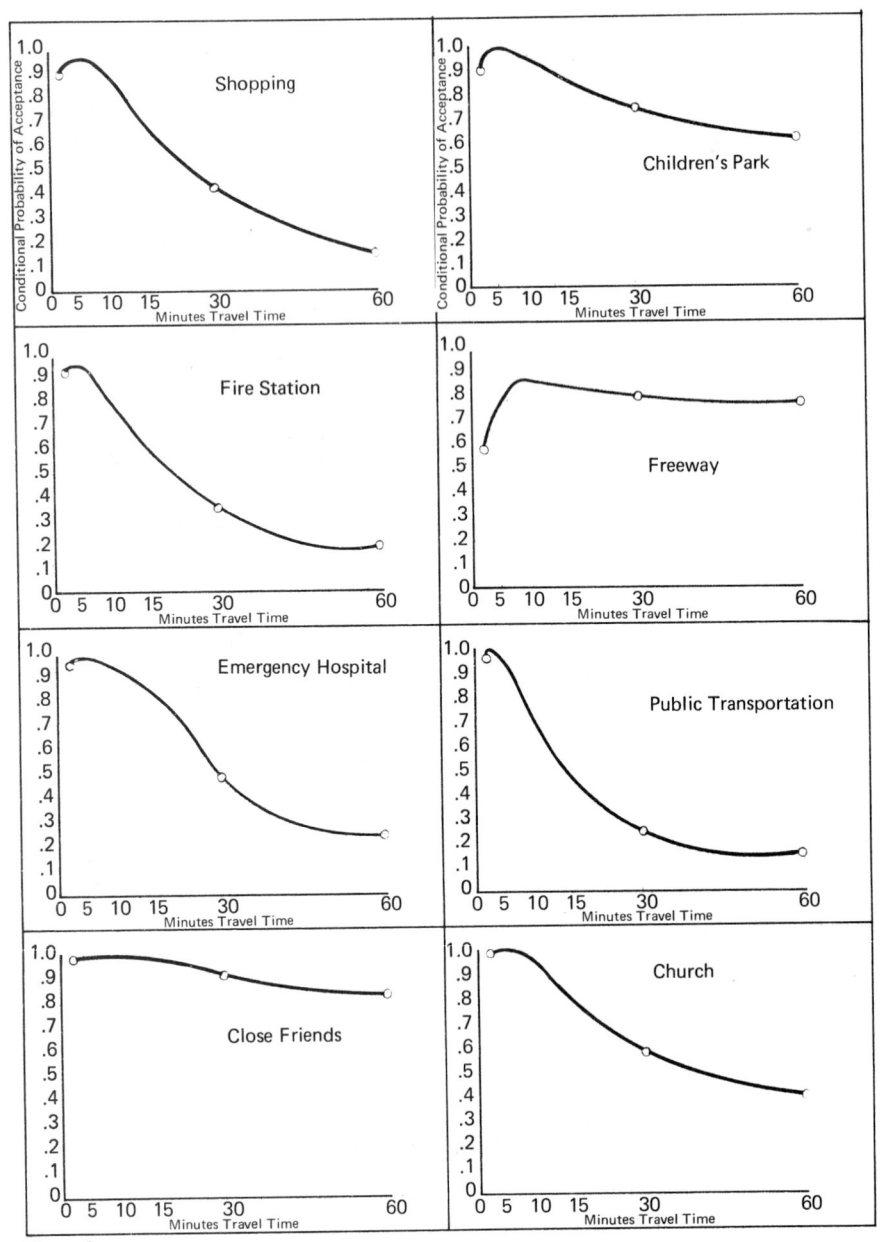

Figure 2

ferent to the location of all eight services. The curves do not appear to approach a probability of zero acceptability as distance increases, but rather some non-zero assymptote, suggesting indifference to the distance or travel-time involved. This is most obvious for *public transportation, freeway, children's park*, and *church,* and is least obvious for shopping. Similarly, the non-zero intercept on the ordinate for zero travel time suggests an equivalent proportion who are insensitive to proximity. Some services, especially *public transportation,* appear to be more sensitive to changes in accessibility than others. *Close friends,* for example is almost totally insensitive to travel time. Finally, it may be noted that there is substantially more distaste for proximity to a freeway than for any of the other services studied—the vertical intercept in this case indicating roughly 30% rejection of a freeway within 5 minutes of the home. Preference for access to a freeway appears relatively insensitive to travel-times greater than 30 minutes.

Toward a Theoretical Structure

Assume, as outlined above, that an individual responds to changes in the accessibility of a given neighborhood service in terms of two independent objectives which he is trying to satisfy simultaneously. One objective is concerned with an abhorrence of proximity, while the other desires convenience of access. For simplicity, assume further that dissatisfaction relative to the first objective increases at an increasing rate as the level of accessibility increases (i.e. as the travel time between home and service location decreases), approaching infinity at some critical travel time. Likewise, assume that dissatisfaction relative to the second objective increases at an increasing rate up to a similar critical level as accessibility decreases (i.e., as travel time increases). Let these two critical values be defined as "t_c" 7 "t_f" respectively, as illustrated in Figure 3.

For any given population, the values of t_c and t_f may be expected to vary from individual to individual. It will be assumed (but not proved) here that this variation is noncorrelated—i.e., that the values of t_c and t_f are independent. It will be further assumed that intermediate travel times t are perceived in terms of the ratio $\frac{t_0}{t_c}$ or $\frac{t_0}{t_f}$.

Let $z_c = \log t_c$ & $z_f = \log t_f$ be independent random variables with distribution functions $g_c(z)$ & $g_f(z)$ as illustrated in Figure 4.

The probability that a given travel time, t, will not be rejected for being too small is then given by:

$$p\{z_c < z\} = \int_{-\infty}^{z} g_c(z) dz = G_c(z), \text{ where } z = \log t.$$

Likewise, the probability that t will not be rejected for being too large is given by:

$$p\{z < z_f\} = \int_{z}^{\infty} g_f(z) dz = G_f(z).$$

Hence, assuming $g_c(z)$ and $g_f(z)$ are independent, the probability that t will not be rejected for being too small or too large is given by:

$$p\{z_c < c < z_f\} = p\{z_c < z\} \cap p\{z < z_f\} = \int_{-\infty}^{z} g_c(z) dz \int_{z}^{\infty} g_f(z) dz = G_c(z) G_f(z)$$

If z_c and z_f are normally distributed (as suggested by the data of Figure 2), the functions $p\{z_c < z\}$, $p\{z < z_f\}$, and $p\{z_c < z < z_f\}$ will have the form illustrated in Figure 5. Transformed back to the original arithmetic scale for t, these curves will be of the form shown in Figure 6.

If it is further assumed that a certain proportion, "a", of the population is indifferent to the closeness of the service, and that some similar proportion, "b" is indifferent to its remoteness, with z_c and z_f being normally distributed for the remainder, then $p\{z_c < z\}$, $p\{z < z_f\}$, and $p\{z_c < z < z_f\}$ will take the form shown in Figure

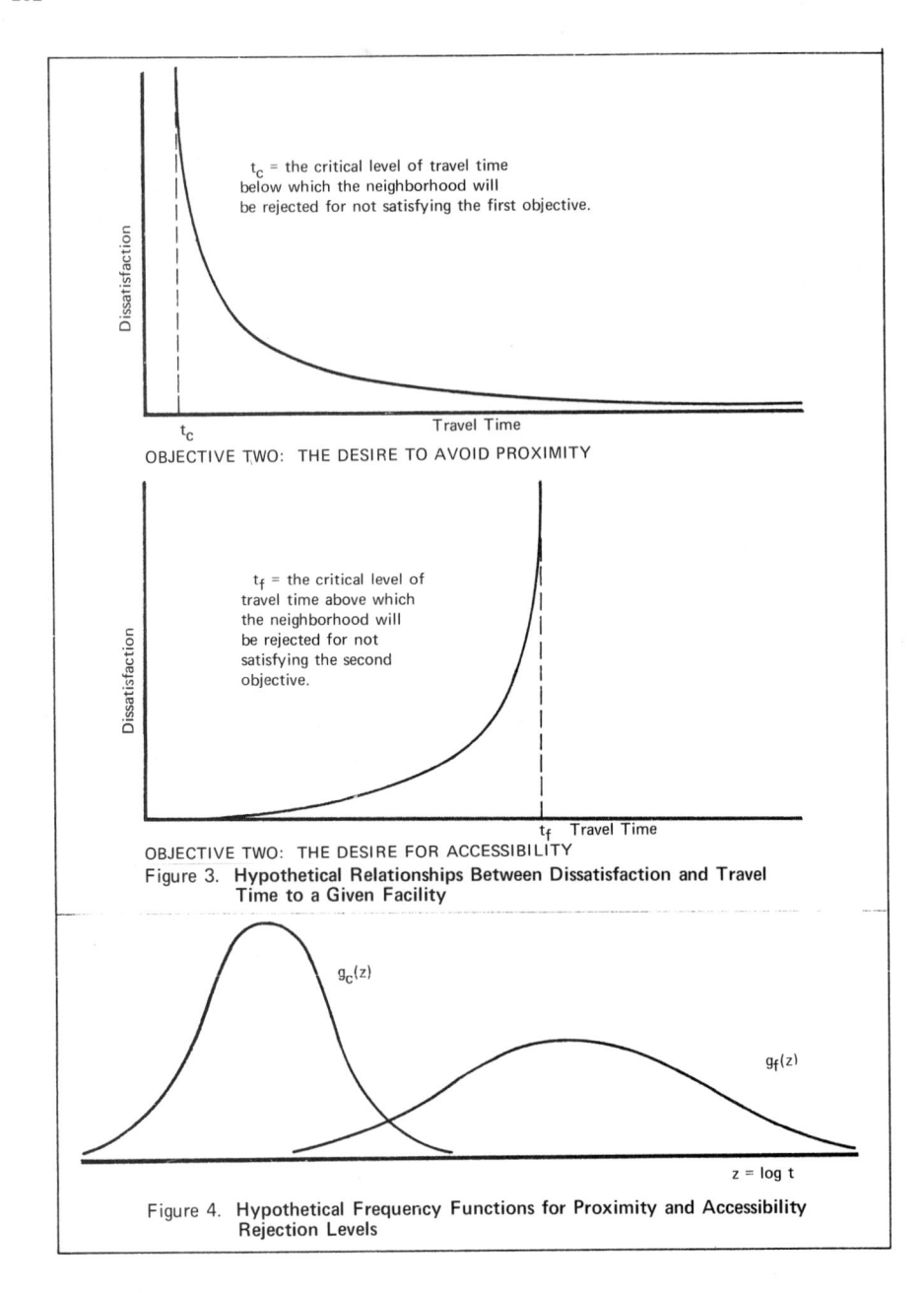

t_c = the critical level of travel time below which the neighborhood will be rejected for not satisfying the first objective.

Dissatisfaction | Travel Time | t_c

OBJECTIVE TWO: THE DESIRE TO AVOID PROXIMITY

t_f = the critical level of travel time above which the neighborhood will be rejected for not satisfying the second objective.

Dissatisfaction | t_f Travel Time

OBJECTIVE TWO: THE DESIRE FOR ACCESSIBILITY

Figure 3. Hypothetical Relationships Between Dissatisfaction and Travel Time to a Given Facility

$g_c(z)$

$g_f(z)$

$z = \log t$

Figure 4. Hypothetical Frequency Functions for Proximity and Accessibility Rejection Levels

7. On the original time scale, these curves will have the appearance of Figure 8.
The resulting form of $p\{z_c\ z\ z_f\}$ in Figure 8 is essentially the same as that of the empirical curves in Figure 2.

A Model of Accessibility Acceptance

The preceding theoretical structure may be summarized formally as:

Let
$$F_c(z) = \int_{-\infty}^{z} f_c(z)\,dz = \frac{1}{1-a}\left[G_c(z) - a\right]$$

$$F_f(z) = \int_{z}^{\infty} f_f(z)\,dz = \frac{1}{1-b}\left[G_f(z) - b\right]$$

be normal distribution functions of the probability that for a given service (all other things held constant) a given travel time $t = \text{Log}^{-1}z$ will not be rejected for being too small or too large, respectively, by that proportion of the population which is not indifferent to proximity in the case of $F_c(z)$, and by that proportion of the population which is not indifferent to remoteness in the case of $F_f(z)$.

Solving for $G_c(z)$ and $G_f(z)$,

$G_c(z) = (1-a)\,F_c(z) + a$, and

$G_f(z) = (1-b)\,F_f(z) + b$.

If $A(z)$ is defined to be the probability that the travel time $t = (\text{Log}^{-1}z)$ will be acceptable, then a general equation for the curves indicated by the data in Figure 1 is given by:

$$A(z) = G_c(z)\,G_f(z) = \left[(1-a)\,F_c(z)+a\right]\left[(1-b)\,F_f(z)+b\right].$$

that is:

$$A(z) = \left\{(1-a)\int_{-\infty}^{z}\frac{1}{\sigma_{zc}\sqrt{2\pi}}\exp\left[-\tfrac{1}{2}\left(\frac{z-\mu_{zc}}{\sigma_{zc}}\right)^2\right]+a\right\}\left\{(1-b)\int_{z}^{\infty}\frac{1}{\sigma_{zf}\sqrt{2\pi}}\exp\left[-\tfrac{1}{2}\left(\frac{z-\mu_{zf}}{\sigma_{zf}}\right)^2\right]+b\right\}$$

Calibration of Model

If it is assumed that the data shown in Figure are in fact the outcome of a process such as that outlined above, with proportions "a" and "b" of the populations being indifferent respectively to proximity and distance, and the residual functions $p\{t_c < t\}$ and $p\{t < t_f\}$ being log-normally distributed, then the two expressions

$$\frac{1}{1-a}\left[\int_{-\infty}^{z} g_c(z)\,dz - a\right] = h(z)$$

and

$$\frac{1}{1-b}\left[\int_{z}^{\infty} g_f(z)\,dz - b\right] = k(z)$$

may be solved graphically for "a" and "b"—i.e., both functions plot as linear expressions of z on normal probability paper.

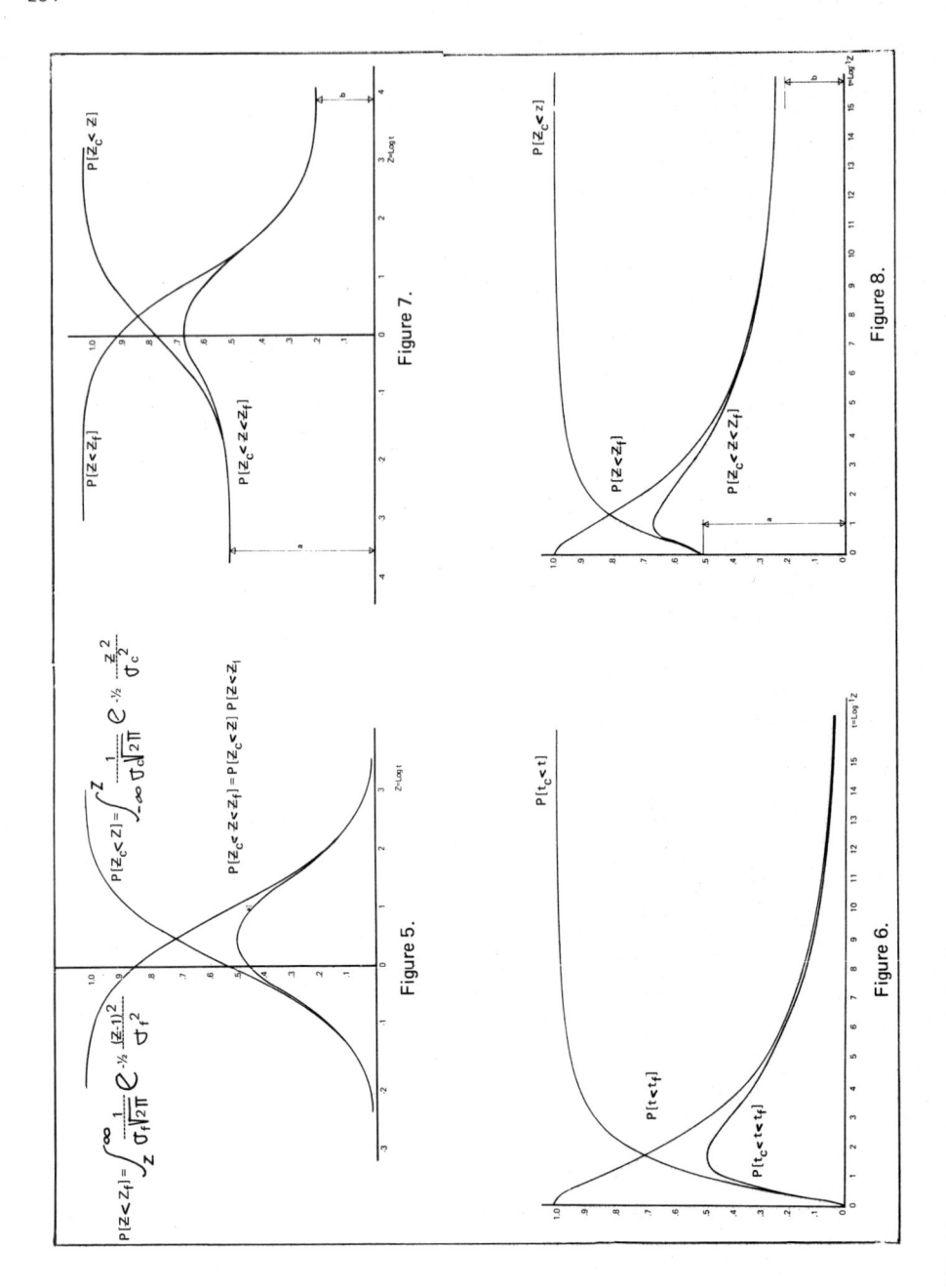

$$P[Z<Z_f]=\int_Z^\infty \frac{1}{\sigma_f\sqrt{2\pi}} e^{-\frac{1}{2}\frac{(Z-1)^2}{\sigma_f^2}}$$

$$P[Z_c\leq Z]=\int_{-\infty}^Z \frac{1}{\sigma_d\sqrt{2\pi}} e^{-\frac{1}{2}\frac{Z_c^2}{\sigma_c^2}}$$

$$P[Z_c<Z<Z_f]=P[Z_c<Z]\,P[Z<Z_f]$$

Figure 5.

Figure 6.

Figure 7.

Figure 8.

Figure 9 illustrates the results of such an analysis for the case of the Fire Station. Note the estimated values of a = 0.625 and b = 0.208 for the two indifference properties.

The remaining parameters may be estimated as:

$$a = 0.625$$
$$b = 0.208$$
$$\mu_{zc} = \text{Log } 1.3$$
$$\sigma_{zc} = \text{Log } 3.2 - \text{Log } 1.3$$
$$\mu_{zf} = \text{Log } 18.5$$
$$\sigma_{zf} = \text{Log } 34 - \text{Log } 18.5$$

Figure 10 compares these values with the plots of the two functions h(z) & k(z) and with the empirical data summarized in Figure 2. Table 3 summarizes the values of h_c and h_z as calculated from the original data and the estimates of a and b, for each of the eight services studies. Table 4 summarizes the parameter estimates for each of the other seven services, for those instances where they could be determined. In each case, the results lend at least tentative credence to the theoretical structure postulated above.

Concluding Remarks

This paper has developed a simple theory of locational preferfene for neighborhood services. It has been suggested that an individual perceives the distribution of neighborhood facilities around his home—shops, parks, fire-stations, transit-stops, etc.—in terms of two competing objectives. One objective is concerned with the need to use the service and desires ready accessibility, the other is concerned with avoidance of irritation and attempts to avoid undue proximity. The result is a tradeoff, the implications of which are exhibited empirically in the form of a log-normal distribution of accessibility acceptance. This distribution may be modeled theoretically as the outcome of two independent, normally distributed response functions, one reflecting the desire for accessibility, the other the desire for insulation from irritation. The empirical validity of this structure has been demonstrated using data from a survey of 240 families in Chicago.

A word of caution, however, is in order. The empirical data were obtained by using an experimental instrument designed for purposes other than that of testing the immediate hypothesis developed here. Although they tend to support rather than negate the hypothesis, they should be subjected to additional external verification. The next logical step would be to design and implement a suitable experiment that would permit a more rigorous examination of the postulates outlined in this paper.

Attention should be equally directed toward the notion of substitutability between services which was ignored here. In this case, the procedure was to locate optimally all but one service and then to vary the location of that one to determine thresholds of acceptability. This results in a conditional rather than an absolute distribution of acceptance for each service, and is clearly dependent on the optimal location of the others. The procedure establishes a threshold of acceptance for each service based on the condition that the others are located so as to provide maximum satisfaction. However, it is conceivable that this threshold would be different if other outside factors came into play. If one or more of the other services were not located optimally, it is possible that this threshold would be different. An alternative approach might be to locate all but one service at minimal acceptance levels, and vary the location of the remaining one until it, too, was just acceptable. The threshold of acceptance thus obtained might well be different from that established here.

Additional research may also usefully be directed into the notion of tradeoffs between accessibilities. The data in this study seem to suggest that there are limits on accessibility that the resident will judge to be minimally acceptable. If this hypothesis can be verified, a next step might be to examine combinations of accessibility to a mix of services, leading ultimately to establishment of acceptable combinations for each that collectively define a satisfying environment rather than intriguing, but unattainable, optimum.

206

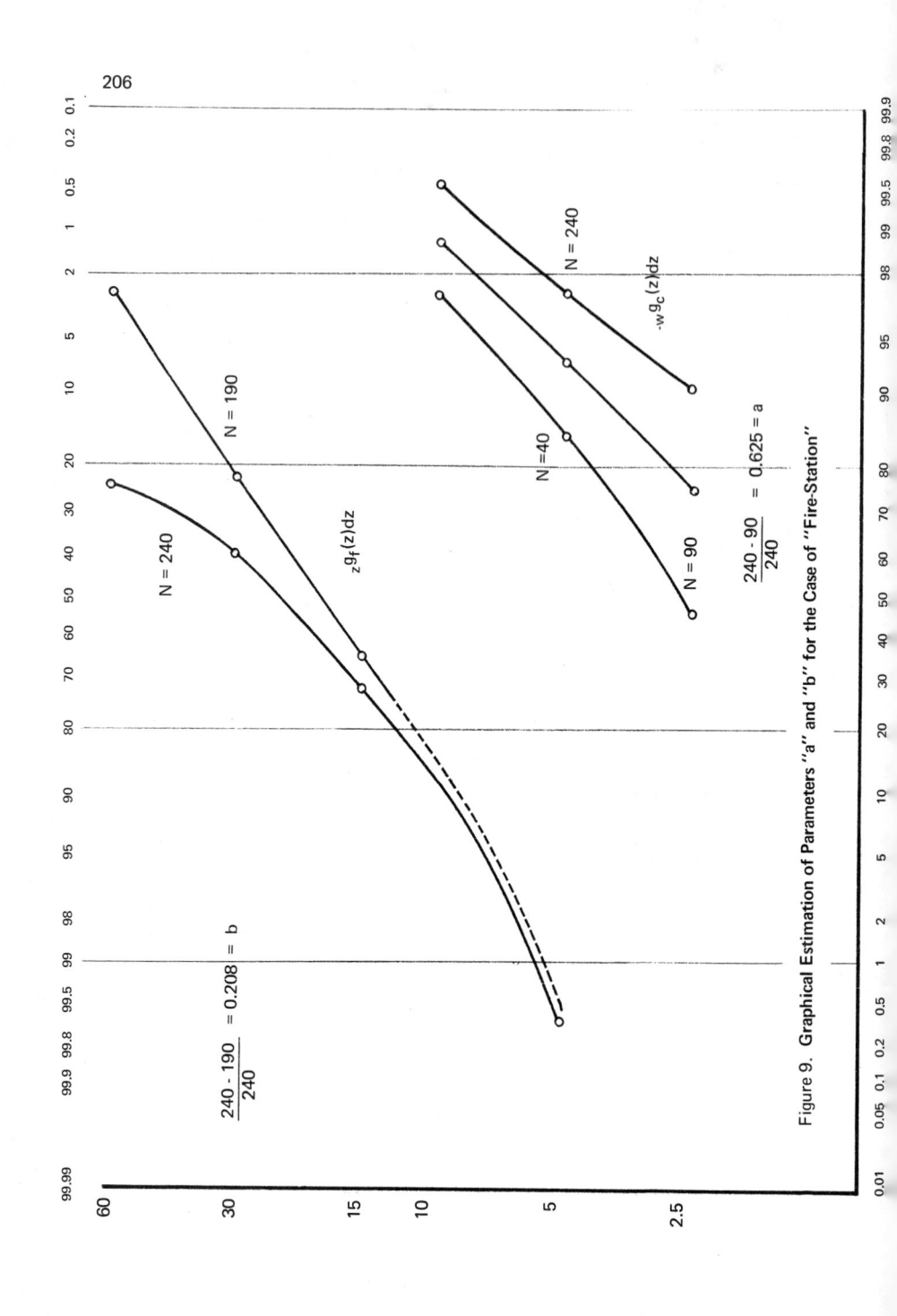

Figure 9. Graphical Estimation of Parameters "a" and "b" for the Case of "Fire-Station"

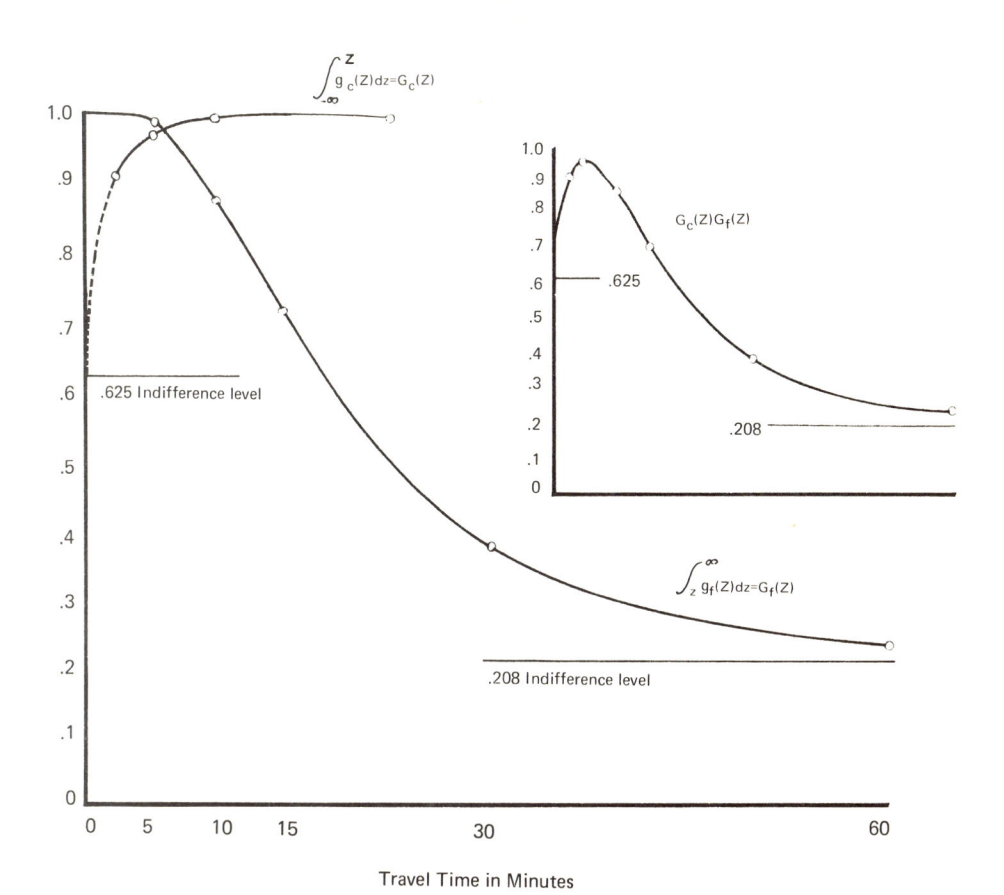

Figure 10. **Fire Station**

	Travel Time "t" (min.)					
	2.5	5	10	15	30	60
Service "i"		Values of $f_a(t)^i$				
Shopping	218	232	222	188	103	44
Fire	219	233	208	171	90	55
Hospital	224	234	226	206	108	60
Park	220	234	231	212	169	146
Freeway	152	177	213	223	198	178
Public Transportation	234	231	179	122	49	37
Friends	237	238	239	235	225	198
Church	236	238	234	209	144	106

TABLE 1. NUMBERS OF HOUSEHOLDS ACCEPTING GIVEN ACCESSIBILITY—$f_a(t)^i$

	Travel Time "t" (min.)					
	2.5	5	10	15	30	60
Service "i"		Values of $p_a(t)^i$				
Shopping	.908	.966	.925	.748	.429	.183
Fire	.913	.971	.866	.713	.375	.229
Hospital	.933	.975	.942	.858	.450	.250
Park	.917	.975	.962	.883	.704	.608
Freeway	.633	.737	.888	.930	.825	.742
Public Transportation	.975	.962	.746	.508	.204	.154
Friends	.987	.922	.996	.979	.937	.825
Church	.983	.992	.975	.871	.600	.442

TABLE 2. PROPORTIONS OF HOUSEHOLDS ACCEPTING GIVEN ACCESSIBILITY—$p_a(t)^i$

			Travel Time t					
			2.5	5	10	15	30	60
Service		(N.a)	Values of Integrals 1 and 2 Defined Below					
Shopping	1	------	.908	.975	.992	1.00	1.00	1.00
	2	*						
Fire	1	------	.913	.975	.996	1.00	1.00	1.00
	2	150	.767	.933	.989	1.00	1.00	1.00
Hospital	1	------	.933	.979	1.00	1.00	1.00	1.00
	2	**						
Park	1	------	.916	.975	.992	1.00	1.00	1.00
	2	*						
Freeway	1	------	.633	.737	.900	.958	.992	1.00
	2	0	.633	.737	.900	.958	.992	1.00
Public Transportation	1	------	.979	.996	1.00	1.00	1.00	1.00
	2	**						
Friends	1	------	.987	.992	1.00	1.00	1.00	1.00
	2	**						
Church	1	------	.983	1.00	1.00	1.00	1.00	1.00
	2	**						

TABLE 3a
Summary of Results of Graphical Estimation of a

*: Negative Indifference level. Linearity requires a<o.

**: Insufficient data

1: $\int_{-\infty}^{z} g_c(z)dz$ for N = 240 (total Sample)

2: $\frac{1}{1-a} \left[\int_{-\infty}^{z} g_c(z)dz - a \right] = \int_{-\infty}^{z} f_c(z)dz$ (estimated maximum linearity)

Travel Time t

Service		(N.b)	2.5	5	10	15	30	60
				Values of Integrals 1 and 2 Defined Below				
Shopping	1	------	1.00	.991	.933	.784	.429	.183
	2	30	1.00	.990	.924	.752	.348	.067
Fire	1	------	1.00	.996	.871	.713	.375	.229
	2	50	1.00	.995	.837	.637	.211	.026
Hospital	1	------	1.00	.996	.941	.858	.450	.250
	2	52	1.00	.995	.926	.819	.298	.043
Park	1	------	1.00	1.00	.971	.883	.704	.608
	2	142	1.00	1.00	.928	.714	.276	.041
Freeway	1	------	1.00	1.00	.987	.971	.833	.742
	2	176	1.00	1.00	.953	.890	.375	.031
Public Transportation	1	------	.996	.966	.746	.508	.204	.154
	2	36	.996	.961	.701	.422	.064	.0049
Friends	1	------	1.00	1.00	.996	.979	.937	.825
	2	240	1.00	1.00	.996	.979	.927	.825
Church	1	------	1.00	.992	.975	.871	.600	.442
	2	100	1.00	.993	.957	.778	.314	;043

TABLE 3 b
Summary of Results of Graphical Estimation of b

1: $\int_z^\infty g_f(z)\,dz$ for $N = 240$ (total sample)

2: $\frac{1}{1-b}\left[\int_z^\infty g_f(z)\,dz - b\right] = \int_z^\infty f_t(z)\,dz$ (estimated maximum linearity)

	a	b	zc	zc	zf	zf
Shopping	*	.125	------	------	Log 23.3	Log 1.90
Fire	.625	.208	Log 1.3	Log 2.46	Log 18.5	Log 1.84
Hospital	**	.216	------	------	Log 22	Log 1.79
Park	*	.592	------	------	Log 21	Log 1.82
Freeway	0	.734	Log e	Log 2.67	Log 25.8	Log 1.57
Public Transportation	*	.150	------	------	Log 13.5	Log 1.72
Friends	*	0	------	------	Log 195	Log 3.4
Church	*	.417	------	------	Log 23.8	Log 1.70

TABLE 4

Summary of Graphically Estimated Parameters for

$$A(z) = \left\{(1-a)\int_{-\infty}^z \frac{1}{\sigma_{zc}\sqrt{2\pi}} \exp\left[-\tfrac{1}{2}\left(\frac{z-\mu zc}{\sigma_{zc}}\right)^2\right] + a\right\} \left\{(1-b)\int_z^\infty \frac{1}{\sigma_{zf}\sqrt{2\pi}} \exp\left[-\tfrac{1}{2}\left(\frac{z-\mu zf}{\sigma_{zf}}\right)^2\right] + b\right\}$$

* insufficient data
** negative indifference level (normality requires a or b<0.)

Raymond G. Studer
Man and Environment Division
Pennsylvania State University

WORKSHOP ON HUMAN RESPONSE TO THE ENVIRONMENT

Each paper deals in some way with behavioral consequences of spatial organization; each has implications for environmental organization; each is a "probe" rather than a complete study; each has strengths and weaknesses of method and possible application which are somewhat complementary.

There is a ponderosity and ambivalence in Finrow's paper which at times makes the investigation difficult to follow. The presentation often vacillates between two issues which operate at vastly different scales: 1) the characteristics of intra-apartment contact systems, and 2) the characteristics of regional contact systems. To assess the effects of a specific environmental component in a residential setting is one thing; to extrapolate to an entire region on the basis of these data is quite another (although a work-related residential ordering seems logical enough).

What one does not readily find in this paper is a well articulated framework within which the research issues in question could be more clearly identified and bounded. The equivocal mode of argumentation, especially as this relates to the data, was at some points rather frustrating, and there emerges a tendency to move too quickly beyond the modest data presented.

The underlying design issue itself embodies certain difficulties. There is some question as to whether or not more intensive human contact is either good for human systems or desired by the participants. While this assumption was briefly questioned, the overriding implication was "the more the better." More important, perhaps, is the quality and composition of these contacts relative to the goals of the human organization in question. Not only is "human contact behavior" too general a class to embody meaningful design directives, but certain of these in certain contexts can be both stressful and dysfunctional with respect to the setting as a whole. A less biased and more refined statement of the behavioral issues might have sharpened the analysis and delivered more incisive information for designers.

As the investigator noted, the first study was less convincing than the second. For example, the experimental and control groups were not really comparable since they included 60% and 30% married subjects respectively. Considereing the size of the sample, this non-homogeneity could be significant when assessing the data. One senses that married and unmarried individuals surely have different contact patterns which might account for certain outcomes in Study 1. A comparison can be made here with Michelson's investigation in which systematic analyses were made to factor out possible socioeconomic or demographic variables which might account for certain environmental responses. Finrow's second study seemed to correct several difficulties of the first and together, these studies seem to verify the general hypothesis that human contact patterns are indeed intensified with the introduction of community facilities.

The problem addressed by Michelson was considerably broader than that by Finrow, but in contrast, he immediately sought an appropriate research framework which was more likely to focus upon relevant variables. The study was problem-oriented from the outset, which precipitated a particular research strategy made even more incisive because of the constraints of time. The general point to be made here is that the problem orientation seemed to have the effect of giving structure and boundaries to an empirical investigation which made it more relevant to the application of its findings. The samples in this study were well drawn, the data effectively analyzed and the conclusions quite convincing. It was indeed a notable example of effective field research techniques.

What emerges, of course, is the endemic problem of applying sociological findings to specific design decisions. This realization is particularly discouraging in this case since the study was specifically designed to provide design specifications. What makes the utilization of these findings to physical design so difficult? The first problem is that the findings are, of course, quite general. How shall we interpret environmentally the result that 33% of townhouse residents intend to move? The second, and overlapping, problem is that these findings are not in a form which makes them readily applicable in the physical design process. The units of environmental information are: 1) quite large and non-quantitative with respect to topographic implications, 2) they are described in conventional environmental units (e.g. building types) and 3) they set forth no positive and precise design directives. To be utilized, then, these findings require transformation into elements of an *environmental program*, i.e., a set of specifications describing environmental quantities, qualities and relationships. These elements may necessarily be in the form of probability statements or "if-then" statements, but such directives seem an implicit aspect of a specific design-related investigation. If such studies do not conclude with a testable environmental hypothesis, little wonder that so few sociological data find their way to actual environmental arrangements. The sociological consultant contends that such hypotheses are outside his sphere of responsibility. But who is better equipped to deliver them? Environmental designers must constantly put their reputations on the line in these matters without the tools available to the social scientist. Should not this responsibility for specifying appropriate environments at least be shared?

One monor assertion by this investigator requires at least a passing comment because it embodies a fairly serious epistemological issue, i.e. that satisfaction or dissatisfaction can be assessed independent of the activities involved. It is difficult to see how a satisfaction state can be detected without observing some form of behavior. More generally, the problem of assessing attitudinal responses with respect to any class of phenomana arises from the fact that the social context of the attitudinal probe involves a different set of contingencies than the subsequent behaviors (ostensibly correlated with them). I do not think that a monolithic position is justified here.

In many ways the Peterson, Worrell and Redding study was the most speculative, but it was also the most relevant to environmental design processes. The analytic tools used were developed for another purpose, i.e., transportation planning. The various elements of this analysis thus forced oversimplification of the (human) problem under investigation (transportation models developed to date generally have this effec in any event). Before one would be confident in applying the findings of such a study it is apparent that 1) more extensive and reliable behavioral data are required, as well as 2) a considerably richer conceptual mode Since the investigators identified both these problems themselves, it is pointless to dwell upon them or the specific conclusions they generate. More important is the direct relevance this approach has to environment al design and management problems. In this respect there emerges an interesting contrast with the other two studies. Like Michelson's, this investigation was defined in a specific, real problem context. Beyond this, these data were defined in (design-relevant) spatial-temporal units and these units admit quantification. Finally, the study concluded with a precise and testable hypothesis (a predictive model) for an aspect of environmental design. Because of these characteristics, not only the findings, but the model itself could be readily integrated into a design process and systematically tested in a real setting.

General Comments

Among other things, such empirical studies as these reveal certain fallacies in the fundamental beliefs held by designers regarding man-environmental causality. That is, designer intuition concerning these matters is often *reversed* when empirical findings are systematically developed. This is ostensibly the purpose of a scientific analysis and it is gratifying to see these techniques at work in the environmental design community.

All three of these investigations, while obviously producing findings which are either explicitly or implicitly relevant, contain a fundamental difficulty with regard to their implementation in design decisions.

Each of these sets of findings deal with behavioral responses to conventional environmental components, e.g., existing residential building types, fire-houses, hospitals. Herein lies a fundamental dilemma for researchers in this area. One could hardly investigate (empirically) behavioral responses to environments which do not exist. But the design problem specifically involves the invention and realization of environments which do not yet exist (otherwise there would simply be no problem). How can our knowledge of these human responses be generalized and extrapolated to entirely new settings yet to be invented? For example, the problem is not to locate n fire-houses, but to swiftly extinguish or prevent destructive fires as these occur within a given demographic and/or geographic boundary. "Fire-houses" are but one method of responding to this problem and not a particularly innovating one at that. The question, then, is what relevance has a population', "fire-house location preferences" to do with the essential problem under analysis. A similar comment could be made about each of these studies. What aspect of the findings are actually relevant to inventing a new system? In order to use these data to invent new systems, they must be transformed, I feel, to another level of discourse which reveals the underlying causal structure.

To carry the above argument a bit further, we may look at environmental design as an attempt to resolve conflicts between things as they are and as they ought to be (relative to the goals of the population in question), and to move a system from the *is* state to the *ought* state. To do this, environmental designers must understand and respond to three levels of requirements: 1) human wants, 2) human needs, and 3) human system equilibrium (a synthesis which realizes collective well-being)—using behavior as an index. These three studies essentially deal with the first of these, i. e., human wants with respect to conventional components. The implication of the Peterson, Worrell and Redding study was that human preferences (wants, capriciously conditioned) can be directly extrapolated to an alternative setting. These preferences are in fact but one of several dimensions affecting the characteristics of the required configuration.

The above comments concerning applicability may seem unjustified since it may be argued that these investigations were intended only to produce empirical findings, not to explicate their application. Be this as it may, the problems of design research (the subject matter of this conference) necessarily involve problems of application. These issues are thus very difficult to overlook.

In responding to this conference generally, a comment by a psychologist exposed here for the first time to design research, seems particularly relevant. He expressed surprise that the research issues pondered in some of these sessions were settled years ago in psychology. Many persons from the design community are moving for the first time into areas of empirical research. While one must begin where he is, there is clearly an obligation to become knowledgeable in the relevant disciplines being applied. Furthermore, we must realize a more critical attitude in this regard when assessing research findings. The basic disciplines generally embody one characteristic lacking within the design disciplines, i.e., a *research consensus* within particular areas of investigation. As a rule, the scientific disciplines have evolved a language and a format for reporting research which is highly effective in communicating their intentions and results. There are clearly advantages in such a research milieu.

Obviously, we must better understand the resources of the basic disciplines from which design research must draw, and we must come to adopt more critical and rigorous modes of doing research. It does not follow, however, that we should adopt the precise research format of these basic disciplines. If their individual approaches were sufficient to resolve the research issues we in the design community now face, there would be no need for this kind of conference nor the organization behind it. Rather, we should seek a format which more clearly reflects the means of applying the results.

In order to understand how man functions, he had to be taken apart. Basic disciplines evolved to study the various sub-systems. To understand the environmental requirements of man, we must put him together again. Most information needed for environmental design purposes is fragmented, incomplete, and in a form which makes its use most difficult to apply in physical problem-solving. One must be skeptical regarding the direct application of existing modes of doing research in the basic disciplines. What is suggested is that research projects should be formulated in the same dimensions as the environmental design problems to be solved. To carry the proposition still further, each new design problem represents an opportunity for relevant environmental research. The Michelson Study, for example, represents in many ways a missed opportunity for it stopped short of formulating and testing an explicit hypothesis concerning environment-behavior causality. I have argued elsewhere for a *behavior-contingent paradigm* in which environmental design problems are formulated in terms of empirically testable hypotheses concerning human outcomes. I would argue here that one and the same paradigm produces a reasonable format for approaching environmental research. To generate empirical findings independent of actual problem situations may further fragment our knowledge and move us further away from an appropriate technical language of environmental design.

Stuart Rose
School of Design
North Carolina State University

TOWARD A STIMULUS-RESPONSE THEORY OF ENVIRONMENTAL DESIGN

When approximations of building costs are generated over the life of a building, the initial cost of construction appears to be about two percent of the total; maintenance costs about six percent of the total; and, cost of salaries, etc. of the building occupants about ninety-two percent of the total. To date most criteria for building design decisions have related to initial maintenance costs plus whatever general and very apparent human considerations appeared to require satisfaction (based upon the judgment of the designer and client). The effect of the quality of the space upon the occupant of that space has either been disregarded, dismissed due to budget "limitations," or subjectively dealt with by the designer. Yet, while such considerations have received little attention, their implications in terms of cost alone—if not in terms of aiding to fulfill human life functions—would suggest a criterion rating forty-six times that of decisions related to initial building costs, and fifteen times that of decisions related to maintenance costs. Until designers of space can demonstrate that the effects of spatial attributes upon the behavior of occupants is significant in terms of helping occupants to fulfill their life activities, and that the process by which they arrive at behavioral types of decisions is measurable and predictable, designers will remain in the ornamental fringe of society.

This study represents an initial step in an exploratory attempt at understanding an aspect of human responses to the physical environment. While the sample size was small and the data lacking statistical significance, the results suggest the kind of data that might be achieved with a larger and more representative sample. The purpose of this study was to determine the effect that variations in the qualitative characteristics of a space have upon the behavior of its occupants,[1] who are performing a specified task. The variations in the qualitiative spatial characteristics included size, shape, color and texture of the space establishing elements[2] (i.e. walls, floor and ceiling). The behavior of the occupants consisted of small group discussion oriented toward achievement of a specified, programmed and measurable set of tasks. Attitudinal response toward the spatial characteristics, considered as the environmental stimulae, was the independent variable; two spaces were constructed in accordance with two patterns of responses toward the stimulae, one congruent and one dissonant (or opposite) with attitudes toward the described activity. Behavior of the occupants in the two spaces was measured by task achievement, quality and quantity of interaction, and attitude toward the activity. The study hypothesized that the space whose characteristics were attitudinally dissonant (or diametrically opposite) to the attitude toward desired behavior would encourage and produce the desired behavior.[3]

The interaction between man and his physical environment was examined from a stimulus-response point of view. The stimulus was considered to come externally to man from the physical environment, while the response was considered to come internally to man in the form of behavioral motivations. In the interaction the desire for comfort is assumed, such that if the environmental stimulants vary from a point of comfort in one direction, the behavioral response would vary from comfort in the other direction to produce a comfortable balance.

Viewing the qualitative attributes of space establishing elements as environmental stimuli the following process suggests a means by which the use of such stimuli may be directed to help facilitate desired behaviors.

In this study, the semantic differential attitude measuring instrument was employed to record the profile of the meaning attached to the behavior desired.

Assuming the neutral axis of the semantic differential as an approximation of the comfort balance, a mirror-image of the desired behavior profile was constructed. The latter was assumed to be a balancing profile to which the meaning attached to the environmental stimulants should conform.

Employing stimuli that conform to both the profile of desired behavior (i. e. "Dissonant Stimuli"), the space designed by incorporation of the balancing "Dissonant Stimulants" should produce the desired behavior.[4]

Four terms critical to the understanding of the study are listed below:

1) *Profile "A"* or the *Behavior Profile,* which is the profile of the behavior desired.

2) *Profile "B"* or the *Balancing Profile,* which is the mirror-image of the Behavior Profile.

3) *Space "A"* or the *Consonant Space,* which is the space designed by the incorporation of Consonant Stimulants (or environmental stimulants consonant with the Behavior Profile).

4) *Space "B"* or the *Dissonant Space,* which is the space designed by the incorporation of Dissonant Stimulants (or environmental stimulants dissonant with the Behavior Profile and consonant with the Balancing Profile).

Characteristics of the Spaces

Space "A", the Consonant Space:

designed according to closeness of response to small group discussion activity concept, (i.e. to the Behavior Profile).

Space "B", the Dissonant Space:

designed according to closeness of response to "mirror-image" of small group discussion activity concept, (i.e. to the Balancing Profile).

Qualitative Attribute	Space "A"—Consonant Space	Space "B"—Dissonant Space
Size	15 X 18 feet, 8½ foot ceiling	8½ X 11½ feet, 8½ foot ceili
Shape	rectangular	oval
Contrast	high	low
Textures	rough	smooth
Colors		
Walls	bright yellow bright orange	medium grey
Ceiling	white with black trim	muted purple
Table	4½ foot diam.—white	4½ foot diam.—muted green
Chairs	white upholstery with silver metal trim	muted blue or green upholst with medium grey metal trir
Floor	bright blue	muted green

Given a similar series of student groups performing identical educational activities, the following hypotheses were asserted:

1) Task achievement will be greater in Space "b", the Dissonant Space, (a space that evokes an atitude opposite to the occupant's attitude toward the activity).

2) Interaction will be greater in quality and quantity in Space "B", the Dissonant Space, (a space that evokes an attitude expression opposite to the occupant's attitude toward the activity).

3) Space "B", the Dissonant Space, will tend to shape attitudes toward the activity in the direction of Profile "A", the Behavior Profile, (i.e. the attitudes toward the small group discussion activity will tend to be shaped in a direction opposite to that of the attitude expression evoked by the space).

4) Attitudes toward the small group discussion activity will be closer to the attitudes toward the subject matter at the completion of the activity than at the start of the activity.

Limitations of the Study

Five basic limitations were evident during the pre-test phase of the study. The first was the quality of the attitude measurement instrument which had been tested thoroughly by Charles Osgood[5] in a variety of situations, but had never been employed with the particular purposes for which it was used in this study. The second limitation during the pre-test was the number of participants, which was just above the minimum required to insure reliability of results. The third limitation was the quality of simulation of the concepts; such qualities as color accuracy were limited by photographic technology. The fourth limitation in the pre-test was the time of exposure; the impact of a color over a period of only a minute or so. The fifth limitation was the number of concepts employed, which was the result of the fatigue that was found to occur after a length of testing; two or more sittings were not possible to arrange.

Six basic limitations were evident during the treatment phase of the study. The first was the treatment of the qualitative characteristics of the space establishing elements as a composite; because of time and fund limitations, each characteristic could not be treated independently as the permutations expanded too quickly. The second limitation was the flexibility of construction of the spaces; fund limitations and restrictions imposed by the physical plant department of the University provided for spaces that could not be specially constructed for the treatment, and only for temporary modifications of existing spaces. The third limitation was the number of participants available for each group, which was six rather than the preferred seven. The fourth limitation was the number of groups that were formed, which was also limited to availability of participants; while five groups per space was assumed to be sufficient to provide reasonable reliability of results, several times that number would have been preferred. The fifth limitation was the amount of time, in terms of the numbers of group meetings, that was able to be scheduled; a result of the schedules of the student participants and the amount of funds available to pay for extended participation. The sixth limitation was the mode of observation employed to record the interaction of the groups. Because the meeting rooms could not be specially constructed for the study, observation booths were not available and interaction measurement had to be done from tape recordings.

The major limitation involved in the post-test was the quality of the attitude measurement instrument, as discussed previously. In all, the most critical of the limitations mentioned were the small number of subjects and the treatment of the attributes of space establishing elements as a composite.

By employing measurement of attitude expression as an intermediate linkage or common medium between the internal forces that lead to behavioral outputs and the external forces that result from physical outputs and the external forces that result from physical environmental stimulae, the following process was established in which the design of the space hinged upon behavioral input and in which the designed space would cause predictable behavioral output.

1) Measurement of the attitudes of occupants toward a given activity, the content of the activity and an array of alternative attributes of space establishing elements, (e.g. various colors, textures, etc.).

2) Determining and recording of the attitude profile toward the activity, and the opposing, or balancing attitude profile toward the activity, (i.e. the Behavior Profile and Balancing Profile.)

3) Determining and recording the attitude profile toward each of the alternative attributes of space establishing elements.

4) Matching the profiles of the attitudes toward the alternative attributes of the space extablishing elements to the Behavior Profile and the Balancing Profile.

5) Determining the particular alternative attributes (i.e. the particular colors, textures, etc.) whose profiles most closely match the Behavior Profile and the Balancing Profile.

6) Creation of two spaces, (Space "A" and Space "B"), each containing those attributes whose profiles were found to match the two activity attitude profiles, (the Behavior Profile and the Balancing Profile.)

7) Conducting of the (small group discussion) activity in the two spaces, half of the occupants in one space and half in the other.

8) Measurement of behavioral outputs, in terms of activity achievement, content achievement, and attitude shaping.

9) Determining whether or not statistically significant differences occurred between behavior in the two spa and whether or not those differences were in the direction predicted, (the greater achievement being predicted occupants in Space "B", designed according to the Balancing Profile.)

Sources of Data

The sources of data for this study were student subjects consisting of 22 students for the pretest phase ar five groups of 6 students in each of the two spaces.

Treatment of Findings

The pre-test data were employed to determine 1) which colors, forms, values (i.e. light-to-dark gradient), degrees of contrast and scale evoked responses most nearly the same as the attitude responses evoked by the s group discussion activity concept (i.e. which matched the Behavior Profile most closely), and 2) which were n nearly the same as the opposite of the attitude response evoked by the small group discussion activity concept (i.e. which matched the Balancing Profile most closely).

The semantic differential instrument contained twelve polar opposite adjectives (Figure 1). The mean score of the twenty-two responses was determined for each variable. For concept responses that were to be co pared to the small group discussion activity concept response, the differences between each of the variable

Figure 1. Attitude Profiles Toward Activity

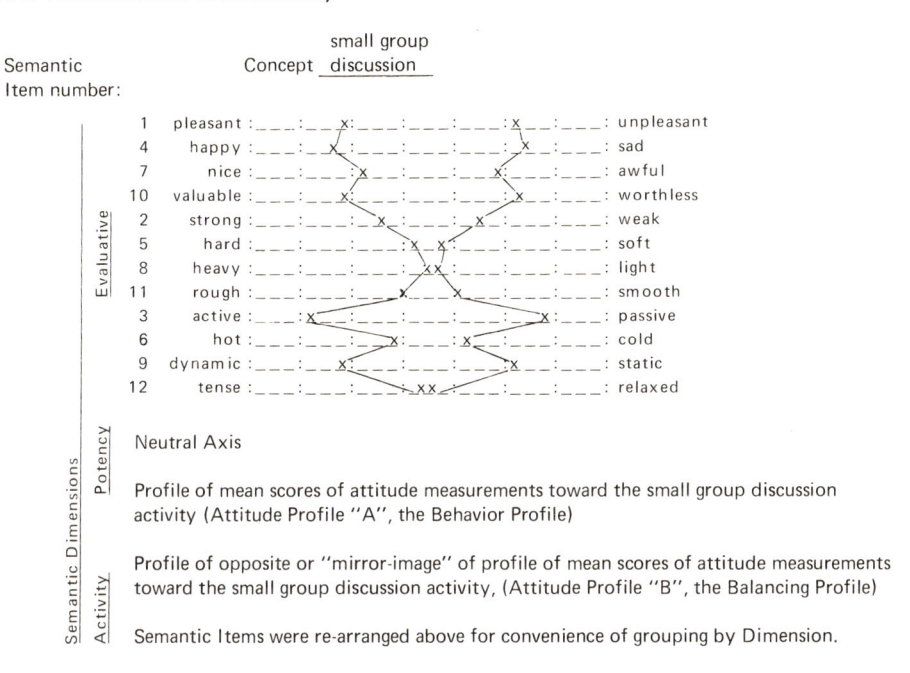

```
                                      small group
Semantic                      Concept  discussion
Item number:
                  1   pleasant :___:___x:___:___:___x___:___: unpleasant
                  4   happy    :___:__x:___:___:___x___:___: sad
                  7   nice     :___:___:_x:___:___:_x:___:___: awful
                 10   valuable :___:___x:___:___:___x___:___: worthless
         Evaluative
                  2   strong   :___:___:_x:___:_x:___:___: weak
                  5   hard     :___:___:___:x_x:___:___:___: soft
                  8   heavy    :___:___:___:xx:___:___:___: light
                 11   rough    :___:___:___x:___x:___:___: smooth
                  3   active   :___:x:___:___:___:___x:___: passive
                  6   hot      :___:___:___:x:___:x:___:___: cold
                  9   dynamic  :___:__x:___:___:___x___:___: static
                 12   tense    :___:___:___xx:___:___:___: relaxed
```

Semantic Dimensions: Evaluative, Potency, Activity

Neutral Axis

Profile of mean scores of attitude measurements toward the small group discussion activity (Attitude Profile "A", the Behavior Profile)

Profile of opposite or "mirror-image" of profile of mean scores of attitude measurements toward the small group discussion activity, (Attitude Profile "B", the Balancing Profile)

Semantic Items were re-arranged above for convenience of grouping by Dimension.

scores of the two compared concepts were summed; the color concept response with less difference than other color concept responses was considered most similar to the small group discussion activity concept response, etc. Thus the colors, textures, forms, values, contrast and scale degrees with the lowest differences were considered to evoke the most similar responses to the small group discussion activity response.

The opposite response to the small group discussion response was determined by subtracting the mean score for each of the twelve bi-polar adjectives from the number four, the middle or "neutral" score on the instrument, and adding that difference to the number four. In this way the "mirror-image" response about the neutral axis was determined and was considered as the opposite or balancing response to the one evoked by the small group discussion activity concept. Comparison to other concepts was accomplished by the same process as described above in relation to the response to the small group discussion activity concept.

The treatment data were employed to determine 1) the quantity and quality of interaction, and 2) achievement on a series of programmed educational tasks. The quantity of interaction was determined by counting the number of separate responses initiated. The quality of interaction was examined by means of a category system developed by Bales.[6] Due to instability in group organization during the early meetings, data were considered for both the nine day period of measurement and for the last seven meetings only. The achievement on the programmed series of education tasks was determined by a summation of the scores attained on each task, as described in the task instruction sheet.

The post-test data were employed to determine 1) the effect of the treatment on attitudes toward the educational psychology subject matter and the small group discussion activity concept, and 2) the similarity between the responses evoked by the spaces and the responses, (Behavior Profile) and the opposite responses (Balancing Profile,) toward the small group discussion activity concept determined at the pretest. The mean of each variable in the responses concept, and the meeting place was determined for the participants who met in each space. The Fisher-Student "t" test for statistical significance was employed to determine whether or not the differences, if any, between the responses of the participants in the two spaces was reliable.

Conclusions

The conclusions were derived from the study findings, related to the hypotheses for purposes of substantiation, and finally employed to examine the implications of this study and to suggest avenues for additional research.

The first conclusion that was observed in relation to the findings was that none of the findings was statistically significant, in terms of demonstrating reliability of the findings for purposes of prediction. Because of the small number of groups employed in the study, the variance in the data generated by the groups in each space would have to be exceedingly small: such was not the case. Thus, all ensuing conclusions must be regarded with an eye towards their lack of predictional reliability.

Task

The groups in the Consonant Space (A) achieved a mean score of 170.80, while the groups in the Dissonant Space (B) achieved a mean score of only 105.40. Thus, the groups in the Consonant Space (A) were concluded to have had a higher task achievement, (about 62% higher,) than the groups in the Dissonant Space (B).

The conclusions varied according to the particular interaction characteristic. The following conclusions were based upon the findings of the interaction from days three through nine (3-9) in preference to the findings from days one through nine (1-9) resulting from instability of group activity during the first day or so. The groups in the Consonant Space (A) achieved, in their quantity of interaction, a mean of 4048.80 combined interactions initiated per day, while the groups in the Dissonant Space (B) achieved a mean of 3704.80 combined interactions initiated per day. Thus, the groups in the Consonant Space (A) were concluded to have had a higher quantity of interaction, (about 9.3% higher,) than the groups in the Dissonant Space (B) (Figure 2).

Two factors were examined in relation to quality of interaction: 1) those interactions which were fundamentally related to the task and 2) those related to socio-emotional (or non-task oriented) characteristics. The groups in the Consonant Space (A) had a mean of 56.74 percent of their interactions related to the task area, while the groups in the Dissonant Space 'B) had a mean of 53.05 percent of their interactions related to the task area. Thus the groups in the Consonant Space (A) were concluded to have devoted a higher percentage of their interactions, (about 7.0 percent higher) the the task area; this would tend to support the higher achievement of the task score. The groups in the Consonant Space (A) had a mean of 43.24 percent of their interaction related to the socio-emotional area. Thus, the groups in the Dissonant Space (B) were concluded to have had a higher percent of their interaction, (about 8.6 percent higher,) devoted to the socio-emotional area.

The conclusions concerning attitude shaping were related to the three attitudes measured, 1) the educational psychology subject matter, 2) the small group discussion activity, and 3) the space in which the group met.

Attitudes toward the subject matter of educational psychology were found to vary within the three dimensions. The groups in the Consonant Space were concluded to have a greater liking or higher regard for the subject matter of educational psychology, (about 3.7 percent more) than the groups in the Dissonant Space (B).

The groups in the Dissonant Space (B) were concluded to feel that the subject matter of educational psychology was more potent, (about 17.0 percent more potent) than the groups in the Consonant Space (A).

Figure 2. Group Interaction in Each of the Spaces

Interaction Characteristics	Consonant Space (A)	Dissonant Space (B)	"t"	Level of Significance
Days 1-9				
Qualitative category: (%)				
I Pos. Socio-Emot. Area	35.89	36.79	0.0349	0.9730
II Pos. Task Area	39.13	39.64	0.0180	0.9860
III Neg. Task Area	18.73	15.11	0.2981	0.7733
IV Neg. Socio-Emot. Area	6.24	8.46	0.4859	0.6954
Quantity:				
Number Interactions Init.	4941.60	4615.20	0.0952	0.9265
Days 3-9 only				
Qualitative category: (%)				
I Pos. Socio-Emot. Area	36.86	38.05	0.1907	0.8535
II Pos. Task ARea	38.38	38.11	0.2412	0.8154
III Neg. Task Area	18.36	14.94	0.6669	0.5236
IV Neg. Socio-Emot. Area	6.40	8.90	0.4338	0.6759
I and IV (Socio-Emotional)	43.24	46.95	0.1408	0.8535
II and III (Task Areas)	56.74	53.05	0.3711	0.7202
Quantity:				
Number Interactions Init.	4048.80	3704.80	0.1237	0.9046

The groups in the Dissonant Space (B) were concluded to feel that the subject matter of educational psychology was more active, (about 15.2 per cent more active) than the groups in the Consonant Space.(A).

Attitudes toward the small group discussion activity were consistent within all three dimensions, the groups in the Dissonant Space (B) feeling a higher regard, greater potency and greater sense of activity twoard the small gropp discussion activity than the groups in the Consonant Space (A).

Attitudes toward the space in which the group met were consistent in all three dimensions, with the groups in the Consonant Space (A) feeling a higher regard, greater potency and greater sense of activity toward the space in which they met than did the groups in the Dissonant Space (B). While the responses toward the two spaces were not identical to the pre-test response and mirror image response (i.e. Behavioral Profile and Balancing Profile) upon which the design of the two spaces was based, the direction intended was identical, (i.e. a feeling of greater liking, potency and activity was intended to be evoked by the space designed consonant with the Behavior Profile). Thus, despite designer influence in combining the elements selected in the pre-test, and a likely difference in response resulting from measurement based upon thirty second exposure versus two week exposure, the spaces were concluded to have evoked the desired direction of response.

Summary

In attempting to explain the above findings an assumption was employed on order to determine whether or not the findings would be consistent with that assumption, and would, therefore, be more clearly explained by means of that assumption. The assumption was, merely, that the subject matter of educational psychology was not the subject matter of greatest concern to the participants, (i.e. given a free choice of subjects in which to become involved, educational psychology was not likely to be freely selected a majority of the time.) If the groups met in a space in which their desire to become involved in their small group discussion activity was greater than their desire to pursue an assignment of programmed tasks dealing with educational psychology, then a higher percent of their interaction would relate to the non-task, or socio-emotional area. In addition, their personal involvement might tend to cause the groups to respond with greater feelings of potency and activity, but a somewhat lesser liking for the assigned educational psychology subject matter. This description appeared consistent with the behavior of the groups meeting in the Dissonant Space (B). Conversely, if groups met in a space in which they sensed little desire for involvement in discussion with one another, they might tend to lean more directly upon the programmed task material presented to them, and relate their interaction more exclusively to that material. Such groups might tend to achieve more on the tasks, feel a greater regard for the subject matter fo the task, have a greater percent of interaction related to the tasks, but feel less liking for the activity and generally a lesser sense of potency and activity. This description appeared consistent with the behavior of the groups meeting in the Consonant Space (A).

The quantity of interaction which was greater for the grops in the Consonant Space (A) than in Dissonant Space (B), did not appear to have been explained by the above assumption. If interaction related to the task material generally involved a greater number of short comments, while interaction related to subject selected by the group involved lengthier comments (and, hence, a fewer munber of comments initiated per unit time,) the data related to quantity of interaction would be explained. Informal comments made by persons listening to the taped recordings of the group meetings for purposes of interaction analysis suggested that the groups in the Consonant Space (A) appeared somewhat quiet or timid, and that a noticeable amount of "dead air" between comments was common, but that the comments appeared to occur in short "flurried", (e.g. a large number of one or two word comments offered by many of the group members). Such comments also suggested that the groups in the Dissonant Space (B) appeared somewhat more "heated" in their discussions and had a more continuous flow of conversation, although individual interactions were often quite lengthy. Also, persons employed to start and stop the gropp discussions, start and stop the tape recorders, etc. indicated that groups in the Consonant Space (A) ceased discussion and began their exit almost as soon as the person entered the space in order to stop the tape recorder; groups in the Dissonant Space (B) tended to remain after the tape recorded was stopped, often continuing until the next group entered the space.

On the basis of the above conclusions regarding attitudinal differences, quality of interaction and achievement, (and employing the stated assumption as a clarifying model), the groups in the Dissonant Space (B) were concluded to have had a "better" small group discussion activity. The Dissonant Space (B) was therefore concluded to have been the more suitable space for small group discussion for those particular occupants.

Hypotheses

Employing the above conclusions as a basis for substantiating or rejecting the stated hypotheses of this study, conclusions were formulated in relation to each of the four hypotheses.

1) Task achievement will be greater in Space "B" the Dissonant Space, (a space that evokes an attitude opposite to the occupants' attitude toward the activity).

This hypothesis was concluded to be unsubstantiated; the rejection of this hypothesis was qualified, however, as a result of the possible difference in behavior that may exist between a situation in which the subject matter is assigned, and one in which it is freely selected by the participants in the group.

2) Interaction will be greater in quality and quantity in Space "B", the Dissonant Space, (a space that evokes an attitude expression opposite to the occupants' attitude toward the activity).

Interaction was concluded to have been "better", overall, in the Dissonant Space (B), which supports the hypothesis.

3) Space "B", the Dissonant Space, will tend to shape attitudes toward the activity in the direction of profile "A", the Behavior Profile, (i.e. the attitudes toward the small group discussion activity will tend to be shaped in a direction opposite to that of the attitude expression evoked by the space.)

The hypothesis was concluded to be substantiated.

4) Attitudes toward the small group discussion activity will be closer to the attitudes toward the subject matter at the completion of the activity than at the start of the activity.

This hypothesis was concluded to be substantiated with regard to two of the three dimensions of attitude measurement, potency and activity, but unsubstantiated with regard to liking (i.e. evaluation) of the subject matter. The conclusion was qualified to refer only to those subjects imposed upon the group, as opposed to subject areas selected by the group.

References

1 Rajendra K. Srivastava, *Human Movement as a Function of Color Stimulation.* Topeka, The Environmental Research Foundation, 1968.

2 James Gibson, *The Perception of the Visual World.* Boston, Houghton, Mifflin Co. 1950. p. 216.

3 Douglas H. K Lee, "The Role of Attitude in Response to Environmental Stress." *Journal of Social Issues,* Vol. XXII, No 4. October 1966, p. 87.

4 Leon Festinger, *Theory of Cognitive Dissonance,* Stanford, Stanford University Press, 1957.

5 Charles Osgood, George Suci, and Percy Tannenbaum, *Measurement of Meaning,* University of Illinois Press, 1957.

6 Robert Bales, "A Set of Categories for the Analysis of Small Group Interaction," *American Sociological Review,* Vol. 15, 1950.

Gary Winkel
Environmental Psychology Program, City University of New York

Roger Malek
Arthur D. Little, Inc., San Francisco

Philip Thiel
Department of Architecture, University of Washington

A STUDY OF HUMAN RESPONSE TO SELECTED ROADSIDE ENVIRONMENTS

The field of environmental research is still relatively young. There is a small but growing body of empirical evidence focused on various aspects of the relationships assumed to be operating between the physical environment and behavior. It will be one of the premises of this paper that the area of environmental research will expand only if a set of methodologies are developed which demonstrate a sophistication equal to the complexity of the environments we wish to investigate. In the early portions of this paper, I intend to review some representative studies which have been addressed to the problem of environmental quality. The methodological implications of the investigations will be considered and a proposal will be made for an alternative approach to quality problems.

Research on problems of environmental quality must be interpreted in light of other investigations of environmental problems. It is possible to detect at least three general areas of research activity:

1. studies of behavior in on-going environments whether natural or man-made. Investigations are designed to link behavior to specific features or patterns of physical elements in various environments (studies of patient behavior in psychiatric hospitals, children in school, campers in wilderness areas, or families in various housing units would fall under this heading)

2. studies relating to the beliefs, meanings, values and attitudes of individuals or groups concerning various environments. Investigations of the quality of environments, the relationship between wilderness values and the physical character of wilderness areas, the symbolic significance of neighborhoods, or even cities are representative issues under this heading.

3. studies of the behavior of those who have varying responsibility for the management of the physical environment (hospital, clinic, prison, and school personnel would be relevant individuals for study in this area)

Greatest emphasis has been placed upon the second category listed above. There appear to be both historical precedents and practical reasons for this. As Glacken (1966) and Lowenthal (1968) have indicated, there is a rich history of speculation concerning the symbolic significance of the physical environment. The majority of this literature is concerned with the changing meaning of the natural environment as a function either of man's adaptation to it or domination of it. A related theme concerns the relative values which are placed upon the natural environment in comparison to the man-made or artificial environment. As Glacken suggests, however, little empirical research exists which would shed light on the extent to which environmental attitudes are shared by those who experience environments but do not write about them. This literature does provide a number of hypotheses concerning environmental meaning and a series of prescriptions for environmental appreciation. One of the objectives of environmental research would be to test these assumptions and hypotheses in a more systematic manner.

As a matter of fact, a number of studies have appeared which try to discover the determinants of environmental meaning. In these studies, attitudes, values, and images are taken to be variables inside the organism which mediate overt response to the external environment. In a brief review of the literature, it will be seen that many of these intervening variables are subsumed under the heading "preference". Experimental efforts are then made to discover those factors in the physical environment which influence preference. Those portions of the experiments having to do with the development of attitude surveys are reasonable well worked out. But dealing with attitudes toward the large-scale physical environment presents some rather unique problems. These have to do with the way in which the environment is represented to the individual whose preference is to be determined. It has only been recently that experimental approaches have been taken to this problem (Winkel and Sasanoff, 1966). One technique has made use of visual simulation devices. These have been developed largely for experimental reasons because they allow the investigator a degree of control over the environment that he might not otherwise possess. Most of these simulation systems make use of drawings, photographs or physical models of the environment being studied.

Early studies of preference for photographic simulations were understandably unsophisticated in their control over the physical variables which might be linked to preference ratings. Wilson (1956) made use of a series of three photographs (a builder's development, a high density residential area and a street scene in Durham, North Carolina). The respondents were asked to rate each of the photographs on the basis of a series of words and phrases which were descriptive of selected qualities which might be imputed to the neighborhood pictured. The words and phrases used by Wilson were as follows:

>   Spaciousness
>   Beauty
>   A character which is good for children
>   Exclusiveness
>   A country-like character
>   Closenses to nature
>   Privacy
>   Greenery
>   Homeyness
>   Quietness
>   Cleanliness
>   Newness
>   Friendliness
>   Crowdedness
>   Dirtiness

No attempt was made to define the physical correlates of these qualities since Wilson was only interested in the extent to which the verbal phrases were meaningful to the respondent, their ranking indicating the relative importance of various qualitative aspects of neighborhood development.

In a similar study, Peterson (1967) prepared a shorter list of attributes including:

>   Preference
>   Greenery
>   Open space
>   Age
>   Expensiveness
>   Safety
>   Privacy
>   Beauty
>   Closeness to nature
>   Quality of the photograph

Peterson intercorrelated and factor analyzed judgments of housing attributes and found four factors which accounted for the majority of the variance in the system. These factors were tentatively identified as:

(1) Harmony with nature
(2) Physical quality of the housing
(3) Quality of the photograph
(4) Noise

Selecting those variables which loaded highest on each factor, Peterson was able to develop a multiple regression equation which related housing attrigutes to to preference in the following way:

Preference= -0.54 (Age) + 0.47 (Closeness to nature)
+0.54 (Quality of the Photograph)

While this study was more sophisticated than that of Wilson there are still difficulties with this approach. Peterson himself indicates that the multiple regression model which he developed is based upon "observed correlations among psychologically obtrusive measurements and are not accompanied by a theory which explains the relationships. . .nothing has been proved or demonstrated by the fact that the numbers fit the model except that the numbers fit the model. Several meaningful models might fit equally well. Because of the nature of the measurements, it must be asked if the subjects like some appearances better than others *because* the hypothesized factors exist in greater perceived quantities. . .This question has not been asnwered (Peterson, 1967).

The question of what factors "cause" the respondents to express greater or lesser amounts of preference will not be satisfactiorily answered until research shows what physical factors contribute to ratings of spaciousness, age, privacy, safety and so on. Aside from the age of the structure, amount of open space and greenery the remaining variables have a decidedly abstract quality about them. It would be difficult to adequately specify the particular grouping of physical variables which were responsible for the judgments given by Peterson's observers.

An alternative procedure which is designed to identify the physical elements which contribute to preferences has been developed by Shafer, Hamilton and Schmidt (1968). One hundred black and white photographs of various natural environments were taken around the United States. The sample included mostly wildland views. The landscaped scenes included mountains, meadows, water and various combinations of these elements. The photographs were divided into ¼ inch squares and were assigned a number corresponding to the physical characteristics present. The variables included the amount of sky, vegetation, non-vegetation (including rocks, exposed soil, grass or snow covered areas, streams, waterfalls, and lakes). The observers were asked to rank the photographs in terms of preference. After factor analyzing the elements in the photographs, Shafer *et.al.* were able to work out a rather complicated second-degree polynomial expression which linked preference to the area and perimeter of certain elements in the photographs.

Shafer *et. al.*'s efforts are to be applauded since they were among the first researchers to scale the photographs which the observers were subsequently asked to rate. But there are a number of difficulties with their techniques. The most obvious problem is that it is not clear whether the observer is rating the photograph or the view which the photograph is supposed to represent. The photograph itself represents a judgment concerning the representativeness of the view presented to the observer. It would be surprising if Shafer and his colleagues could predict user preference for the intact environment based on the photographs. In studies of this sort it would be advisable to take a range of photographs and ask a group of observers to make judgments concerning that photograph or those photographs which are most representative of the scene under consideration. It would also be helpful if studies existed which demonstrated how people build up representations of environments as a function of successive experiences with them. It seems difficult to believe that one view of an environment would be sufficient to give an accurate picture.

The second difficulty with Shafer, *et. al.*'s approach is the rather arbitrary method employed in categorizing the photographs. It would have been instructive if Shafer and his colleagues had included variables which were not assumed to have any bearing on preference. For example, Shafer might have taken measurements of the grey values of the elements in the photographs, the number of distinctly articulated groupings of elements or the rated degree of perceivable detail of groupings of elements in the photographs. Any one of these variables might have contributed to a prediction of preference. It should be noted in passing that Peterson (1967) did include a rating of the quality of the photographs used in his study. While he found that this factor did not significantly alter the estimated multiple correlation coefficient it might have, and he was justified in the inclusion of that variable in the study.

This brief review of three approaches to environmental preference points to one of the chief obstacles to further progress on environmental perception: the development of experimantal techniques which will allow the experimenter to specify the physical elements which influence preference judgments. I would suggest that experiments of the type considered here create more difficulties than they solve because of the problems of interpreting the findings. It is often difficult to sort out the causal factors involved in correlation coefficients. Consider a simple example from Peterson's study. He found that the age of the houses in his study was negatively correlated with preference. But, it may be asked whether age or the condition of the housing is the critical factor which "causes" the changes in preference. It is probably true that age of the house is correlated with its physical condition but it is not clear which of these two factors are the more important in producing change in the ratings. And it is only under special conditions that the relevant factors can be identified.

Future studies of environmental preference should aim to give the experimenter a greater degree of control over the variables which are entering into the observer's judgment. Ideally, the experimenter should be able to select photographs which show every combination of the relevant physical variables while maintaining a constant pattern for the non-relevant physical variables. For instance, suppose an investigator is interested in the role of different landscaping schemes, different house styles and different degrees of housing maintenance as they affect preference ratings. If using photographs, the investigator should have available pictures showing every combination of relevant physical variables. It is not sufficient to simply select a set of photographs because they happen to be available and are varied. Even so, it is the present author's opinion that it is next to impossible to develop such sets based on naturalistically occurring variation in the physical environment. Under these conditions another alternative suggests itself. This is the use of photographic retouching in which elements may be added and subtracted almost at will.

The environment selected for study were two examples of roadside development on an urban arterial. The elements which were removed from the photographs were those which many critics felt were the most offensive to trevelers: overhead wires and telephone poles, proprietary and institutional sighs, and billboards. The technique was designed to study how these factors influenced user evaluations while holding all other physical factors on the road and roadside constant.

Method

Observers

A total of 80 observers was used in the study. The group was divided into 40 females and 40 males. The total group of observers was further divided on the basis of whether the observer was a student or not. Tables 1 through 3 contain a summary of the demographic information on the observers used in this study.

The pool of 843 observers was developed by placing advertisements in the "Help-Wanted" columns of two newspapers in Seattle, Washington, for one week.

Every attempt was made to minimize the observer's possible fear about participating in the study, while not giving him information about what was to be expected of him. They were given no information other than that the task was quite simple and involved looking at a series of pictures and expressing their evaluation of them.

Test Location

Two offices were used for the study. The offices were in a private office building near the University of Washington in Seattle. One room was used as a reception and paper and pencil testing area. The second room was a dark room containing the Mackworth eye-movement camera, projectors, and other supporting equipment.

Location and Description of the Routes

Two different segments of the roadside along Aurora Avenue in Seattle, Washington were selected as the routes to be used as stimulus materials for this study. One segment was characterized as the "commercial route" and the other as the landscaped route.

The commercial route is a typical commercial strip of medium density. The route contains a variety of elements which are characteristic of this type of development and the urban roadside generally, such as utility poles and overhead wires, billboards, proprietary and institutional signs, buildings and trees and planting.

These two aspects of the roadside were selected as test displays because they exhibited variable backgrounds and a common overlay of the kinds of physical objects typically placed along urban roadsides. One of the primary goals of our study was to detect changes in observer evaluation which might occur as the existing environment is modified through the elimination of objects in that environment. We were interested in the differences in evaluation that might occur when similar objects were removed from environments that had dissimilar background characteristics.

Slides of the Routes

One of the major purposes of our research design was to simulate major modifications of visual scenes by retouching photographs of the visual environments which they represented. Initially, we considered using either a movie display or colored still photographs for the stimulus materials. These were precluded for technical-economic reasons. We decided that modifications of the environment through retouching was more important to our research purpose than perfect environmental simulation. Although a relative degree of realism would be lost in the displays, it was not the intention of the research to focus on perfect environmental simulation, but rather to observe changes in response to modifications in the visual environment. Therefore, we decided to use black and white still photography for the photographic displays. Observer familiarity with black and white photography in the mass media was considered to be sufficient to ensure realistic acceptance by the observers. In fact, it was felt that the quality of the photographic retouching would be a more important consideration than a more perfect environmental simulation. Once the observers accepted the experimental ground rules we felt they would probably be more sensitive to imperfections within that frame of reference than they would have the tendency to test the realism of the simulation.

Figure 1. No Modifications

Figure 2. Route Minus Utility Poles and Overhead Wires

Figure 3. Route Minus Utility Poles, Overhead Wires and Billboards

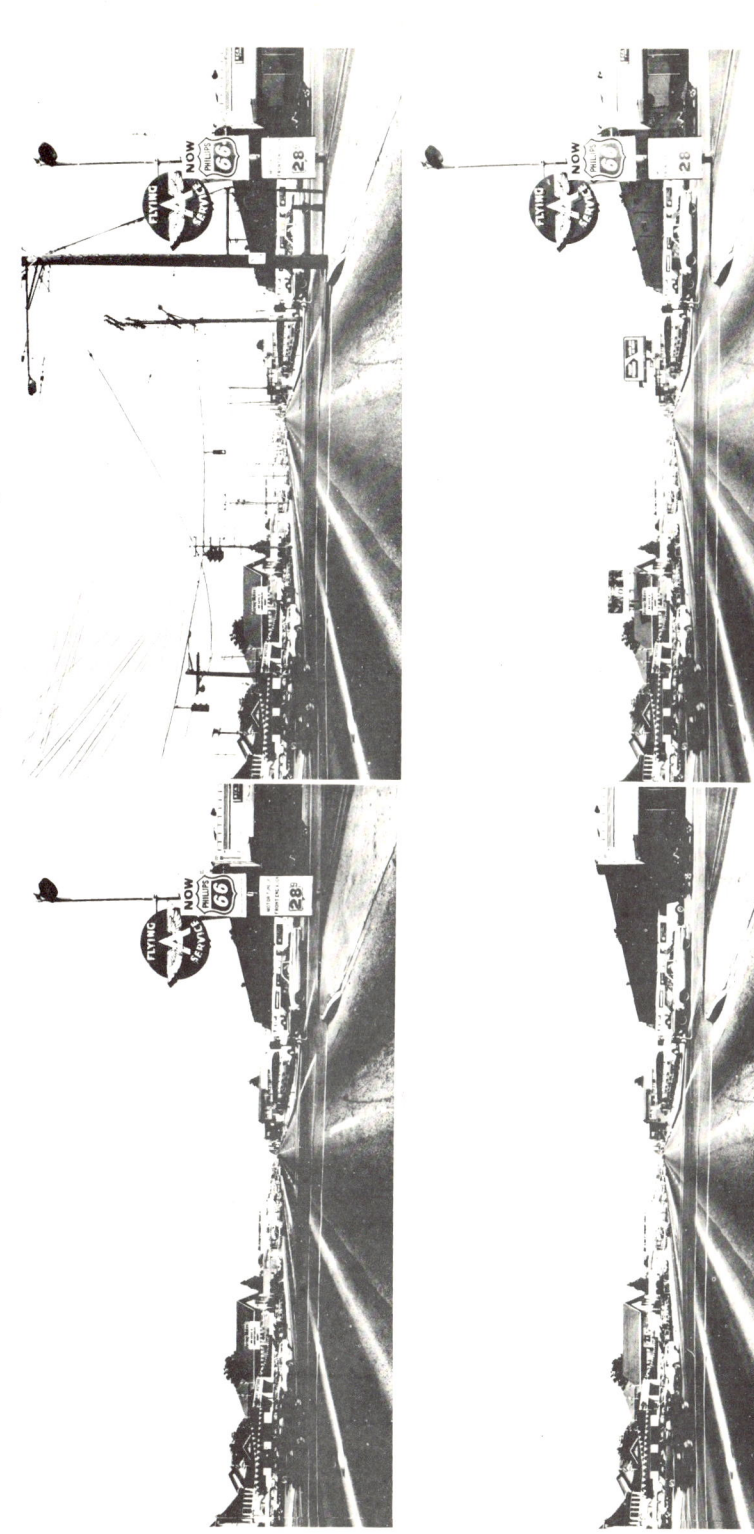

Figure 4. Route Minus Utility Poles, Overhead Wires, Billboards and Other Signs

Ten photographs were taken of each route (the "route as it is") at intervals of approximately 300 feet. Transformations of each route were made by eliminating objects from the roadside environment. Each observer was shown ten slides of the "route as it is" and ten slides each of the route minus all billboards; minus utility poles; and other sighs. Each observer was shown a complete set of slides for one route; the original and four transformations, fifty slides in all. Figures 1 through 4 show a typical slide from the commercial route under various alterations.

The Semantic Differential

The Semantic Differential Scale as a device which measures the meaning of an object to an individual has been described by Osgood, Suci and Tannenbaum (1957). In this technique, the observer is asked to rate a given concept on a series of seven-point bi-polar rating scales. The center point of the scale indicates an indifferent response, and directionality from the centerpoint indicates the emphasis and direction of the meaning intended by the respondent. Any concept can be rated by this technique.

The observers in this study were asked to rate or describe a series of slides of the urban roadside on such adjective pairs as: *simple-complex, usual-unusual, desirable-undesirable, safe-dangerous, useless-useful, organized-disorganized,* etc. These adjectives were selected from Anselm Strauss' *Images of the American City* and Peter Blake's *God's Own Junkyard,* from articles in the *Progressive Architecture, Architectural Record* and *Architectural Forum* and other critical and topical literature. Figure 5, the adjective questionnaire, lists the adjectives used in the final study, including the format and the instructions given to the observers.

Procedure

Each observer was initially briefed on the nature of the test and the functioning of the eye-movement camera. After being acclimated to the camera, the observer was given instructions to take the point of view of a passenger in an automobile riding down the route which was to be shown to him in the slides. He was asked to pay close attention to the characteristics of the roadside and informed that he would later be questioned about the roadside and asked to evaluate it in terms of some adjectives. After the observer was given preliminary instructions in the use of the bi-polar adjective scale, the first set of ten slides, always of the route "as it is," were shown. The observer was then asked to rate the route on the semantic differential scale.

After completing this task, the observer was again seated in front of the screen and instructed that he was going to be shown another set of slides of the same roadside taken in the same place as before. He was told that some changes had been made but that we were not particularly concerned with whether or not he could detect what specific changes had been introduced. Rather our concern was whether, in his opinion, the set he was about to see was similar or different in overall character to the first series of slides he had seen. If the observer noted a difference in the slides, he was asked to indicate what seemed to make the difference. This generally took the form of identifying those elements which had been removed. The observer was then asked if in his opinion the difference was sufficiently large to warrant a change in his adjective ratings of the route. If so, he was asked to change his adjective ratings. The observer was allowed to change his adjectives only if he correctly indicated what it was about the slides which made them appear different from the first set he saw.

Each observer was shown five sets of slides: the original route and all transformations of the route. The order of viewing of the transformed routes was altered so that each transformation had an equal opportunity of appearing after the slides of the route "as it is."

natural	unnatural
worthless	valuable
hazardous	harmless
varied	monotonous
refreshing	tiresome
irregular	regular
revolting	glamorous
effective	useless
confusing	clear
fascinating	dull
similar	different
planned	unplanned
useful	useless
uncluttered	cluttered
rational	irrational
dull	exciting
offensive	pleasing
bad	good
personal	impersonal
disorganized	organized
disliked	liked
weak	powerful
unattractive	attractive
informative	uninformative
safe	dangerous
dirty	clean
delightful	disgusting
changing	lasting
unimaginative	imaginative
unpleasant	pleasant
impressive	unimpressive
organized	disorganized
random	asymmetrical
annoying	agreeable
effective	useless
accept	reject
depressing	stimulating
confused	clear
balanced	unbalanced
complex	simple
logical	chaotic
effective	ineffective
desirable	undesirable
dishonest	honest
neat	sloppy
orderly	disorderly
harmonious	discordant
unimportant	important
diverse	similar
acceptable	unacceptable
interesting	uninteresting
unnecessary	necessary
expressive	unexpressive
beautiful	ugly
disturbing	satisfying
illogical	logical
lively	deadly
symmetrical	asymmetrical
weak	powerful
consistent	inconsistent
usual	unusual
rich	poor
degrading	uplifting
sick	healthy

Figure 5. **Polar Adjective Questionnaire**

Data Analysis

In the process of constructing the bi-polar adjective scale it seemed unrealistic to expect that the sixty-four adjectives finally included represented sixty-four different aspects of judgment. The more likely expectation was that many of the adjectives would tend to cluster or group together and form units of meaning. To test this, the data from the bi-polar adjective scales were factor analyzed using a principal axis solution. The resulting facgor matrices were then rotated to Varimax criterion (Horst, 1965).

After the principal factors had been defined, a procedure was developed to allow us to understand how each observer rated the route "as it is" and how those ratings changed when a retouched sequence was viewed.[1] Two computer programs (SCORE and COMPARE [Little, 1968]) were developed which would allow comparisons between ratings of the route "as it is" and pre-evaluations of the transformed route. Essentially the programs allow judgments of the actual ratings compared to a hypothetical observer (referred to as Mr. Bland) who consistently used the middle reference point on the bi-polar adjective scale (i.e., a scale rating of four).

Use of this system allowed the investigators a simple average score for each factor. The sign attached to this score gives an indication of the direction of the rating compared to the neutral Mr. Bland.

Results

Observer Characteristics

Tables 1 through 3 summarize relevant demographic information describing the observers.

Perception of Changes in the Simulated Environment

On Table 4 is a summary of the percentages of observers who were able to detect differences in the simulated environments under various changes. As might be expected, the relationship is an increasing function of the number of elements removed from the photographs by retouching.

A consistent finding for both routes is that the elements least noticed upon removal were the billboards. It should also be noted that in the third elimination sequence (utility poles, overhead wires and billboards removed) the clues most often used to detect a difference was the absence of utility poles and wires. This finding was more applicable to the commercial route than to the landscaped route. For elimination sequence 4, the clues used most often for both routes were the absence of utility poles, overhead wires, and/or on-premise signs.

Analysis of the Adjectives

The results of all of the separate evaluations made by the observers in the experiment were intercorrelated and factor analyzed. The data used in the analysis included all of the adjective ratings of the route "as it is" (the initial display), as well as all of the adjective ratings of the route transformations. A separate factor analysis was done for the total sample of observers of both routes, the commercial route observers and the landscaped route observers. Tables 5 and 6 summarize the factor loadings on the normalized rotated factor loading matrix.

Bi-polar Factors, Total Sample—Both Routes

Factor 1 was tentatively designated as a Monotony-Depression scale insofar as it measures response to the urban roadside (as it might be similar to the examples used in this pilot study).. The highest factor loadings occur for adjective pairs such as *Impersonal-Personal, Deadly-Lively, Regular-Irregular, Weak-Powerful, Similar-Different.*

Substantial loadings occur for the following adjective pairs: *Unimaginative-Imaginative, Dull-Fascinating, Monotonous-Varied, Depressing-Stimulating, Tiresome-Refreshing, Usual-Unusual, Uninteresting-Interesting, Weak-Powerful,* and *Revolting-Glamorous.*

Adjectives which reflect an aesthetic and moral acceptance or rejection of the routes also appear on this factor, but have lower factor loadings than those listed above. These include adjective pairs such as *Unattractive-Attractive, Delightful-Disgusting, Unpleasant-Pleasant, Degrading-Uplifting* and *Sick-Healthy.*

The adjective pairs included in this factor have components that reflect the capacity of the visual displays to incite feelings that stimulate and refresh or depress; that describe the roadside environment as having the projected human quality (pathetic fallacy) of being personal, lively, interesting and expressive or impersonal, deadly, uninteresting, regular and similar or irregular and different. Although there is considerable variety in the adjective pairs, there seems to be a consistency in describing the visual environment in terms which either incite in the observer or project onto the environment itself feelings which excite or depress, while at the same time more objectively describint the environment in terms of the degree of diversity and regularity it exhibits. Because the adjective pair *Varied-Monotonous* has both an objective-descriptive component and a feeling component we have used it along with the stronger *Depressing-Stimulating* pair to characterize the factor.

Factor II is more clearly related to a description of the compositional and planning characteristics of the routes being viewed. We tentatively defined this group of adjectives as a *Simplicity-Complexity* scale.

The highest factor loadings occur for adjectives such as *Simple-Complex, Consistent-Inconsistent, Balanced-Unbalanced, Symmetrical-Asymmetrical, Orderly-Disorderly, Neat-Sloppy, Uncluttered-Cluttered, Organized-Disorganized.* The factor also reflects tendencies to accept or reject the route based on its compositional or planned characteristics. This is evident from adjective pairs which have lower factor loadings, such as, *Offensive-Pleasing, Bad-Good, Unpleasant-Pleasant, Disliked-Liked* and *Desirable-Undesirable.*

The factor is not completely denotative in character since the two pairs which express the observer's concern with the safety features of the roadside have relatively high loadings in this factor, i.e., *Harmless-Hazardous* and *Safe-Dangerous.*

Factor III is clearly related to the degree of utility and usefulness which the route is seen to have. We tentatively designated it as a *Useless-Effectiveness* scale. There are very high factor loadings for adjective pairs such as *Useless-Effective, Useless-Useful, Unnecessary-Necessary, Unimportant-Important, Ineffective-Effective.*

Similar to factors I and II there is a tendency for other connotative adjective pairs to appear on factor III. These adjectives have to do with the rationality, naturalness and informativeness, as well as the acceptability, value and impressiveness of the route. Their factor loadings are, however, considerably lower.

Bi-polar Factors, Commercial Route Observers

Factor I is similar to both factors I (*Monotony-Depression*) and III (*Uselessness-Effectiveness*) scale of the adjective analysis of the total sample. In this analysis, there were high factor loadings for adjective pairs such as *Varied-Monotonous, Fascinating-Dull, Dull-Exciting, Depressing-Stimulating, Effective-Ineffective,* as well as for *Effective-Useless* and *Useful-Useless.* This factor also contains high loadings for such adjective pairs as *Offensive-Pleasing, Bad-Good, Disliked-Liked, Unattractive-Attractive* and *Beautiful-Ugly.*

It is apparent that for this group factor I reflects a wide range of different evaluative adjectives operating together to define observer attitudes toward the commercial route.

Factor II describes the compositional character of the roadside and can be described as a Simplicity-Complexity scale similar to factor II for the total group. There are high loadings for such adjective pairs as *Disorganized-Organized, Random-Asymmetrical, Confused-Clear, Balanced-Unbalanced, Complex-Simple and Orderly-Disorderly*. This factor also contains high loadings for *Hazardous-Harmless* and *Safe-Dangerous*, similar to factor II for the total group.

Factor III is somewhat more difficult to describe for this group. Very few adjectives load heavily on this factor. There are high loadings for adjectives such as *Irregular-Regular, Similar—Different, Personal-Impersonal and Diverse-Similar*. Concern is expressed for objective-descriptive characteristics of the route; however, because of the small number of significant adjective pairs, the data is difficult to interpret.

Bi-polar Factors, Landscaped Route Observers

Factor I is similar in character to factor II for both the commercial route observers and the total sample of observers, since it refers to the planning and compositional characteristics of the landscaped route. High factor loadings appear for adjective pairs such as *Confusing-Clear, Uncluttered-Cluttered, Disorganized-Organized, Changing-Lasting, Balanced-Unbalanced, Orderly-Disorderly and Simple-Complex*. Similarly there was a concern for the safety features of the route because of high loadings for the adjective pairs *Hazardous-Harmless* and *Safe-Dangerous*.

Factor II for this group refers to the stimulating qualities of the roadside, similar to the Monotony-Depression scale of factor I for the total group of observers. High factor loadings appear for adjective pairs such as *Refershing-Tiresome, Fascinating-Dull, Depressing-Stimulating and Lively-Deadly*. Adjectives with strong expression of aesthetic feeling such as *Revolting-Glamorous, Offensive-Pleasing, Bad-Good, Unattractive-Attractive and Beautiful Ugly* also appear.

Factor II for this group is a Uselessness-Effectiveness scale as reflected in high loadings for such adjective pairs as *Effective-Useless, Useful-Useless, Effective-Ineffective*.

Throughout the analysis of the bi-polar adjectives for the different observer groups, it is possible to detect the existence of a factor which describes, from the standpoint of the observers in this experiment, the Monotory-Depression and Pleasing-Offensive aspects of the roadside. The second major factor refers to the compositional and organizational aspects of the roadside as reflected in such adjectives as *Simplicity-Complexity* and *Symmetry-Asymmetry*. The third factor appears related to the degree of efficiency or utility the roadside is thought to have.

Initial and Modified Evaluations of The Routes

After the principal factors had been defined by the factor analysis of the separate observer evaluations, we wished to know how the observer ratings changed when a retouched sequence was viewed. The Score and Compare programs were devised to quantify and assess the directionality of the observer ratings as defined by the principal bi-polar adjective factors.

The factors for the total sample of observers for both routes were used in defining the directions and values resulting from the Score and Compare analysis.

The Initial Evaluations

The only significant differences in the evaluations of the various observer groups occurred between the student and non-student group scores on bi-polar factor II, the Simplicity-Complexity scale. The students generally saw the routes as simpler than the non-students, who tended to see the routes as alightly toward the complex side of the scale. (An estimated t=2.61, for 78 degrees of freedom.)

The commercial route observers evaluated the commercial route as more depressing (factor I= 14.38) than the landscaped route observers (factor I=9.84) evaluated the landscaped route; less simple (factor II=I.06) than the landscaped route observers (factor III=3.20) than the landscaped route observers (factor III=0.10) evaluated the landscaped route. The statistical significance of the differences are open to question, however, as the values are 1.05 or lower.

Compared to Mr. Bland, who would be completely indifferent and neutral, the students and the non-students, the commercial route observers, and the landscaped route observers rated the routes as more monotonous and depressing (factor I). The student observers rated the commercial and landscaped routes as slightly more simple (factor II) than Mr. Bland. The non-student group, however, rated the routes as slightly more complex than Mr. Bland. All of the observers rated the routes as slightly more useless (factor III) than Mr. Bland.

Changes in Evaluation

Table 7 summarizes changes in bi-polar factor scores as a function of the removal of elements from the two routes.

The changes in evaluation which resulted from the removal of elements from the routes, again by comparison with Mr. Bland, were most pronounced when *utility poles* and *overhead wires were* removed from the commercial route; and when *utility poles* and *billboards* and *utility poles, billboards* and *other signs* were removed from the landscaped route.

When utility poles and overhead wires were removed from the commercial route, there was a substantial shift in the evaluation of the route as more personal and varied (-33.55), much simpler (26.72) and much more effective (-23.10) than Mr. Bland's evaluation.

When *utility poles* and *billboards* and *utility poles, billboards* and *other signs* were removed from the landscaped route, it was evaluated by the re-evaluators as much more personal and varied (-31.89, -21.24), much simpler (25.59, 27.34), and much more effective (-19.79, -17.90) than Mr. Bland who would have been indifferent or neutral in his ratings.

One comment should be made concerning the single individual who chose to re-evaluate the commercial route when overhead wires and billboards had been removed. As Table 9 indicates, this person did not respond as might have been expected. Rather than approving of the removal of the utility poles and billboards, he objected strongly to this step believing that the urban roadside should be junky and cluttered. His ratings reflected this attitude.

The second inconsistency to be noted in Table 7 occurs for the five observers of the landscaped route who re-evaluated when utility poles and overhead wires were removed. Our expectation was that they would respond positively to this step. Based on subsequent interviews it was true that they did. But when they rated the route on the adjectives, they were responding to the billboards on the route which were now quite visible rather than to the removal of the utility poles and overhead wires.

Another one of the surprising results of the elimination sequences was the change in evaluation made by the thirteen commercial route re-evaluators when billboards, utility poles, overhead wires, and other sighs were eliminated from the commercial route. The route was rated as only very slightly more personal and varied (-0.78), somewhat more simple (7.09) and more useless and ineffective (I.90) than Mr. Bland would have rated the route. A much slighter but similar change in the Factor I score occurred when billboards, utility poles, overhead wires, and other signs were removed from the landscaped route (-21.24) as compared with evaluation of the same route (-31.89) when only billboards and utility poles were removed from the same route. The interview material contained a partial answer for this finding. The re-evaluators of the commercial route with three sets of elements removed were surprised at the dull and uninteresting quality of the resulting environment. Some expressed surprise that the resulting environment would look appalling. The same finding was true for the landscaped route but it was less pronounced here.

Conclusions

A careful analysis of the findings from the present study suggest that the technique of introducing simulated environmental change under experimental control can be potentially useful in studies of environmental quality and preference. The major advantage of this technique is that the experimenter is in a position to know what changes have been made in the environmental display and the observers can comment not only on the changes which have been introduced but on what remains after the experimental alterations. It will be recalled that this experimental feature was considered essential in photographic simulations.

The technique also allows the investigator to examine the ability of observers to detect changes in various enivronmental elements. Within the constraints of the present study, for instance, it is clear that the process of detecting change is far from random. Unfortunately, the findings from the present study cannot be readily generalized to the real world without the introduction of some of the richness of experience involved in actually traveling the road. For example, future studies would have to make use of color rather than black and white photographs, vary the rate at which the roadside elements were shown to the observer, and the range of roadside developments would have to be expanded beyond the two examples which made up this study.

A second finding of potential relevance concerns the manner in which observers organize their evaluations of the environment. On the basis of a pre-test (Little, 1968) observers were asked to give open-ended evaluations of the roadsides subsequently investigated. It is significant to point out that their evaluations were quite similar to those of professional critics of the environment. Many of the adjectives included on the semantic differential scale were checked against the evaluations given by the pre-test observers to insure some representativeness. A second point about the evaluations is that environmental preference is probably not a single dimensioned concept. The factor analysis of the adjective ratings indicate at least three different aspects of the evaluative process. And it is clear that the assignment of positive or negative adjectives to the roadside environment is partially a function of the rated complexity of the environment as well as the content of the elements included on the roadside. Ultimately, of course, it is necessary to see whether the same factors will be identified in similar studies and more importantly, whether there are behavioral correlates to these evaluations. For example, would the observers have been willing to pay for the removal of telephone poles and overhead wires since their absence from the photographs significantly changed the evaluation given? Or would the observers hold out for more extensive environmental changes rather than using a piece-meal approach? It must be remembered that even though the observers felt that the removal of the overhead wires was an improvement, their overall rating of the roadside remained negative.

In a consideration of the future of this technique, it is obvious that greater attention must be given to accurately recreating the experiences involved in actually traveling the highway. In this case, physical models of the environment seem like a reasonable option. If models could be developed which accurately predict response to the real environment, environmental change at the discretion of the de signer would be considerably simplified. The designer would then possess a tool for the evaluation of environments before they were constructed. This would represent an important improvement in the design process.

Table 1

Age of Observers

Students			Non-Students	
Age	Number		Age	Number
18	7		20-24	8
19	9		25-29	4
20	10		30-34	8
21	6		35-39	7
22	4		40-44	6
23	2		45-49	4
24	1		50-54	1
25	0		55-59	2
26	1			40
	40			

Table 2

Family Income

Income	Students	Non-Students
Under $3,000	0	1
3,000 – 5,000	1	4
5,000 – 7,500	9	11
7,500 – 10,000	11	12
10,000 – 20,000	13	10
Over 20,000	6	2
	40	40

Table 3

Years of Education

Years in School	Students	Non-Students
8	0	1
9	0	0
10	0	1
11	0	1
12	6	9
13	11	7
14	12	8
15	8	6
16	1	2
17	1	1
18	0	2
19	0	0
20	1	2
	40	40

Table 4
Summary of the Percentage of Observers Who Noticed
A Difference in the Character of the Route
Or Re-evaluated the Modified Route

Route	Elimination Sequence	Percent Who Noticed the Difference	Percent Who Noticed the Wrong Difference	Percent Who Re-evaluated
Commercial (37 Observers)				
	1	32.4	13.5	8.1
	2	59.5	8.1	16.2
	3	64.9	18.9	2.7
	4	83.8	8.1	40.5
Landscaped (43 Observers)				
	1	37.2	2.3	11.6
	2	53.5	16.3	11.6
	3	67.4	4.7	34.9
	4	83.7	7.0	23.3

Table 5
Rotated Factor Loading Matrix (Normalized)
Bi-polar Adjectives
Commercial Route

Adjective Pair	Factor I	Factor II	Factor III	Adjective Pair	Factor I	Factor II	Factor III
1	0.819	-0.185	-0.542	33	0.361	0.918	-0.162
2	-0.958	0.127	-0.255	34	-0.840	0.540	-0.027
3	-0.251	0.861	0.441	35	0.980	-0.153	0.119
4	0.945	-0.043	0.324	36	0.931	-0.362	0.043
5	0.838	-0.455	0.297	37	-0.880	0.256	-0.399
6	0.174	-0.054	0.983	38	-0.553	0.826	0.098
7	-0.771	0.622	-0.134	39	0.028	-0.999	0.014
8	0.991	0.030	0.127	40	0.099	0.901	0.420
9	-0.606	0.527	0.551	41	0.576	-0.806	-0.129
10	0.879	-0.134	0.455	42	0.988	-0.139	0.062
11	-0.442	-0.059	-0.894	43	0.871	-0.448	0.196
12	0.358	-0.914	0.186	44	0.419	0.623	-0.659
13	0.922	0.128	0.364	45	0.209	0.977	0.022
14	0.103	-0.972	0.208	46	0.277	0.973	0.019
15	0.906	-0.354	-0.230	47	0.531	0.804	0.265
16	-0.897	0.145	-0.416	48	0.976	0.112	0.184
17	-0.819	0.563	0.108	49	0.320	0.367	0.873
18	-0.885	0.429	0.176	50	0.890	-0.445	-0.086
19	0.558	-0.072	0.826	51	0.893	-0.270	0.359
20	-0.385	0.914	-0.125	52	-0.925	0.295	0.235
21	-0.902	0.423	-0.077	53	0.802	0.100	0.587
22	-0.979	0.038	-0.198	54	0.821	-0.474	0.315
23	-0.759	0.516	-0.395	55	-0.824	0.536	-0.048
24	0.796	0.382	0.467	56	-0.583	0.707	0.398
25	0.134	-0.941	0.310	57	0.650	-0.069	0.756
26	-0.211	0.976	0.024	58	0.282	-0.958	0.025
27	0.843	0.537	0.004	59	-0.946	0.206	-0.249
28	0.665	0.631	0.397	60	-0.171	-0.906	-0.386
29	-0.807	-0.041	-0.599	61	0.124	0.209	-0.969
30	-0.868	0.488	-0.084	62	0.860	-0.409	0.301
31	0.738	0.026	0.673	63	-0.768	0.615	-0.175
32	0.399	-0.940	0.011	64	-0.785	0.587	0.193

Table 6
Rotated Factor Loading Matrix (Normalized)
Bi-polar Adjectives
Landscaped Route

Adjective Pair	Factor I	Factor II	Factor III	Adjective Pair	Factor I	Factor II	Factor III
1	0.634	0.390	0.666	33	-0.588	0.155	-0.793
2	-0.571	-0.652	-0.497	34	-0.593	-0.607	-0.528
3	-0.903	-0.425	-0.058	35	0.314	0.370	0.873
4	-0.300	0.770	0.561	36	0.620	0.477	0.621
5	0.153	0.914	0.373	37	-0.409	-0.825	-0.389
6	-0.649	0.737	0.182	38	-0.881	-0.241	-0.405
7	-0.480	-0.875	0.057	39	0.864	0.163	0.474
8	0.301	0.282	0.910	40	-0.979	-0.952	0.195
9	-0.963	-0.032	-0.266	41	0.824	0.077	0.560
10	0.078	0.888	0.452	42	0.391	0.425	0.815
11	0.501	-0.848	-0.169	43	0.572	0.768	0.284
12	0.460	0.276	0.843	44	-0.895	0.106	-0.431
13	0.266	0.281	0.921	45	0.796	0.432	0.431
14	0.838	0.434	0.327	46	0.796	0.218	0.564
15	0.765	0.147	0.626	47	0.850	0.422	0.312
16	-0.091	-0.857	-0.505	48	-0.182	-0.567	-0.803
17	-0.642	-0.696	-0.319	49	-0.554	0.820	-0.138
18	-0.684	-0.673	-0.280	50	0.559	0.465	0.685
19	-0.046	0.831	0.553	51	0.088	0.838	0.538
20	-0.844	-0.331	-0.419	52	-0.262	-0.152	-0.952
21	-0.652	-0.706	-0.273	53	0.236	0.874	0.423
22	-0.561	-0.741	-0.367	54	0.620	0.703	0.345
23	-0.622	-0.740	-0.252	55	-0.711	-0.693	-0.114
24	-0.218	0.403	0.888	56	-0.769	-0.159	-0.617
25	0.959	0.242	0.143	57	0.338	0.939	-0.048
26	-0.612	-0.544	-0.573	58	0.780	0.624	0.027
27	0.670	0.700	0.242	59	-0.313	-0.901	-0.299
28	-0.991	0.132	0.013	60	0.953	-0.028	0.300
29	-0.300	-0.874	-0.381	61	0.038	-0.866	0.498
30	-0.644	-0.696	-0.316	62	0.719	0.676	0.155
31	0.203	0.784	0.585	63	-0.448	-0.748	-0.487
32	0.890	0.268	0.366	64	-0.572	-0.808	-0.135

Table 7

Changes in Bi-polar Factor Scores Between
The Original and Transformed Routes

Route	Observers	Number	Change in Factor Score Factor I	Factor II	Factor III
Commercial Total: 37 Observers	2	1	-6.97	14.86	-2.93
	6	2	-33.55	26.72	-23.10
	1	3	18.28	0.49	11.83
	13	4	-0.78	7.09	1.90
Landscape Total: 43 Observers	5	1	-3.12	1.94	0.66
	5	2	6.98	-16.91	10.84
	15	3	-31.89	25.59	-19.79
	10	4	-21.24	27.34	-17.90

References

1. Glacken, C. (1966) Reflections on the Man-Nature Theme as a Subject for Study in F. Darling and J. Milton (eds.) *Future Environments of North America.* Garden City, New York: The Natural History Press.

2. Horst, P. (1965) *Factor Analysis of Data Matrices,* New York: Holt, Rinehart and Winston.

3. Little, A. (1968) *Response to the Roadside Environment,* San Francisco, Arthur D. Little, Inc.

4. Lowenthal, D. (1968) The American Scene *Geographical Review, 58,* No. 2.

5. Osgood, C. et. al. (1957) *The Measurement of Meaning,* Urbana: The University of Illinois Press.

6. Peterson, G. (1967) A Model of Preference: Quantitative Analysis of the Perception of the Visual Appearance of Residential Neighborhoods, *Journal of Regional Science, 7,* No. 1.

7. Shafer, E. et. al. (1969) Natural Landscape Preferences: A Predictive Model *Journal of Leisure Research, 1,* No. 1.

8. Wilson, R. (1962) Livability of the City: Attitudes and Urban Development in F. Chapin and E. Weiss (eds.) *Urban Growth Dynamics,* New York: John Wiley and Sons.

9. Winkel, G. and Sasanoff, R. (1966) *An Approach to an Objective Analysis of Behavior in Architectural Space, Architectural Development Series No. 5,* Seattle: University of Washington, Department of Architecture.

Dan Carson
Man and Environment Division
Pennsylvania State University

WORKSHOP ON ENVIRONMENTAL PERCEPTION

There was some question whether these studies were actually dealing with perception of the environment. In the first instance a quantitative method virtually independent of the observer, except for the shape of visual field, was constructed to measure environments. In the second instance, ideals rather than perceptions were used to construct the real environments used later. And in the third instance, perhaps the closest to perception, stimuli were limited to visual materials which were simulations of passing through an environment.

In addition, the three studies had little in common: one used no real environment, one used slide projections of two environments and the third used two real environments. In one, the method was entirely rational, in one the method of evaluating change was rational and the rest somewhat intuitive, and in the third, the rational method was nearly divorced from the experiment, being used only to set up a pattern from which the design was inferred. Despite these disparities, two central issues were brought up in the general discussion: problems relating to methodology and concern for relevance of this research to real world problems.

Environmental Perception

Questions of relevance arose from some of the intrinsic limitations imposed by the necessary economy of choosing too few different environments on the one hand,,or of choosing only the visual sense on the other. There were somewhat inconsistent criticisms which, while they are imbedded in methodology, bear directly on relevance. For example, the audience took one researcher to task for changing too many things from one real environment to another, and they also simultaneously challenged another researcher for the use of single variables without trying to assess the interactions between variables. How many environments are sufficient in an experimental study to establish the hypothesis, or to provide normative data for design criteria? This question, while again deriving from methodological considerations of experimental constraints, was important to many as a question of relevance for designers or for evaluating behavior in many different types of environments. Most telling of all questions asked for information to tell designers how to decide instead of telling them how to go about telling how to decide (how to decide how to decide).

Apart from the standard questions about choice of populations, choice of criteria and points about statistical inference, two important points were considered under methodology. The first dealt with the proposition that if we pay too much attention to statistical and experimental control, we shall lead ever farther from real world situations. Misplaced precision of this sort at this time appeared to be inherent in each of the studies, according to one vocal group. This concept derives from excessive experimenter constraint in designing his experiment, and is defensible on the grounds of internal consistency and ability to generalize within those studies which adhere to the same rules of development. The second point dealt with the appropriate choice of mathematical machinery and the kinds of models used to explain experimenta phenomena. A corollary point is the choice of test used in statistical inference, These are questions of the assumptions behind the methods and whether they are met by the specific conditions of the experiment or the theory.

These papers were clearly the point of departure for a penetrating and generally sophisticated discussion about not only the specifics covered by each researcher, but about topics important and relevant for the field of environmental perception.

Wolfgang F. E. Preiser
College of Architecture
Virginia Polytechnic Institute

BEHAVIORAL DESIGN CRITERIA IN STUDENT HOUSING

In order to create more habitable environments, research is necessary to understand users' attitudes and preferences about environments in which they dwell. An integral part of the behaviorally defined design process is the evaluation of existing environments in regard to user perception and performance within a set of given environmental factors. For the qualitative assessment of these factors an evaluative tool has been developed which measures verbalized response to physical environment on a relative scale of importance. This method is based on Thurstone's scale of appearing intervals.

Description of the Method

To measure the attitude construct of the degree of acceptance, rejection or indifference toward features of physical environment (dormitories), a method consisting of two major experiment stages was devised. Stage A was designed to have subjects construct attitude statements about dormitories. Then, another group of subjects was asked to place these statements on a ranking scale continuum, thus determining the relative importance of certain features of the environment.

Stage B was intended to validate the results of Stage A by differentiating various categories of environments (dormitories) qualitatively. This evaluative concept of assessing verbalized response to physical environment was to yield user relevant information of points of conflict and priorities of concerns in a particular environment. The sequence of steps involved in this method is as follows:

Stage A	1.	Construction of Attitude Statements
	2.	Correction and Selection of Statements
	3.	Scaling of Attitude Statements
Stage B	1.	Statistical Evaluation (Means, Standard Deviation, Frequency Distribution) of Scaled Statements
	2.	Selection of "Best" Scoring Statements
	3.	Validation Procedure

The Subjects

The subjects who participated in the experiments for Stage A were students attending Virginia Polytechnic Institute whose majors included Engineering, Business, Architecture, Psychology, etc. Although most subjects were Sophomores, 19 to 20 years old, other undergraduate class levels were represented as well. About 15% of the subjects were female which is higher than the University average of 13%. The quality credit average of all subjects was 2.3. All subjects had experienced dormitory life and 95% of them were living in dormitories at the time of the experiments.

The subjects for Stage B were male students of all 4 undergraduate class levels who lived in three different dormitories at Virginia Polytechnic Institute. 157 subjects participated in Stage A1, 191 in A3 and 215 in B3.

The Procedure

Stage A

1. Construction of Attitude Statements

The firse experimental session served the purpose of having attitude statements on dormitories constructed by the students living in dormitories. The 157 subjects of the session (in Seitz Auditorium, VPI, October 14, 1968, 2 p.m. were not informed of the nature of the investigation in order not to bias their responses.

Blank sheets of paper were distributed to the subjects who were asked to write brief statements which reflected favorable, indifferent or unfavorable attitudes toward features of dormitories and dormitory life. The subjects were asked to write statements which covered the entire range from desirable to undesirable features. The session lasted 30 minutes, after which time the statement sheets were collected for evaluation.

2. Correction and Selection of Statements for Scaling

More than 2,000 statements about dormitories and dormitory life were given by the subjects in the session for attitude statement construction. They expressed favorable, unfavorable and indifferent attitudes toward dormitory environment.

The statements were categorized and grouped into 100 distinct content categories which represented a cross-section of all attitudes expressed in the frequently mentioned topics of statements.

Major criterion for the inclusion of statements, among the 100 which were to be scaled, was that a statement should clearly reflect conditions in dormitories as they were experienced by the inhabitants. Unsuitable statements were discarded if they did not refer to dormitories and their impact upon the users.

For presentation in the scaling sessions an equal number of positive and negative statements in random order was included.

3. Scaling of Attitude Statements

The first objective of the scaling procedure was to measure the relative importance of certain features of dormitories at VPI as described in 100 statements which had been made by the users of dormitory environment.

The second objective was to investigate whether there is substantial agreement among a number of respondees in judging the relative importance of features of the environment.

Scale items were administered to five different groups of subjects. The subjects received instructions repeatedly as each set of 25 statements was completed and had been judged by the subjects. They were given approximately 15 seconds to understand and judge each statement in regard to which position on the 11-point scale best reflected the desirability or undesirability of the particular feature contained in the statement. The 100 statements which were read to the subjects are listed in Table 1; this list also contains part of the results, i.e. means and standard deviations for each statement computed for all 5 sessions combined.

The subjects were instructed to place each statement on an 11-point scale with equal appearing intervals ranging form -5 to +5. The negative half of the continuum was to represent statements which reflect undesirable features of dormitories, whereby -5 represented extremely undesirable features, and -3 fairly undesirable features. Accordingly, the positive hapf represented desirable features, with +3 standing for fairly desirable and +5 for extremely desirable features. The scale positions around the midpoint (Zero=0) were to reflect indifferent attitudes.

Many statements were hypothetical and did not apply to the subjects' personal environment and therefore, disregarding their own situation, the subjects were asked to judge objectively the degree of desirability or undesirability of the features contained in the statements.

Stage B

1. Statistical Evaluation

In the five scaling sessions mentioned above, a total of 19,100 statements had to be scaled by the 191 subjects. Before a statistical evaluation could be made, it was necessary to detect and eliminate the scaling results of those respondees who obviously disregarded the instructions and those who did not willingly cooperate and therefore would cause unproportional distortions in the results. For each of the scales of the remaining 148 subjects, the program BMD 01D Simple Data Description—Version of June 6, 1966 Health Sciences Computing Facility, UCLA—was used to compute means and standard deviations of all scale values marked by the subjects. First, means and standard deviations were obtained separately for each of the five scaling sessions and then the values of all sessions combined were computed. They are included in Table 1 for each of the 100 statements of the scaling sessions.

Statements were sorted into categories according to their mean values as indicators of importance of the feature contained in the statement. In order to determine the degree of scalability of the statements, their standard deviations were examined.

Profiles of marked scale positions were obtained for each statement in form of frequency tables. The frequency curves gave a measure for relevance, agreement and general scalability of the content. The mean value for each statement was included in the respective graphs. The greater the mean value of a statement was and the more it approached the extreme scale positions in a positive and negative sense, the more urgent the matter described was. The graphs thus served as clear indicators for priorities in abolishing dysfunctional aspects of a behaviorally defined environment.

2. Selection of Statements for Validation

For a final validation procedure, 25 statements were selected, about half of which represented positive and the other half negative features of dormitories. Only statements with low standard deviation, i.e., which had achieved considerable agreement among the judging subjects, were included and ordered according to mean values. The objective was to choose statements that would cover the entire attitude continuum form -5 to +5 in approximately .5 intervals, which was not always possible.

3. Validation Procedure

The validation procedure was to prove that the method of measuring verbalized response to physical environment can be used as a reliable evaluation tool. Once a relative scale of priorities of features of a particular environment had been established it would be possible to differentiate environments of different qualities, such as "good" dormitories from "bad" dormitories. As described above, 25 statements which represented a contiuum of attitudes toward dormitory features ranging from extremely undesirable to extremely desirable were selected to be included in a final validation questionnaire. The statements were presented in random order.

The questionnaire was distributed to a 20% sample of students living in three different types of dormitories. According to the varying sizes of dormitories, the sample sizes varied also (Dorm A—170, Dorm B—68 and Dorm C—57 respondees). Random number tables were used to determine the room numbers of those students who should receive a questionnaire. In case a room number occurred twice, both roommates would receive a questionnaire. The instruction sheet with the room number was not returned with the questionnaire to make identification of respondees impossible and to facilitate unrestricted judgment of their dormitories. The respondees were to agree or disagree with the 25 statements contained in the questionnaire.

Results and Discussion

In this chapter, results will be discussed in the same sequence as the experiments were conducted.

Following a few pilot studies, the session for construction of attitude statements about dormitories was held. The results yielded information about features of dormitories which were relevant and mattered to their users, in the language spoken by dormitory inhabitants. The most important subcategories into which the statements could be grouped were:

1) Physical-environmental features
    a) Rooms and immediate environment of students
    b) Total dormitory environment and vicinity

2) Psychological-behavioral features
    a) Performance criteria
    b) Life "enjoyment" criteria

3) Sociological features
    a) The student's social interaction with friends and neighbors
    b) The network of relationships within the entire dormitory and among dormitories

4) Administrative features and regulations

Although experiment I was unstructured in that the subjects were free to express things which they were concerned about, the results showed that most statements dealt with physical features of dormitories. Several positive features were mentioned, such as sinks within the dormitory rooms, water fountains, candy and coke machines, elevators, the small size and the appearance of the exterior of dormitories. Social aspects such as "bull sessions," making new friends, sense of togetherness and meeting different types of people were among the issues reoccurring most often. Desire to liberalize and to change dormitory regulations was frequently expressed.

Among the negative statements, physical features were again most frequently mentioned. These included extremely poor acoustics, monotonous and poor design in general, too long and noisy hallways, institutional character of dormitories, drab and depressing appearance of materials used, lack of recreation facilities and adequate study spaces, insufficient sizes of rooms and therefore lack of privacy and features such as inconvenient bunk beds, bad lighting, insufficient shelf and closet space. Complaints about regulations pertained to lack of freedom to choose a roommate, entertaining females in men's dorms, the prohibition of alcoholic beverages, hot plates, refrigerators and air conditioners in dormitories. The unavailability of different kind s of housing for undergraduate students was not appreciated as well as the lack of university housing for graduate and married students.

A list fo the features most frequently mentioned in experiment I can be found on Table 1 in the 100 statements which were used in the scaling session. The 157 subjects wrote approximately 2,000 statements about dormitories. The range was 15 to 25 statements per subject. The experiment lasted 30 minutes, i.e. as long as the subjects were willing to write statements.

A listing of mean categories shows the number of statements which were placed into the respective categories 1-10. It is interesting to note that no statements were placed in the indifferent region around scale point 6.

Original Scale	-5	-4	-3	-2	-1	0	+1	+2	+3	+4	+5
New Scale Mean Category	1	2	3	4	5	6	7	8	9	10	11
Number of Statements	1	7	9	6	0	0	4	12	20	1	-

Data were examined to determine whether the mean categories coincided with categories of contents of statements. Several subjects were asked to sort the statements into natural groupings. There was considerable agreement among these judges and the groupings chosen were essentially those described above under construction of attitude statements.

Table 1 containing the standard deviations for the 100 scaled statements lead to the construction of a frequency table which further clarified the distribution of judgments for each statement. Although in most cases, a clear concentration of judgments around one scale point can be observed, there are in most cases a few "wrong" judgments on the opposite side of the scale. A number of examples was chosen from the frequency table to be represented graphically (see Figure 3-4). The slope of each frequency distribution curve indicated the amount of agreement among the judges and the location of the peak of the curve showed the degree of relevance of the feature contained in the respective statement.

The examination of frequency distributions resulted in the selection of 25 statements with low standard deviations for the final validation procedure. It was intended that these statements would cover the whole range of the attitude continuum form -5 to +5 (1-11) (Table 5). The final validation questionnaire contained these statements in random order. It was distributed to a 20% random sample of 3 dormitories of different size and construction date.

The results of the validation questionnaire showed no considerable differences in the qualities of the three dormitories.

3. Validation

The results of the validation questionnaire were converted into percentages in such a way that "agree" and "disagree" responses for each statement would total 100%, except for cases where more than two subjects had failed to judge the statement.

The fact that there was little differentiation in the tendencies of responses toward features of the dormitories A, B and C confirmed the initial assumption that the dormitories were essentially of the same type (double-loaded corridors with two man rooms) except for the number of floors which the contained.

For each dormitory the number of responses regarding a particular statement was plotted. The points thus obtained were connected and the areas under the resulting graphs then represented the degree of acceptance of the respective dormitories. By this definition dormitory B, Vawter, appears to be "better" than A, Monteith, or C, Lee. However, there were instances where certain favorable features seemed to occur in one of the dorms A and C, but not in B. There are two explanations for this:

a) The dormitory is really deficient regarding the particular feature
b) The feature is more important to one kind of student and not so much to others

Some of the statements represented in Figure 5 (Comparative Profiles of Qualitative Judgments of 3 dormitories) shall be discussed here. Proceeding from the extreme scale positions -5 and +5 to the scale midpoint 0, the degree of importance of the feature decreases. At the same time, the amount of differentiation among the dormitories increases.

1) Statement No. 98: "From the outside, my dorm looks like a jail" differentiates clearly 82% of the students in Dorm B believe it does not look like a jail, whereas 62% in Dorm A agree.

2) No. 33: "The floor material of my room looks very ugly." In the oldest dorm, A, 92% agree, in the second oldest, B. 57% agree and in the newest and largest dorm, C, only 40% believe that their floor is ugly.

3) No. 46: "In my dorm orientation is poor" applied to 42% of the students in Dorm C (Lee) and 46% believe the corridors are too long in that dorm.

4) No. 11: "The halls in my dorm create a sense of unity among the residents." 78% in Dorm B (Freshmen only) agree with this statement; however, 74% of the residents of the largest dorm, C, (over 700 students) cannot agree.

5) No. 2: "The exterior of my dorm is very pleasing to the eye" was to check the responses of statement No. 98. In Dorm A only 5% agree, in Dorm B, however, 67% and in Dorm C, 48%.

6) Almost everybody in all 3 dorms agrees to Statement No. 59: "Without personal decorations, the barrenness of my room is distasteful."

7) According to Statement No. 100 "There is a lack of entertainment facilities in my dorm," all three dorms need major improvements regarding this issue.

8) No. 8: "The lounge of my dorm is attractive" was disagreed with by almost everybody in all 3 dorms.

9) No. 86: "Many rooms on a corridor lead to frequent disturbance through friends dropping in" was agreed with by about 75% of all students.

10) Almost everybody agreed that the furnishings in all dorms were very dull (No. 29) and very few students found the interior of their dorms look pleasant (no. 85).

Behavioral Design Criteria in Student Housing at VPI

The following list of recommendations contains design criteria which have evolved from the results of this study on behavioral response to student housing. The most important ones are:

1. The student's room as living-learning unit

Students' needs are basically the same as those of any other segment of the adult population, i.e. they deserve appropriate living conditions providing for privacy, security, the satisfactory social interaction to feel "at home", motivation to work and study, life enjoyment and above all, sensory and spiritual stimulation.

- a) At least 25% of all students express the need or a single room, if necessary at higher cost, in order to achieve satisfactory study and living conditions.

- b) Two-man rooms are satisfactory for most fresnmen because of the greater need for socialization, but single rooms should be available for those who need one, with approximately 120 square feet per student.

- c) Students' needs change as they advance to higher class levels and so does the use of facilities which are provided for them. Therefore, a variety of types of accomodations should be made available in order to give students opportunity and freedom of choice of their personal environment.

- d) High quality "luxury" rooms providing more than the bare minimum of nocturnal storage should be made available for those who are willing to pay for them.

- e) Flexibility in furniture arrangement is highly important to the users of dormitory rooms. It demands adequate room sizes and moveable furniture except for the closets. Provisions should be made to allow students to personalize rooms with their own decorations and objects, i.e. posters pictures, drapes, rugs, easy chairs, stereo sets, personal book shelves, etc.

- f) Physical features of rooms need to be improved, i.e. a careful choice of building materials should be made. The drabness of concrete block walls and their lack of sound proofing capacity should outweigh economical considerations. In 2-man rooms, the window sizes are not sufficient to provide enough daylight for 2 study desks. One "naked" bulb in the center of the room provides very unpleasant artificial lighting.
Each student should be able individually to control his thermal environment; heat controls in every room are necessary as well as satisfactory ventilation. Color schemes generally criticezed as being too institutional should be varied from room to room to enable students to identify themselves with their environment within a large number of otherwise identical units.

2) The arrangement of students' rooms

- a) With the exception of freshmen, most students would like to live in groups not larger than 8-12 individuals. This figure coincides with the number of friends of the average student. "Groups" of 60 students might be reasonable from an administrative or supervisory point of view; however, usually only casual contact is possible among such a large number of students.

- b) At least one third of all students prefer suite-like arrangements of rooms to the conventional double loaded corridor type with all its disadvantages like noise problems, institutional appearance and sources of disturbance.

- c) These smaller groups of room arrangements lead to greater flexibility in terms of changing uses. They should contain shower and bathroom facilities, a kitchenette, a common congregation area, and possibly a study area separately from the individual rooms. Independent access should be given to each group of rooms.

3. Study areas

- a) Study rooms should be small, close to the students' individual rooms and large enough in number to accomodate working places for at least 50% of the students.

- b) To avoid frequent disturbances, they should not be directly accessible from general circulation areas.
- c) Provisions should be made for different kinds of study activities like space for drawing boards, typewriting or reading and writing only.
- d) Places should be provided where students can be alone for studying.

4. Lounges and recreation areas
    - a) The barren furnishings of lounges and the poor choice of material in general cannot create a homey atmosphere for the users. Existing lounges are far too small to accomodate all students who would like to watch TV or engage in other activities.
    - b) Spaces for small group activities like card playing should be incorporated into the groups of living-learning units.
    - c) Spaces for informal and casual gatherings should be in the immediate vicinity of the student's rooms.

5. The arrangement of dormitory complexes
    - a) Dormitories with 1,000 and more students each are too large if they aren't broken down into smaller parts. High density low-rise walk-up dormitories are preferred to high-rises standing isolated within large open spaces. The size of 250 to 300 students per building is accepted.
    - b) High-rise dormitories are appreciated, provided an appropriate grouping of rooms and activity areas has been achieved, with easy access.
    - c) Apart from the properties described above, excessively long corridors (200-300 feet) create an institutional feeling among students. Although lighting is mostly sufficient, an impression of sterility is evoked in the person walking through them. Orientation often is difficult within the dormitories.
    - d) The architectural appearance of dormitories at VPI leaves much to be desired. Although impressive stonework has been used, they give the impression of jails, of huge storage spaces.
    - e) Indoor-outdoor relationships of dormitories to their surroundings need to be carefully analyzed. Roof tops might be utilized for sun-bathing and other activities. Landscaping including benches and quiet places needs to be provided for socializing, recreation and outdoor studying.
    - f) Segregation of students of different class levels and majors should be avoided. A healthy mixture of all kinds of students whoul be desirable for the purpose of intensified stinulation.
    - g) Co-ed dormitories are highly desirable among male students, less so among female students. They constitute an important step forward toward a "normalization" of university life at VPI. There is a great need for attractive activity spaces where co-eds can be entertained within the dormitory complexes.

h) The incorporation of classrooms in dormitories is very unpopular among students. They want a clear separation of formal teaching and living spaces.

6. The overall size of the university campus in relationship to the supporting community will be limited to the extent in which the town can sustain cultural, shopping, recreation and other facilities necessary for a life of fulfillment.

Conclusions and Recommendations

The following conclusions can be drawn from the experiments described above:

1. User-relevant information regarding response to physical environment can be obtained through unstructured experiments for attitude statement construction.

2. The degree of relevance of features of the environment as perceived by individual users can be measured by using the scaling technique outlined above.

3. Response to physical environment as perceived by larger groups of users can be measured. Thus a hierarchy of priorities regarding acceptance or rejection of environmental features can be established.

4. Complex environmental concepts can be analyzed by testing verbalized descriptive elements as they relate to the total concept, using the method of relevance scales.

5. Any design criteria obtained through the method of relevance scales are characteristic of the particular environment and subjects under investigation and cannot be generalized.

Recommendations for future in-depth research into the topic area are:

1. For the purpose of further validation, the method of measuring verbalized response to physical environment might be tested in more differentiated environments than were available at VPI. Then, distinct patterns of response could be obtained.

2. A comparison and further investigation of the scalability of the various items contained in the attitude continuum regarding a particular environment might give insight to the investigator as to how subjects respond on different levels of perception to different categories of content.

3. In order to measure the adaptability of users to constructed environment over a period of time, a series of experiments might be conducted, using identical scale items.

4. The influence of dependent variables, such as the socio-economic background of the subjects, upon the results of various scaled categories of features might be of interest to the investigator.

5. Shortcut procedures might be devised to make the method universally applicable, e.g. for the testing and and evaluation of prototypes for housing developments through prospective buyers or users.

Experimental Data (Tables)

Table 1. Means and Standard Deviations of 100 Statements

The following is a list of the 100 statements in identical sequence as they were read to the subjects of the scaling sessions. Listed are also the mean value and standard deviation of each statement for the combined scaling sessions A-E. Table 2 and Table 3 list the values for means and standard deviations for each individual scaling session in order to allow comparisons among the various sessions. For computational purposes, the original scale from -5 to +5 had been converted into a scale from +1 to +11. All mean values refer to the new scale.

Original Scale	-5 -4 -3 -2 -1 0 +1 +2 +3 +4 +5
New Scale	+1 +2 +3 +4 +5 +6 +7 +8 +9 +10 +11

List of Statements

Statement Number	Mean (of 148)	S.D.	Statement
1	3.3767	1.3500	The size of my double room makes it difficult to maintain order
2	7.8571	1.7886	The exterior of my dorm is very pleasing to the eye.
3	8.6986	1.8841	The construction of my dorm is very good and solid.
4	2.7466	2.1101	In my dorm, the individual rooms have no heat control.
5	8.5510	2.6617	My dorm is close enough to classes.
6	3.6122	2.6251	There is hardly space in my room to accommodate private shelves, an easy chair, etc.
7	3.2345	2.8137	In my dorm, I have to lock my room because thefts have occurr
8	8.4354	1.9796	The lounge of my dorm is attractive.
9	8.5374	1.7449	The lighting in the corridors of my dorm is good.
10	3.3630	2.3426	In my room there is hardly enough daylight coming in through the window.
11	7.5724	1.7390	The halls in my dorm create a sense of unity among the resident
12	3.4422	2.3353	In my room there are no antenna plugs for radio and TV.
13	3.9864	2.1707	It is forbidden to have air-conditioning in the window of my roc
14	4.0340	2.7057	The cinderblock walls in my room give a drab appearance.
15	9.1027	2.7082	In my dorm I have a place for studying where I can be alone.
16	8.1701	1.7994	The appearance of my dorm fits with the rest of the campus.
17	3.7823	2.4425	The view from my window is lousy.
18	7.2635	1.7742	The arrangement of our dorm rooms along a corridor is satisfac
19	2.3082	2.2639	The soundproofing in my dorm is very poor.
20	8.3401	2.5762	The dormitory room arrangement is really suited to the student needs.
21	3.6081	2.7569	When the wind is blowing outside, there is a strong draft from t window of my room.
22	4.2721	2.2862	There is no variation in the wall colors of my dorm.
23	4.5646	2.0071	The bunk bed in my room is hard to make up.

Statement Number	Mean (of 148)	S.D.	Statement
24	3.9662	1.8532	My sink is placed badly.
25	4.6014	2.0726	In my dorm, the arrangement of rooms along a corridor reminds me of a barrack.
26	9.1824	1.4524	A feature of my dorm is that I am permitted to put rugs in my room.
27	9.1622	1.6332	The lounges of my dorm are comfortable because they contain drapes and wall-to-wall carpeting.
28	9.3469	1.7582	Our recreational lounges provide a good place to relax.
29	3.1014	1.7330	The furnishings in my dorm are very dull.
30	2.3378	2.1932	There are no washer and dryer facilities in my dorm.
31	8.9595	2.1185	The halls in my dorm are cleaned sufficiently.
32	3.6081	2.3230	Overcrowded shower facilities make communal bathrooms in my dorm unpopular.
33	3.5946	1.8620	The floor material of my room looks very ugly.
34	9.6284	1.4347	In my room I can arrange the furniture according to my personal ideas.
35	2.6824	2.4690	The closet in my room is too small to hang up all clothes.
36	9.3699	2.3282	My room has large enough closet space.
37	9.0270	1.6780	In my dorm rules are set up by the students.
38	9.7162	1.2673	I am allowed to put up personal decorations on the walls of my room.
39	3.9932	2.1523	In my dorm there is lack of community spirit.
40	2.7568	2.1080	There is no soundproofing against slamming doors in my dorm.
41	4.6892	1.9858	The stone facade of my dorm is not very attractive.
42	9.0270	2.1695	My room has some place to lock up valuables.
43	2.9932	2.6548	My dorm is too far from the parking lot.
44	9.5270	2.0978	The study room of my dorm has excellent soundproofing.
45	8.0000	2.4826	The sink in my room has one outlet for mixing hot and cold water.
46	4.2770	1.8874	In my dorm orientation is poor.
47	2.6824	2.1823	Our study room is at times very hot and stuffy.
48	9.8649	1.7444	The bathroom in my dorm is kept fairly clean by the janitors.
49	4.6014	1.8470	Among the various floors of my dorm, there is lack of communication.
50	2.8649	2.2213	My room does not provide real privacy.
51	4.0608	2.3567	The telephone in my hall is a place for congregation and talk.
52	3.8844	2.0856	The sameness of rooms make my dorm depressing.

Statement Number	Mean (of 148)	S. D.	Statement
53	9.0952	1.9561	If I need academic advice a great many people are within reach in my dorm.
54	10.2905	0.9494	In my dorm I can choose my roommate.
55	9.4932	1.9743	My dorm is within 10 minutes walking distance of classes.
56	4.5541	2.3102	Because of their impersonal character, lounges are not used in my dorm for bull sessions, etc.
57	9.1149	1.5497	Within my dorm, informal dress makes you feel more at home.
58	9.1429	1.9515	In a single room I am free to do what I want.
59	2.2770	1.6488	Without personal decorations, the barrenness of my room is distasteful.
60	4.5473	1.7275	The corridors in my dorm are too long.
61	7.6014	1.7133	Living in a small dorm makes it easier to feel at home.
62	2.7551	2.1054	Open telephones in my dorm do not allow for private conversation
63	8.4932	1.9881	My dorm has light walls and bright colors.
64	9.5811	1.8881	In my dorm I have a place where I can be alone for studying.
65	9.1622	1.2727	The fluorescent lamps in the halls of my dorm provide good lighting.
66	3.2568	2.0804	The TV lounge in my dorm is overcrowded.
67	8.5890	2.0932	The planning of rooms in my dorm is well handled.
68	4.7770	1.7254	My hall is too big to let me leel at home.
69	7.0135	2.6367	Our dormitory rooms are the most popular place for bull-sessions and meetings.
70	9.3311	1.5359	My dorm has satisfactory sink facilities.
71	3.0541	1.9510	According to regulations, I cannot have a refrigerator in my room
72	2.9388	1.9523	The rooms in my dorm are unvaried monotonous boxes with little stimulation.
73	8.8571	1.6427	The drink and candy machine in my dorm is convenient.
74	8.5850	1.7511	Students have participated in the design of my dorm.
75	4.5203	2.3019	The uniform wall colors of my dorm make it look very institutio
76	8.6757	1.3259	The ceiling height of my room is adequate.
77			
77	3.8919	1.8957	There are no single rooms in my dorm.
78	3.0405	2.1089	In my dorm there is no place for warming up soup, tea, etc.

Statement Number	Mean (of 148)	S. D.	Statement
79	8.8446	1.4787	In my dorm I have the opportunity to meet all kinds of people from different parts of the country.
80	3.9865	1.7103	In my dorm there are few group activities.
81	2.1892	1.6177	There are very few places within my dorm where studying can be done.
82	3.1757	1.7756	The walk from my room to the communal bathroom is convenient.
83	9.0408	1.8390	My dorm offers a variety of room types and sizes to fit different needs of students.
84	2.1351	1.9679	In my room noise can be heard from two or three doors away.
85	9.1622	1.5255	The inside of my dorm looks very pleasing.
86	4.1554	1.9049	Many rooms on a corridor lead to frequent disturbance through friends dropping by.
87	9.4595	1.4351	There is sufficient shelf space in my room.
88	9.4354	1.2609	In my dorm I am able to have friends close by.
89	3.2313	1.5397	In my dorm the dreariness of most rooms does not provide a homey atmosphere.
90	9.6939	1.5243	I have enough drawer space in my room.
91	8.5608	1.7112	The atmosphere of my dorm is very personal.
92	2.7905	1.6833	Recreational facilities in my dorm are insufficient.
93	8.6486	1.5424	The color scheme in my dorm is very appropriate.
94	9.8176	1.2886	The size of my room is adequate.
95	8.2973	1.4212	My dorm has an attractive stone exterior.
96	7.7619	2.5437	The fact that my dorm is set up in suites of 4 or 5 rooms is desirable.
97	7.7365	2.0975	The closeness of other student's rooms leads to a lot of social interaction in my dorm.
98	2.7838	1.5974	From the outside my dorm looks like a jail.
99	1.6486	1.2337	There is no ventilation in my room.
100	2.4459	1.2469	There is a lack of entertainment facilities in my dorm.

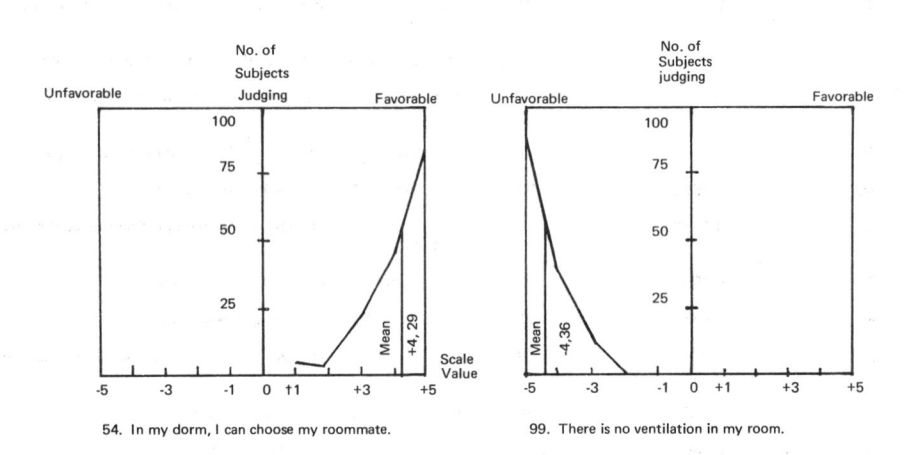

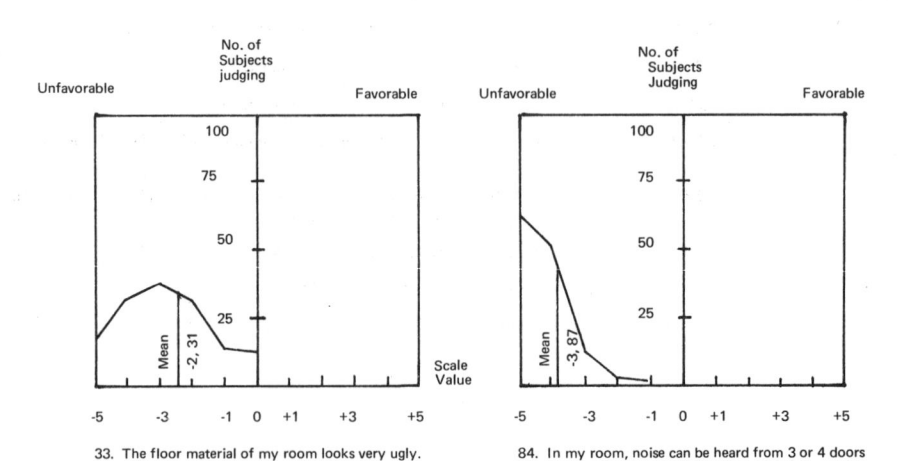

Figure 1. Frequency Distributions for Selected Statements

54. In my dorm, I can choose my roommate.

99. There is no ventilation in my room.

33. The floor material of my room looks very ugly.

84. In my room, noise can be heard from 3 or 4 doors away.

Figure 2. Frequency Distribution for Selected Statements

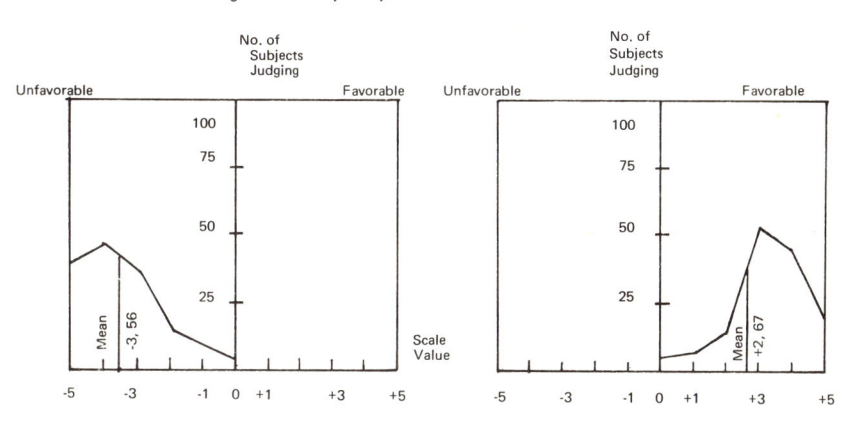

100. There is lack of entertainment facilities in my dorm.

76. The ceiling height of my room is adequate.

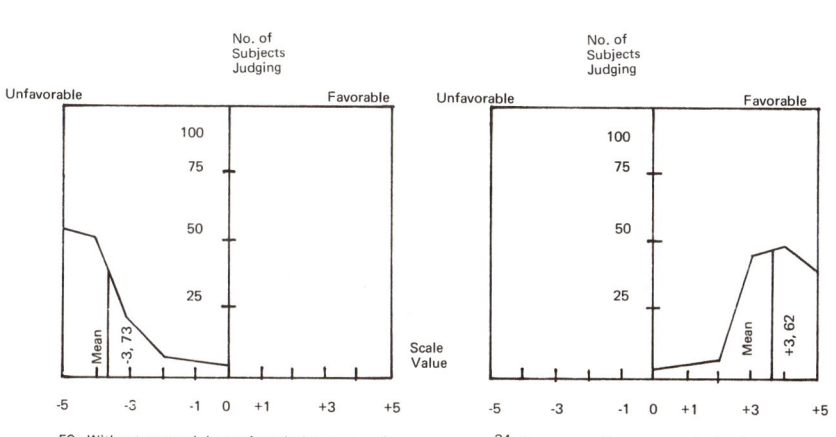

59. Without personal decorations the barrenness of my room is distasteful.

34. In my room I can arrange the furniture according to my personal ideas.

258

Figure 3. Comparative Profiles of Qualitative Judgments of Three Dormitories

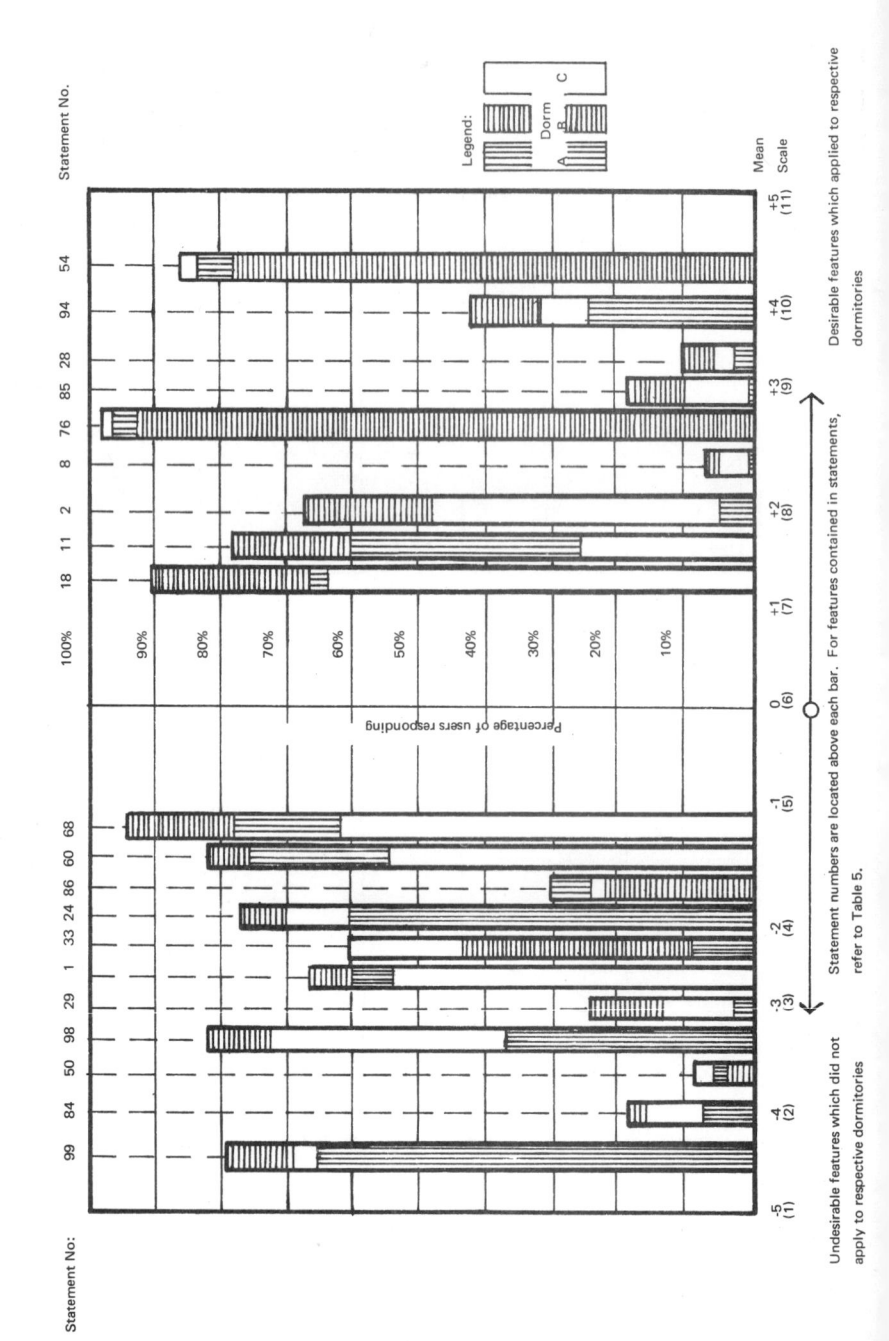

References

1 Bruner, T. S. and Goodman, C. "Value and Need as Organizing Factors in Perception", *Journal of Abnormal Psychology,* XLII, (1947).

2 Brunswik, E. *Wahrnehmung und Gegenstandswelt,* Wien, 1934.

3 Fechner, G. T. "Zur Experimentalen Aesthetik," *Kaiserlide Saechsische Gesellschaft der Wissenschaften,* Abhandlungen, IX, 6, (1871).

4 Guilford, J. P. *Psychometric Methods,* McGraw Hill, New York, 1954.

5 Hsia, V. "Residence Hall Environments; An Architectural Psychology Case Study", University of Utah, Salt Lake City, 1967.

6 Likert, R. "A Technique for the Measurement of Attitudes", *Archive of Psychology,* 140, (1932).

7 Oppenheim, A. N. *Questionnaire Design and Attitude Measurement,* Basic Books, New York, 1966.

8 Osgood, C. E., Suci, G. J. and Tannenbaum, P. H. *The Measurement of Meaning,* University of Illinois Press Urbana, 1957.

9 Osgood, C. E. "The Cross Cultural Generality of Visual Verbal Synestetic Tendencies", *Behavioral Scientist,* 5, (1960), 146-169.

10 Thurstone, L. L. and Chave, E. J. *The Measurement of Attitudes,* University of Chicago Press, Chicago, 1929.

11 Thurstone, L. L. *The Measurement of Values,* University of Chicago Press, Chicago, 1959.

12 Torgerson, W. S. *Theory and Methods of Scaling,* Wiley, New York, 1962.

13 Van der Ryn, S. and Silverstein, M. *Dorms at Berkeley, An Environmental Analysis,* Educational Facilities Laboratories, New York, 1967.

14 Wright, J. H., and Hicks, J. M. "Construction and Validation of a Thurstone Scale of Liberalism-Conservatism", *Journal of Applied Psychology,* L. 1, (1966).

M. Mohan Sawhney
Department of Sociology and Anthropology
North Carolina State University

WORKSHOP ON ATTITUDINAL RESPONSES TO THE ENVIRONMENT

Two papers presented in this workshop dealt with the problem of measuring students' responses to and evaluations of the dormitories in which they were living. The explicit or implicit objective was to establish criteria for designing better student housing on college campuses.

An evaluative summary report on any event like this workshop is generally presented in relation to the expectations of the reporter. If one were to assign a high place to rigorous research design and broadly applicable, valid and reliable research findings it is easy to share the feelings expressed by Stuart Silverstone and a few other workshop chairmen. There was indeed a tendency to seek simplistic answers to very complex problems of correspondence between physical and perceptual or psychological spaces. However, in view of the present evolutionary stage of the development of what could be referred to as environmental social psychology, I feel more justified in making an inventory of pluses and minuses of the workshop papers than an overall objective evaluation.

On the plus side it was significant that the two speakers and other workshop participants shared the strong conviction that planning of environment could not be left any longer to personal insights and hunches. Rather, the planning process should be based on empirical research dealing with social, psychological and cultural dimensions of the physical environment.

On the minus side there was quite a gap between what the two authors claimed to have sought and what in fact their studies produced. Gerald Davis wished to deal with "the relationship between architecture and happiness" but ended up asking college students rather routine evaluative questions about the dormitories in which they lived. Happiness being a complex psycho-emotional state dependent on a large number of variables, the designer can, at the most, regard physical environment as one of the variables which could become the limiting factor, not the determining factor of happiness. Wolfgang Preiser after expressing his concern about social disorders in slums, crime rates in suburbia and student riots, comes down to the study of what V.P.I. students regarded as relevant dimensions or aspects of their dormitories. In fact, both the researchers were trying to find the degree to which the students found their physical envirnoment satisfactory (or desirable) for performing the activities they wished to perform in the physical environment. After examining the findings of the two reports, one wonders whether he now has any definite criteria for better dormitory design, much less for increasing student happiness.

The above remarks are not, however, intended to undermine the significance of the two reports. Rather, the remarks are intended to put the studies in a more appropriate perspective. It is these studies with many mor future studies with broader frameworks and improved methodologies that would provide an accumulated body of knowledge from which broadly applicable prescriptive criteria would emerge. In this respect the papers do have an important contribution to make.

Also, methodologically these studies and the group discussions during the paper presentation made some very useful conclusions for guidance of future research. Wolfgang Preiser and Gerald Davis made a convincing case for the use of verbal responses to get valid and realistic information about users' attitudes and values. Both the researchers outlined some necessary precautions. It was strongly suggested, especially by Preiser, that in order to draw valid inferences from verbal responses there is a need for using realistic constraints. They expressed agreement with the suggestions made from the floor that the respondents should not be asked to respond to purely hypothetical situations.

Future researchers, especially those connected with universities, were also cautioned against indiscriminately using their own students as research subjects for evaluating campus environment. In these cases, the respondents' evaluations of campus physical environment might be seriously biased by their evaluations of university administrators. In one of the examples cited, it was hypothesized that the students who thought the dormitories looked like prisons (Preiser study), were indeed evaluating their administration.

In summary, one must admit that sociopsychological research in environment has to go a long way. The present phase of analytical fact-finding will have to accumulate more knowledge before entering the synthesizing-generalizing phase. The emphasis will have to move from purely descriptive to exploratory research. The variations in responses of different users will have to be explained in terms of social, demographic, psychological and cultural variables.

Gerald Davis' paper, "Architectural Determinants of Student Satisfaction in College Residence Halls" was not available when this volume was ready for publication.

Wilfred E. Holton, Bernard M. Kramer, Peter Kong-ming New, and Grazia Marzot
Social Science Division
Department of Preventive Medicine
School of Medicine
Tufts University

LOCATIONAL DECISIONS: OBJECTIVES AND CONSEQUENCES

Introduction

Locational decisions play an important role in the interplay between man and his environment. The pattern of land-use in any geographic area results from a large number of separate site choices. The geographic location of social activities has many implications for individual social units and for the social system as a whole. A hospital has no value, for instance, if it is completely inaccessible to its community. Therefore planners are concerned about achieving consonance between the objectives and consequences of locational decisions. Working toward such consonance requires a thorough understanding of the decision making process.

Planners and students of planning recognize certain shortcomings of the rational model. One office planner recently told us, "We are pretty well indoctrinated to the fact that there are a lot of things wrong with the usual approaches to design and that many foolish things get done because of it." This person and many others emphasize the nonrational elements in the activities of professional planners and decision-makers themselves. These *internal* non-rational factors are important, and we include a review of some of them. We focus, however, on certain forces which are *external* to the activities of planners themselves and out of their direct control. Such external forces have important consequences. We have done detailed case studies of the locational decision-making for a number of Community Mental Health Centers. Interviews with many people on-the-scene and the collection of available materials help to reveal the complexities of the locational decisions. In this paper, we use case study data to illustrate the concepts and conclusions. We focus on five external forces: *legal restrictions, fortuitous opportunities, ideologies, veto patterns, and the inability to identify and measure relevant variables.*

There is a considerable body of literature on internal nonrational forces in environmental planning. Braybrooke and Lindblom (1963) studied the process of governmental policy analysis. They concluded that "disjointed incrementalism" is the most common sort of decision-making. This pattern includes consideration of a limited number of policy alternatives, lack of coordination among the several decision-makers, and non-rational interpretation of data. In a later paper, Lindblom (1964) treats the weaknesses of the "rational-comprehensive analysis" model and proposes the "branch" model (with limited intellect and sources of information, impact of values, non-existence of "best" or "correct" choices, and limited comprehensiveness). Banfield (1962) goes so far as to say that most important decisions are the result of accident rather than design. Demone (1965) describes similar non-rational elements in mental health planning, the focus of this paper: "The selection of goals is limited and value-laden, containing unintended consequences." Locational decisions are known often to differ from the "rational" model. Management and Economics Research, Inc., (1967) studied site-selection practices of Indiana Industries. Almost ¾ of the locational decisions were made by top management, usually one man, and only 6 of 70 companies called on outside location experts as consultants. Locational problems of small businesses are treated by Mayer and Goldstein (1961). The conclude: "The strong impression conveyed by these data is that location factors are not taken seriously by the large majority of business owners." Duncan and Phillips (1967) discovered non-rational factors in site-selection for retail stores. "A local drug chain relies largely on the judgement of one of its chief executives who usually appraises a site 'just by taking a ride around a particular area, talking to some of the people living there, and getting a general feel of the site's expansion possibilities.'

I will now turn to a consideration of external forces in locational decision-making. Our studies show that several external factors may introduce important nonrational elements into the decision-making process. These external forces have a direct impact on the congruence between objectives and consequences. When all five of the forces are in full play, the following pattern is the result. *Ideologies* replace well-articulated objectives. Expediency and legal restrictions (of the type which forbid specific solutions) limit the number of alternatives considered. Veto patterns may enter after either of the first two stages. Vetoes cause the decision-making process to evert to an earlier stage. The inability to identify and measure relevant variables weakens the objective comparison of alternatives. Legal restrictions (of the type which alter the order of the planning process) may change the flow between stages. The development of objectives programs, etc.) may not be allowed until after the locational and design choices have already been made. The external force model esults in consequences which seldom meet the original (ideological) objectives.

History of Mental Health Theories and Treatment

The present interpretations and treatments of mental illness represent only one point in a long historical development. A brief review of the traditions and patterns of change puts current mental health planning into context.

The early interpretations of mental illness (in Egypt and Greece, for instance) saw the cause in divine or demoniacal visitations. Deutch (1938) traces theories and treatment of mental illness from early history into the 20th Century. The demon-and-witch theories persisted, with some exceptions, through the Renaissance period almost to 1880. This theory led to "treatments" based on exorcism and corporal punishment. Most mentally ill persons were held in city or town jails and treated as criminals. In the mid-18th century, the rational humanitarianism movement began to advocate less harsh treatment of the mentally ill; in 1751 Benjamin Franklin led a successful fight to build a new and better "Hospital for the Relief of the Sick Poor of this Province (Pennsylvania), and for the Reception and Cure of Lunaticks" (Deutch, page 59). It was at this Pennsylvania Hospital that Dr. Benjamin Rush advanced more humane and intelligent treatment of the insane. Although "moral treatment" was in vogue, large numbers of the mentally ill remained in poorhouses and jails, subjected to extremely bad conditions and punishments. From 1830 and for several decades the theory flourished that mental illness was easily curable in its early stages. The state hospital boom began during this period, spurred also by Dorothea Lynde Dix and other reformers. In this period, the locations of many state hospitals were decided by competitions among towns which were vying for the economic advantages of proximity to a large institution. In 1838 this pattern was followed in New Hampshire: "The state contributed 30 shares of New Hampshire bank stock, the town of Concord contributed $9,500. to secure the location of the asylum within its precincts, and the rest was raised by private donations." The state hospitals represented an improvement over the previous extremely inhumane methods of treatment although their locations were remote and inaccessible. Physical constraint was gradually replaced by greater freedom for patients. In 1909, a former mental patient, Clifford W. Beers, founded the National Committee for Mental Hygiene, which was later to become the present-day National Association for Mental Health. This public sympathetic concern was coupled with modern views of psychiatry and insanity to bring further reforms in institutional treatment.

The concept of "community mental health" is quite new but its roots are traced by Jeanne L. Brand (1968) from about the turn of the century. The first "psychopathic pavillions" in general hospitals were opened in 1902, and outpatient services and aftercare programs (for released mental hospital patients) began to become available. In some cities "psychopathic hospitals" were founded to provide examination and preliminary treatment of mental illness. Dr. Adolf Meyer stressed the role played by social factors in promoting mental problems and urged community-based early care and prevention. Freud's method and theory of psychoanalysis gained followers in America from about 1910 on. Freudian psychiatry has played an important role ever since. In 1928, William Alanson White proposed the marriage of sociology and psychiatry, thereby setting the stage for "social (or community) psychiatry."

Group therapy and milieu therapy grew out of the basic assumption that man is a social being interacting with other people. World War II helped to educate the public in the high cost of mental illness (1,100,000 men were rejected for military duty because of mental or neurological disorders), and in the potential effectiveness of early intensified psychiatric care. Earlier drug and electric "treatments" have proven ineffective, but the present-day tranquilizing drugs allow outpatient and open hospital care of formerly violent people. In the 1950's extensive research was done on day hospitals, milieu therapy, emergency psychiatric services, and rehabilitation programs. The National Institute of Mental Health was established in 1949. The Community Services Branch of NIMH provided consultation and grants-in-aid for the development of local programs. One of the earliest demonstration project grants (1957) was to the Massachusetts Mental Health Center in Boston to explore emergency outpatient services for mental patients. The Final Report of the Joint Commission of Mental Illness and Health was delivered to Congress in 1961. The Commission strongly endorsed new community care concepts: "A national mental health program should set as an objective one fully staffed, full-time mental health clinic available to each 50,000 of the population." Kennedy's presidency brought a strong proponent of mental health programs to the White House. In 1963, after the assassination, Congress enacted the Community Mental Health Centers Act which authorized the appropriation of up to $150,000,000. to finance up to two-thirds the cost of construction of community mental health centers. Two years later new legislation provided grants for three years to assist in the costs of staffing community mental health centers.

The current concept of community mental health goes beyond the provision of scattered clinics. Consultation and cooperation with other agencies (schools, welfare agencies, courts, etc.) is designed to locate emotional problems at early stages and to help other local professionals. Seeking out people who need help ("outreach" in the professional jargon) reverses the hospital you-come-to-me model. Many psychiatrists and other professionals are defining their roles as more closely tied to social issues and problems (poverty, race relations, etc.). This general orientation has obvious consequences for locational solutions. Centralized or isolated sites will weaken such programs as these. A large number of people believe that satellite or multiple center patterns offer the best solution. Holton and Kramer (1969) discuss the issues of centralization and decentralization in mental health planning. We oppose the use of the term "site-selection" because it suggests a single location. This paper employs the concept of "LOCATIONAL SOLUTION" as a substitute.

The Massachusetts legislation reflects a number of the facets of community mental health:

Comprehensive Services

>Children and adults suffering mental and emotional disturbances will be offered a full range of services. Special attention will be given to problems of aging, crime, delinquency, alcoholism, drug addiction, epilepsy, mental retardation and occupational mental illnesses.
>
>Each area program will include:
>* Inpatient services
>* Outpatient services
>* 24-hour emergency service
>* Partial hospitalization for day-care and night care
>* Mental health consultation and educational services to community agencies and professionals practicing in the area
>* Diagnostic services
>* Rehabilitation services

* Precare and aftercare services, including foster homes, halfway houses and home visiting
* Personnel training programs
* Research and program evaluation studies

For the mentally retarded services will include:

* Diagnosis, evaluation and re-evaluation
* Treatment
* Training programs
* Vocational adjustment programs
* Preschool clinical services
* Long and short term, day and night-care, residential services
* Personnel training programs
* Research and evaluation programs
* Preventive services

Methodology of the Research

This research is designed to investigate "factors affecting the location of mental health services." Our interests as sociologists led us to focus on the process of locational decision-making. Many decisions combine to determine the pattern of residences, businesses, industries, services and other land uses in any given geographic area. The importance of locational desicions for planners is asserted in the following statement by Altschuler (1965): "The most important specific decisions that planners are regularly asked to advise upon are those having to do with the location of new public facilities." Goodman and Freund (1968) stress the importance of location for such public buildings as government offices, civic centers, libraries, fire stations, police stations, and municipal garages and yards. Locational decisions are a pivotal point: interests, goals, traditional positions and existing programs go before; consequences of the new locational setting follow. We discovered an added bonus from the researcher's point of view. Because locational decisions are important steps in the life of organizations, participants volunteer information freely. At the same time, questions about location lead naturally to discussion of professional issues, programs, facilities, community issues and related services.

Our case studies to date have covered five mental health centers in Massachusetts. The small sample was chosen to give a cross-section by several criteria: stage in locational decision-making, type of geographic setting, and type of program presently offered. Center A (all names have been changed) has a two-year-old Comprehensive Community Mental Health Center operated by the Massachusetts Department of Mental Health. Two areas (Cities B and C) are in the process of buying land for a new facility. One area (D) is actively working out a decision on a locational solution, and the fifth area (Center E) has done no active locational planning since 1964. D is an inner part of Metropolis, populated mostly by lower middle class whites of foreign stock. City C is a suburban satellite city on the edge of a large metropolitan area. Cities E and A are declining factory cities 50 and 30 miles from Metropolis. City A has inpatient, outpatient and day hospital care for both adults and children; very few services are delivered in other agencies or in the community at large. D has inpatient care in a large teaching hospital, outpatient services integrated in a Medical School Department of Psychiatry, and additional professional services located in social service agencies in the community. City C has outpatient services for adults and children and an extensive system of consultation to area schools and courts. The B City mental health center provides outpatient services for adults and children. In Center E only children are treated and school consultation is an active program.

The interviewing technique is open and free-wheeling. Many persons (11 in City E, 17 in City B, 12 in City C, etc.) who are familiar with the locational decision and/or its consequences are asked to contribute their knowledge and opinions. The respondents are psychiatric professionals, laymen and politicians. There is no set interview format: the research follows whatever leads become evident.

Reasoned thought and the sorting of similar findings has led to the development of concepts such as those presented in this paper. Collection of more case studies will facilitate refinement of the conceptual framework. This roughly follows the method proposed by Glaser and Strauss (1968).

External Forces Affecting Locational Decisions

Five external forces in locational decisions are defined and illustrated. Interview data and newspaper clippings are used (all names are changed, but essential facts are presented). This section concludes by tracing linkages between the external forces and the consequences of the locational solution for the service unit itself and the consequences for the total area environment.

Legal Restrictions

Almost all locational decision-making encounters legal restrictions such as zoning regulations. In the public sector of the economy, the legal system plays a more direct role in shaping the processes of planning Legal restrictions may impose a planning pattern which differs from the ideal "rational" model. In Massachusetts, fo example, money may not be provided for architectural planning until land has been purchased; also, planning of programs and staffing is not supported until after the building is completed. These restrictions run counter to the model of well-articulated objectives formulated before action is begun. The alteration of stages by legal restrictions limits the number of alternative locational solutions by forbidding certain patterns. As an exar ple, in Massachusetts, funds are available only for separate Mental Health Center buildings owned by the state and not for facilities coupled with other institutions.

The mental health activists in City B have been disturbed by the legal restrictions on the order of plannin activities. The director of the outpatient clinic says that the state has never allowed any program planning in conjunction with architectural design. Dr. W, director of children's services, elaborates:

The State does things in the opposite way. We have not talked about the program because they won't let us until the building is there. Nothing has been said about money for the buildings. It is going to be one million or is it going to be six million? We don't really know.

A member of the Mental Health Association mentions that an architect cannot be hired until the land has been purchased. Mrs. B, a leading proponent of the lical mental health plans, complains about the delays in program development and planning which result from these legal restrictions. As a result, several staff mem bers are very frustrated and some may leave the Center.

In City C, legal restrictions seem to have affected the communications between the state-appointed architect and his client, the City C Mental Health Centr. The architect, Mr. L, says:

It has been a funny process. My primary client is the Bureau of Building Construction of the State of Massachusetts and the Department of Mental Health is my secondary client. We knew that the City C Mental Health Center was soing something and has a good program, but we never saw the City C people.

D also encountered the problem that legal restrictions precluded preliminary program planning and archi tectural planning. D had two advantages over the smaller Mental Health Center in B City. Affiliation with a University Department of Psychiatry provided a reservoir of talent for research and program planning, and the Planning Office of the Medical Center had competence in architectural design. D was able to circumvent the State legal restrictions by obtaining federal grants for research and planning purposes. A 5-year NIMH staffing grant allowed provision of new services in the community and evaluation of them. The Planning Office received a grant to study ways of coupling the new Community Mental Health Center system with the existing General Hospital system. Another legal restriction came to light when D considered combining the Mental Health Center facilities in the new Medical Center complex. It was discovered that the State will only build separate buildings on state-owned land. Lawyers are now working on this technicality in an effort to remove the restriction.

Legal restrictions impinge on the planning by requiring land-taking or construction before well articulated objectives are developed, or legal restrictions may limit the number of alternatives considered in the choice of a locational solution. Legal restrictions and regulations may change over time. When change occurs an entirely new situation is created. An example of this is given by Marzot (1969) in her discussion of recent changes in the nursing home field.

Fortuitous Opportunities

Expedient locational decisions may be made, thereby speeding and easing the process of site-selection. We see expediency as an external force as it depends on the existence of fortuitous opportunities which are provided from outside sources. Expediency serves to limit the number of alternative locational solutions which are considered. In this study we find three types of fortuitous opportunities which serve as external forces. Money may be offered for the construction of a Community Mental Health Center. Land may be donated or offered as the site for a Center. Long delays in finding a site may encourage the mental health activists to accept a "path of least resistance", a site which will not be opposed.

About five years ago, Dr. S, former State Commissioner of Mental Health, wanted to establish several model Community Mental Health Centers in Massachusetts. Dr. S was able to obtain the money for construction and searched for sites which could be secured quickly and cheaply. The architect of the prototype building which was constructed in A, explains:

A just happened to be the first site after they decided that the study was effective and that the building could serve a good purpose. That A has turned out to be the Commonwealth's first new mental health center appears to be as much the result of coincidence as it does the result of anyone's design. (Spivack, 1966).

At this time (about 1961) Dr. S also approached the Mental Health Association in Red River. A City got the facility because its General Hospital and Mental Health Association moved more quickly to make land available when the financing was offered. The site is on the hospital grounds, isolated from the poverty areas and from downtown A. Two years later Dr. S approached several more communities with an offer of $1.6 million as bait. City E is among these. Mrs. J, executive secretary of the E area Mental Health center, remembers, "There was talk at that time of 6 or 7 Mental Health Centers in the State." the E City Sentinal made it more explicit under the headline, "$1.6 MILLION FACILITY PLANS—HEALTH CENTER POSSIBLE HERE:"

Plans for the possible construction of a $1.6 million mental health center by the State Department of Mental Health on the grounds of Fernbank Hospital were disclosed by state officials Friday.

State officials have broached the subject to the trustees and superintendent of Fernbank Hospital and the idea has been met with enthusiasm, it was stated.

This fortuitous opportunity for City E spurred a flurry of activity. Local mental health activists cooperated with Dr. S on a public hearing which was held at the Fernbank Hospital. The hearing was designed to muster support and to urge donation of a plot of land by the hospital trustees. The attempt failed.

In A, land was donated to the State Department of Mental Health by the A General Hospital. The administrator of the hospital, with the support of the local Mental Health Association, persuaded the Board of Trustees to donate a 3.5 acre plot of land. This speeded up the process of land-acquisition.

In C City, a plot of land was donated by the Metropolitan Park Authority. Mrs. H, a leading member of the C City Mental Health Association, tells how the donation of park land came about:

> We submitted a bill to the State Legislature for transfer of the land from the MPA, and it looked like the bill would be defeated. Then we found out that the Commissioner of the MPA could transfer the land to us personally, and he did so.

Taking the "path of least resistance" is a third type of expediency stemming from fortuitous opportunities. The mental health activists in C went through three attempts to get a site near the City Hospital, all of which were blocked by neighborhood protests. Finally, it was decided that the City would assume responsibility for finding a suitable site. The Mayor appointed a site-selection committee which included several interested citizens, the director of the C City YMCA, and a local architect. The committee chose a site 2 miles from the downtown area, an 8 acre tract near the shipyard which is now an automobile junkyard. The mental health advocates agreed to this plan as a means of overcoming or avoiding opposition. The assistant director of the Center said that "whether the community would allow the site" was crucial.

A member of the C City Mental Health Association Board stresses the political expediency and compromise in the committee's choice.

> The final site is not most desirable, but then you have politics. It is within 2 miles of a general hospital. Politicians are impressed not of what's good for the community, but who's making the most noise. At the time of the hearings we had the most signatures. Politicians would like them (Mental Health Centers) out in the woods somewhere. Everyone bought this place as a compromise. Most people wanted it out there for their own comfort.

Mrs. H is less bitter about the expedient decision. She believes that because land is scarce in Quarry no better choice could have been made. One of the local State Representatives believes that the acceptance of a fortuitous opportunity resulted in an inconvenient location which is too far from the general hospital to facilitate cooperative treatment of patients. A local lawyer, and member of the site-selection committee, sums up the argument in favor of expediency:

> Even if the South Avenue site weren't as good at least they had it, and there would be no opposition. At least the Mental Health Center will be able to serve their clients down there. Regardless of their program that site won't be too bad. There's enough space and they won't be bothered by the neighbors. I understand that the Mental Health people have fallen in love with the site.

Regardless of their opinions, each person agreees that the South Avenue site was chosen as a "path of least resistance."

Fortuitous opportunities tend to limit the number of alternatives considered. In A there was no real choice. Only one site was even mentioned as a possibility. Thus the likelihood is slim that original objectives will be met.

Ideologies

The objectives of planning may be more or less ideological positions rather than proven program goals. W use Mannheim's (1952) definition of ideology as our starting point:

> *The particular conception of ideology is implied when the term denotes that we are sceptical of the ideas and representations advanced by our opponent. They are regarded as more or less conscious disguises of the real nature of the situation, the true recognition of which would not be in accord with his interests. These distortions range all the way from conscious lies to half-conscious and unwitting disguises; from calculated attempts to dupe others to self deception.*

The treatment ideologies in the mental health field are not "conscious lies" or "calculated attempts to dupe others." Instead they are honest, but unproven, attempts to find a "best" way of treating mental health problems. A fuller treatment of ideological determinants in mental health planning is given by New (1969). At this time the mental health treatment ideoligies fall on a continuum between the traditional psychotherapy model and the community mental health model. The importance of these ideologies is proven by the fact that Baker and Schulberg (1968) are offering a "COMMUNITY MENTAL HEALTH IDEOLOGY SCALE" for sale. They describe the scale as follows:

> *A standardized, 38-item questionnaire for measuring the opinions of professionals about innovative practices in mental health, including manual reporting validity, reliability and item analysis. Scale discriminates between groups highly oriented to community mental health ideology and random samples of mental health professionals.*

Ideologies may be substituted for well-articualted program objectives. When this is the case few alternative locational solutions are considered. The dominant treatment ideology takes precedence in the decision-making process.

We have found ideologies at both ends of the continuum, and several variations in-between. The community mental health treatment ideology is expressed most strongly in Center C. A member of the Mental Health Association describes the present program:

> *We stress community-oriented mental health, developing programs which will go out to the people. We have mental health consultation in the schools (every school has one day a week of consultation). You have to find out what the community needs are and start from where they are, no matter how ridiculous it may seem. We have tried it the other way. We also consult with the Police Departments, the Courts, other agencies, and ministers.*

The head Psychologist in the C City Mental Health Center emphasizes prevention as the basic premise of the community mental health ideology. She believes that early diagnosis and treatment in the school systems is the best type of prevention.

The community mental health ideology also sees value in basic educational and human relations efforts. In C City there is a program of teaching behavioral science in the elementary schools; class lessons on fear, learning and reinforcement are included. This treatment ideology uses many different kinds of professionals in community oriented work. The head nurse at the C City Mental Health Center illustrates this:

> *Caplan stresses primary, secondary and tertiary prevention. Direct involvement in the treatment process is stressed. Our psychiatric nurses have dual roles: (1) direct clinical service, and (2) consultation. Nurses are an integral part of community mental health, not functioning separately.*

It is not clear that the proposed location will be appropriate for the community mental health ideology.

I will discuss the opinions about this in the section on consequences below. In E City the community mental health ideology is present to a lesser degree. The acting director of the E Children's Clinic, says, "thirty percent of the clinic's time and effort is spent in school consultation."

In addition to consultation, the E City clinic uses family-oriented therapy and tutoring programs for disturbed children. These services are delivered in the Clinic at Fernbank Hospital. Mrs. J, Executive secretary of the Mental Health Association, says:

The clinic uses Green School students in a tutoring program. Tom Ballou runs this and it has been quite successful. The 1970 budget has a request included for a supervisor for the tutoring program. The director thinks that this tutoring plan is a very good buy for the money; it gives the kids close contact with a responsible older person.

Usually the clinic deals mostly with the parents and concentrates on attempting to alleviate the home problems which led to the children's difficulty.

One psychiatrist in B City, Dr. W, subscribes to many aspects of the community mental health ideology. His viewpoint illustrates the relationship between ideology and location:

I want to have crisis intervention and that takes a central location. We could use outside property for long term care. We would certainly try to reach out for the poor. I even thought about having a pizza parlor for delinquents.

The traditional psychotherapy ideology calls only for office space in which one-to-one treatment may be given. Dr. W believes that the traditional approach is too conservative and inefficient, requiring too much time and effort. Dr. W is a minority among the psychiatrists in B City, however. The traditional ideology, in its extreme form, leads to an interesting design solution. The following statement by a layman in B City is revealing although probably an exaggeration:

All I remember is when the plans were drawn up, I think Dr. M (head of the Adult Unit) wanted a huge number of offices, something like 140. Now, that seems like a lot of offices. What could they do with that many? But I don't really remember the total figure.

In D several psychiatrists in the University Department of Psychiatry are strongly committed to the traditional psychotherapy ideology. This leads them to advocate a centralized Mental Health Center in the Medical Center.

Dr. A, a psychiatrist at the Medical Center, expresses one of the aspects of this ideology. He advocates the medical model of care in which the patients must seek help from the professionals.

In B City several people hold an ideology which falls between the two extremes. The "milieu therapy" ideology holds that a pleasant environment is a valuable aid to traditional treatment. Dr. H, Director of the Center, is the main advocate of this ideology. He says, "one basic factor is our idea that we needed enough space for a 'total environment' for living." Mr. E, former president of the Mental Health Association, says that Dr. H especially liked the bucolic setting by the lake. Another member of the Association again gives Dr. H's ideology the central role:

The Hillview property is really ideal. Dr. H wants woods and trees and we have lots of it there. He wants a pleasant rural setting.

The Mental Health Center in B City will include inpatient facilities for the mentally retarded. This adds another facet to the milieu therapy ideology. Mr. H, a member of the Association for Retarded Children discusses the implications for design:

Well, we had talked a lot with the architect. We want an architect that can build us something as human as possible. We want a functional building but not too modern. We want it in a serene setting, with a home-like atmosphere. This is the reason we are thinking about cottage type arrangements for the mentally retarded.

The proposed B City site certainly fits the milieu therapy ideology. There will be 100 acres of hills, woods and lake shore.

In A the treatment ideology has been molded by the former Commissioner and Dr. B, director of the Mental Health Center. The two psychiatrists had worked together for many years at the Boston Psychopathic Hospital. There Dr. S developed a mixture of traditional treatment and day hospital services. This ideology requires a sizeable building and does not have to be extremely accessible. The present A City location is appropriate to the ideology; it is a mile from the city, near the General Hospital.

Veto Patterns

Veto patterns arise whenever one or more of the interest groups involved in a decision-making process see fit to block a given decision. We have adapted the concept from Riesman (1953). From a study of power in the U.S. political system he concludes:

Today we have substituted for that leadership (of business interests) a series of groups, each of which has struggled for and finally attained, a power to stop things conceivably inimical to its interests and, within far narrower limits, to start things. By their very nature, the veto groups exist as defense groups, not as leadership groups.

The same sort of veto group occurs in the mental health field. We also find groups which exercise veto power only at a particular point in time; otherwise they are either quiescent or contribute to positive planning. This veto pattern blocks the decision-making process at least temporarily. If locational decisions are to be made, the process must be resumed at a stage prior to that at which the veto occurred.

In C City, the mental health planners were confronted with many vetos, from many quarters. Three sites were considered near the C City Hospital in a middle class residential area. Each time the same neighborhood association was an effective veto group. A local State Representative says:

It's the same old story of build it somewhere else and not in my neighborhood. I was disturbed by the reaction of the people in the MPA Parkway area. It was sidetracked because of the political pressure they put on; you have to listen to the voice of the people and 40 or 50 adamant people are hard to ignore. I haven't been swayed by this kind of reaction as much as some others.

The Mayor remembers that he went out on a limb in favor of a site near the hospital. Mr. M, head of the neighborhood association, led the opposition. Failure was apparent, so the Mayor substituted a bill for a study committee. Mr. C, lawyer for the Neighborhood Association, gives another slant on the story:

The Neighborhood Associates case (blocking building of the Mental Health Centr on the White St. site) was the easiest case I've ever had. Usually I'm fighting 200 neighbors, but then I had 200 screaming people on my side. No. 1, we had the backing of the people and therefore the politicians listened because they have to take account of where the votes are. No. 2, the mental health people broutht in experts from Metropolis to speak for them and no politicians cared about their opinion

The power of the Neighborhood Association veto group must not be underestimated. Mr. L, the architect, points out that the state had every right to build on the first site because it was not under local zone ordinances. The state did not wish to damage its relations with the city of C and backed out. The C City Republican carried a headline, "482 OPPOSE LOCATION OF MENTAL HEALTH CENTER." The opposition view was summed up by a Mr. M: "I think they should get out in the country. Transportation is very easy today. Everyone owns a car."

Mr. C believes that the veto group pattern is endemic in C City:

You have a power structure, unfortunately. If a power structure is good, okay, but here it accomplishes nothing and prevents all change. The politicians have got to go where the votes are. You can't even turn around in C City without some association fighting you. Preserving the status quo is important to many people here.

The mental health advocates in C City were also blocked by vetoes from two other sources—the State Department of Mental Health, and private developers who owned land which was chosen as a possible site. Mrs. H remembers that the Commissioner denied priority to C City in 1964, causing the appropriation bill to fail in the legislature. In 1966, the report of the Massachusetts Mental Health Planning Project placed C City 34th out of 37 on the priority list. Mrs. H believes that these were key factors in the delayed planning for a new mental health center. The final site selection committee preferred a piece of land which was held by a powerful local developer who is on the Mental Health Association Board. Mr. L says:

The Mayor finally appointed a committee to choose a site for the mental health center and they found the best site; at first they considered the top of the hill out there on South Avenue but had trouble with private developers (underlining mine) and now have got the dump an adjacent parcel for $250,000.

Mrs. H says that P refused to sell because he wanted a top price for his land. Mr. C believes that P was motivated by a different economic consideration. He says, "The P Trust (real estate firm) will find it economical to subdivide their property down there after the sewage, water, and electricity have been brought in for the Mental Health Center." In B City vetoes came from two different sources: The Mental Health Association itself, and the State government. One veto affected the choice of an architect while the other delayed acquisition of land. Mr. A, retired medical reporter for the B City Journal who wrote many articles on the incidents, says:

The Mental Health Association affairs were brought to my attention when the mental health board called a meeting about the double cross. They had been told that they would have a voice in the choice of an architect and had had a committee preparing the specifications for a collection of small buildings in an open site. The local group was dismayed when the state Department of Mental Helath announced having selected an architect. The Attorney General gave a talk here saying that local voices would be heard and during the talk that night Mr. E was called to the phone and told that B (an architect in Metropolis) had been chosen. Bob then blew the bomb up in the Attorney General's face, and he then got right on the telephone to Metropolis. You don't do it this way. It's very embarrassing to public officials.

Mr. B, the architect who was appointed by the State, admits that he had received the job because of contributions to political campaigns. He withdrew when the Commissioner of Finance and the Governor were in a tough spot because of the protests from the B mental health activists. So the veto by the Mental Health Association was successful. Or was it? The architect favored by the local people has been appointed since, but many feel that the State has been exercising vetoes on progress as a result of the episode. Mr. A says, "Getting the money (for construction) and buying the land has been crippled, I think, by the original argument over the architect." A local State Representative from B has a theory that, "someone in the State House, some little clerk, may have it in for them because of this architect's deal and...every time that piece of paper comes up (he) just whisks it away and puts it somewhere else."

In City E vetoes were exercised by the local medical interests and by the State Department of Mental Health. Mrs. J comments on both of these actions:

The Department of Mental Health tried to persuade the Fernbank Hospital trustees to give land for a new center. The M.D.'s there had indignation meetings. They thought the center was for their referral; they feared that then the Department of Mental Health would have jurisdiction, and patients would be out of their hands. They don't consider cooperating with psychiatrists as consultants.

Soon a nasty letter came from Dr. S to the Fernbank Hospital saying that there wasn't enough civic interest in E City: therefore the issue was being dropped. Later Dr. B and I cornered Dr. S and he repeated his statement that no one was pushing very hard in E City so it wasn't worth the effort of the Department of Mental Health.

It goes almost without saying that vetoes stem from the values and vested interests of the groups which use them. Vetoes may represent concern for a "greater good" as in the priority scheme of the Department of Mental Health, or as in the architect veto in B City. More often, however, vetoes seem to halt healthy progress in locational decision-making. Whenever vetoes occur site-selection is halted at least temporarily. In E City, there has been no planning for a new facility since the last veto four years ago.

Inability to Identify and Measure Relevant Variables

Demone (1965) stresses the difficulty of measuring important variables in mental health planning. He asks the following pointed question:

The quality of information must be considered. Remember the variables. Every good planning operation, every rational planning operation, should have data. How much hard data is available? What do we really know about key variables?

Objective information is usually lacking for locational decisions in the mental health field. "Need" and "demand" for mental health services are hard to determine. Measure of the current use of existing facilities is like looking at the top of an iceberg. "Felt need" studies (in which people are asked what mental health problems are prevalent in the area) have also been shown to have important weaknesses. Most activists feel that transportation (accessibility) for patients is an important factor for the location of mental health facilities. There is no agreement, however, on how the accessibility issue should be measured or solved. Access to related health services such as general hospitals may be important; yet here again the professionals do not agree. The inability to identify and measure relevant variables severely weakens the attempt to choose a "best" alternative. It may be as difficult as groping in the dark.

The easiest index of the inability to identify and measure relevant variables is the extent to which this is even attempted in the process of locational decision-making. Only one of our cases reveals a strong effort to evaluate relevant factors. In one case measurement seems to have come after the decision. In two cases only proximity to a general hospital was considered, and in the fourth case this step was completely lacking.

In A and B City access to general hospitals was stressed. Spivack (1966) tells the A story:

Dr. S wished the mental health center to be proximate to the hospital. They (local mental health activists) appeared to accept this wish as reasonable and not requiring discussion or debate.

In B City Dr. H stresses that proximity to the Hillview Hospital will allow the Mental Health Center to purchase medical services for its patients. Transportation is currently very poor to the proposed site, but it may improve. A member of the Mental Health Association in B says:

> With the community college moving out that way and with the high school out that way, there may be more bus services. Right now buses go out to that area twice a day. We'll see what will happen with the bus company.

Mrs. O, a social worker, believes that inaccessibility to the city of B will be a serious problem. She says that access to distant parts of the county has been favored at the expense of poor people in the city who won't be able to get out.

The mayor's site selection committee in C City released a study of accessibility with its choice of the shipyard land. Mrs. H says:

> A study was made to determine the "center of population" of the area. There was also a study of transportation patterns, but I am not sure how many ride the buses. The new site is perfect from that point of view; there are buses up from Highland and out from C City. I would guess that most people have their own transportation, but there are several people in the after-care program who do need transportation.

Dr. K is not at all pleased with accessibility to the proposed site; he stresses the distance (about two miles) from the city and the hospital.

In D financing from a federal grant allowed the University Department of Psychiatry (in cooperation with the Planning Office of the Medical Center) to do extensive preliminary studies. A "felt need" survey in 1966 showed delinquency and alcoholism to be the most important community problems. The same survey found that only 11% of the respondents were aware of the existing mental health clinic in the medical center. Another study investigated the use of mental health services by area residents. The community had a very low rate of outpatient treatment, but at the same time it had the State's highest rate of State Hospital admissions: this shows a reluctance to seek assistance until the trouble has become serious. Another study was done of transportation patterns; transportation is difficult from most parts of the area to the Medical Center. Although these data are an important resource there is not agreement on the best locational solution for the D Mental Health Center. Some people question the validity of the data, and many are uncertain of their proper interpretation.

Consequences of Locational Solutions:

We identify two classes of consequences: those at the micro- and macro-levels. Consequences for an individual service unit (micro) can be measured in terms of how programs function in a given location. Consequences for the total area environment (macro) rely on the impact of the advent of a new facility.

Only A has a Comprehensive Mental Health Center whose locational consequences for programs (micro) can be measured with some certainty. Dr. B defends proximity to the hospital, but admits that inaccessibility to poverty areas is a serious problem. Miss F, a social worker at the Center, says:

> I would like to take this whole building and put it down in the middle of A so that more people could get to us.

In C City and B City sites for new Mental Health Centers have been chosen. Local people are voicing opinions on the probable consequences for the service units. In B City Dr. W and Mrs. O believe that the rural site will weaken service to poverty areas. Others, notably Dr. H, minimize the seriousness of this factor. In C City Dr. K and Representative B are disappointed about the site. Rep. B forsees undesirable consequences:

> *I shouldn't think that the South Avenue site was as good. It certainly is not as convenient for many reasons. It is not convenient for those who are to make use of the facilities as patients. There is less public transportation out there. There are some busses but it is really an out-of-the-way area. I think that its distance from the hospital is also a problem because some patients may have both mental and physical ailments and need both kinds of treatment.*

The present outpatient clinic at D demonstrates a consequence of location being relatively inaccessible to the community. Only 16% of the present patients at the Medical Center location are from the catchment area; the others are drawn from the entire metropolitan region.

Consequences on the macro-level are mentioned by people in C City and B City when they talk about the proposed sites. In both places the Mental Health Center is seen as part of a more general improvement of the neighborhood. Previously we have seen reference to the prospects for a new high school and community college near the site in B City. The Mayor in C City cites similar expected environmental consequences:

> *Geographically it is in the population center of the district and not prime industrial land. Therefore we are not loosing much taxable property. It would also be rehabilitation from the neighborhood point of view. That area is going up; there is a senior citizen's home and a nursing center nearby on the main artery. Also, the new community college is planned to go into the dump on the other side of South Avenue from the Mental Health Center site.*

There is no direct relationship between specific consequences and the presence of external forces in the process of locational decision-making. We can only make this statement: When external forces enter into locational decisions to a greater extent, there is greater probability that the consequences will not meet the original objectives.

Analysis

The data of this study allow us to develop a planning model which incorporates the external force model with a simplified version of the rational planning model. The line between the two is dotted to indicate that the models are not mutually exclusive. Elements of each model may be mixed in any combination in reality. The bottom half of the figure represents the extreme form of the external force model.

Ideological positions may replace well-articulated objectives. Choices are then determined by the implications of the ideology. For instance, if the milieu therapy ideology is held, the logical conclusion is an extremely beautiful and isolated site. Ideological goals limit the range of possible choices. Where milieu therapy is the dominant ideology downtown sites are not seriously considered.

Fortuitous opportunities and legal restrictions may limit the number of alternatives considered. When a fortuitous opportunity occurs early in the decision-making process its acceptance will preclude any other alternatives. Legal restrictions eliminate whole classes of possible locational solutions.

Veto patterns may enter into the decision-making process after either the "goal setting" or "consideration of alternatives" stages. Vetoes impose at least temporary delays in planning. The process may resume at an earlier point. Vetoes occur most frequently during the consideration of alternatives; in C City three alternatives were opposed and defeated by a Neighborhood Association.

The inability to identify and measure relevant variables weakens the objectivity of the choice. If there are no criteria for discriminating among alternatives, the choice will be essentially a matter of chance.

Conclusion

Knowledge of the external force model allows the planner to anticipate and account for a number of important factors. Awareness of the external forces introduces a degree of objective detachment from the decision-making process. We hope that this perspective will enable the planner to: 1) recognize and carefully evaluate ideological objectives, 2) consider a large number of alternatives in spite of expediency and legal restrictions, 3) anticipate and overcome dysfunctional veto patterns, and 4) find ways of improving the seeming inability to identify and measure relevant variables. This would increase the probability that the consequences of locational solutions would meet the original objectives.

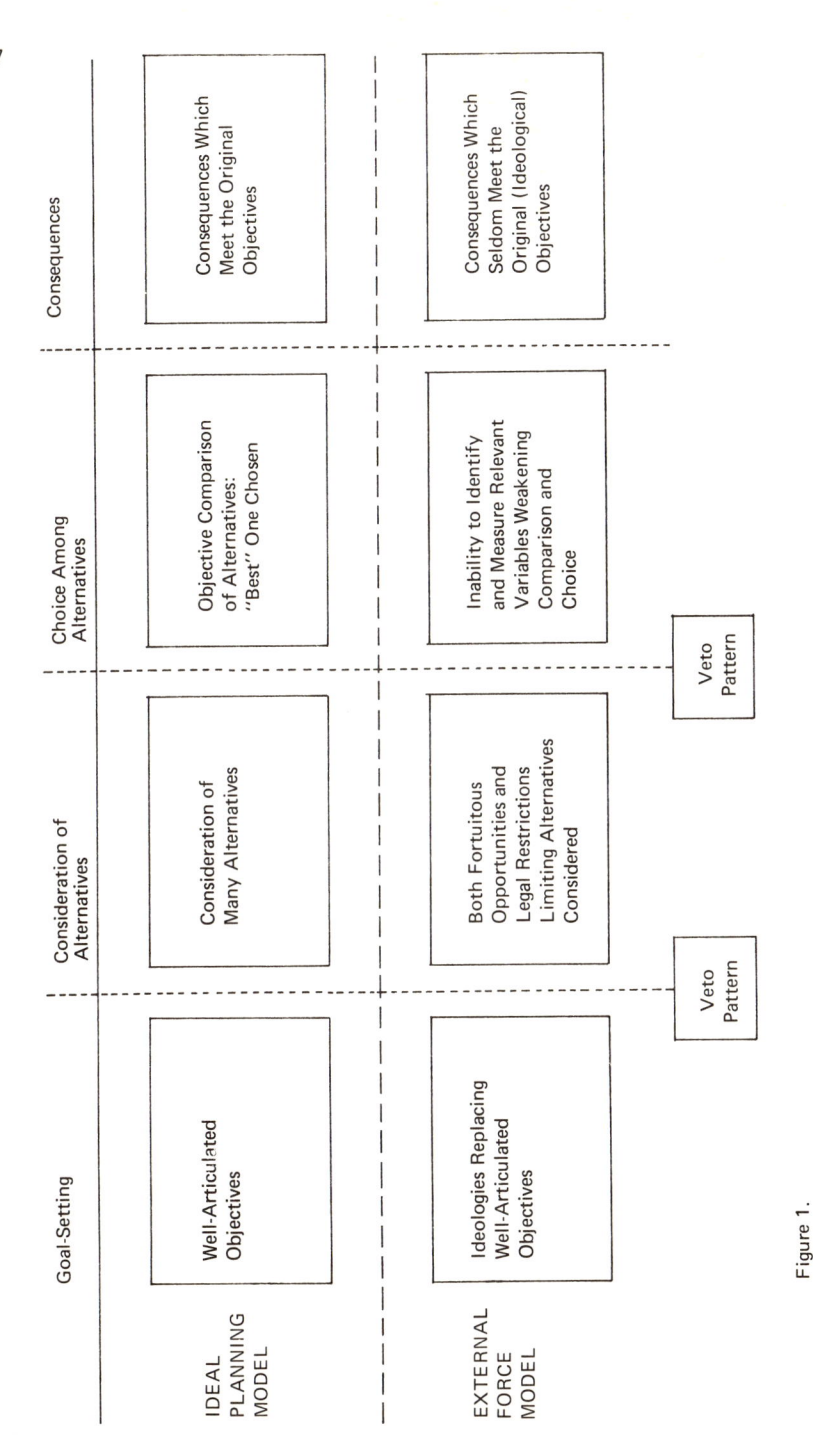

	Goal-Setting	Consideration of Alternatives	Choice Among Alternatives	Consequences
IDEAL PLANNING MODEL	Well-Articulated Objectives	Consideration of Many Alternatives	Objective Comparison of Alternatives: "Best" One Chosen	Consequences Which Meet the Original Objectives
EXTERNAL FORCE MODEL	Ideologies Replacing Well-Articulated Objectives	Both Fortuitous Opportunities and Legal Restrictions Limiting Alternatives Considered	Inability to Identify and Measure Relevant Variables Weakening Comparison and Choice	Consequences Which Seldom Meet the Original (Ideological) Objectives

Veto Pattern (between Goal-Setting and Consideration of Alternatives)

Veto Pattern (between Consideration of Alternatives and Choice Among Alternatives)

Figure 1.

*Legal Restrictions (of the type which alter the order of planning) affect the planning process by changing the flow between the stages of goal-setting, alternatives, and choice.

References

1. Altschuler, Alan A.: *The City Planning Process: A Political Analysis,* Cornell University Press, Ithaca, 1965.

2. Baker, Frank, and Herbert C. Schulberg: *Community Mental Health Ideology Scale,* Behavioral Publication, New York, 1968.

3. Banfield, E. C.: "Ends and Means in Planning", in S. Malick and E. H. Van Ness (editors), *Concepts and Issues in Administrative Behavior,* Englewood Cliffs, N. J., Prentice-Hall, 1962, pp. 70-80.

4. Brand, Jeanne L.: "The United States: A Historical Perspective," in *Community Mental Health: An International Perspective,* edited by Richard H. Williams, Jossey-Bass Inc., San Francisco, 1968, pp. 18-43.

5. Braybrooke, David, and Charles E. Lindblom: *A Strategy of Decision: Policy Evaluation as a Social Process,* Glencoe Free Press, 1963.

6. Demone, Harold W.: "The Limits of Rationality in Planning", *Community Mental Health Journal,* Volume 1, Number 4, Winter, 1965, pp. 378-381.

7. Deutch, Albert: *The Mentally Ill in America,* Doubleday, Doran and Company, Inc., Garden City, New York, 1938.

8. Duncan, Delbert J., and Charles F. Phillips: *Retailing: Principles and Methods,* 1967, Richard D. Irwin, Homewood, Ill., (7th edition).

9. Glaser, Barney G., and Anselm L. Strauss: *The Discovery of Grounded Theory: Strategies for Qualitative Research,* Aldine Publishing Company, Chicago, 1968.

10. Goodman, William I., and Eric C. Freund (editors), *Principles and Practice in Urban Planning,* International City Manager's Association (Municipal Management Series), Washington, D. C., 1968.

11. Holton, Wilfred E., and Bernard M. Kramer: "The Community, the Hospital, and the Mental Health Center Design Implications of Site-Selection," background paper prepared for the Workshop on Environmental Design in Community Planning, Boston Architectural Center, Boston, Mass., May 9, 1969.

12. Lindblom, Charles E.: "The Science of Muddling Through", in William J. Gore and J. W. Dyson (editors), *The Making of Decisions,* Glencoe Free Press, 1964.

13. Management and Economics Research, Inc. (for the Office of Regional Development Planning), *Industrial Location as a Factor in Regional Economic Development,* U. S. Department of Commerce, 1967.

14. Mannheim, Karl: *Ideology and Utopia: An Introduction to the Sociology of Knowledge,* Harcourt, Brace and Co., New York, 1952.

15. Marzot, M. Grazia, and Peter Kong-ming New: "Nursing Homes and Their Administrators: A Study of Their Changing Position in the Community," paper to be presented at the XXII Congress of the Institute Internationale de Sociologie, Rome Italy, September 15-21, 1969.

16. Mayer, Kurt B., and Sidney Goldstein: *The First Two Years: Problems of Small Firm Growth and Survival,* Small Business Administration, Washington, D. C., 1961.

17 New, Peter Kong-ming, Bernard M. Kramer, Wilfred E. Holton, and Grazia Marzot: "Ideological Determinants in the Planning of Mental Health Centers", paper presented at the Society for Applied Anthropology Meetings, Mexico City, April 11, 1969.

18 Riesman, David, with Nathan Glazer and Reuel Denney: *The Lonely Crowd,* Doubleday, Garden City, New York, 1953.

19 Spivack, Mayer: unpublished materials on an NMIH-sponsored study of planning for mental health centers in Massachusetts (1966).

Edward J. Kaiser
Department of City and Regional Planning and
the Center for Urban and Regional Studies
University of North Carolina at Chapel Hill

A DECISION AGENT MODELING APPROACH TO
PLANNING FOR URBAN RESIDENTIAL GROWTH

The center for Urban and Regional Studies for some time has been developing decision agent models to aid in explaining and planning for urban growth.[1] I would like to describe one such model and although much of the work has been of a basic research nature, I will attempt to discuss it in a planning and design modeling context.

First, it should be made clear that this particular decision agent modeling approach fills only one of the several roles required of models in planning for urban residential growth. Its purpose is to help the planner understand and anticipate the impact of local public policies as well as other non-controllable factors on residential location. It does not optimize the distribution of residences or even evaluate the estimated residential distribution which is the model output. It does not help the planner estimate the feasibility of implementing any of the public policies assumed as imput into the model, by including, for example, the public governmental and bureaucratic decision making processes. Although all of these other roles, briefly diagrammed in Figure 1 are required in planning rationally for residential growth, the modeling approach I will describe attempts to fill only one of these roles—the urban growth and change modeling role. (See figure 1).

Our initial research efforts focusing on the decisions agents were "behavioral" in their approach. By behavioral, I mean that we attempted to identify the individual decision agents, their decision processes and the factors influencing those decisions, especially their locational decisions. The initial empirical work was a series of interviews with various decision agents, mostly in the cities of the North Carolina Piedmont. These interviews were conducted with predevelopment landowners, developers, households (white and non-white, buyers and renters), financial intermediaries, realtors, attorneys, and public officials. Essentially we probed the decision behavior by asking how decisions were reached and what factors influenced the decision. Most of these were tape recorded in-depth interviews followed by more structured questionnaires. Other interviews were less formal.

A Conceptual Overview

As a result of exploratory "behavioral" investigations and structured questionnaires, we conceptualized the urban residential pattern as the result of a land conversion process broadly referred to as the residential development process. In this process, non-urban land on the urban fringe is converted to urban residential use by going through a number of intermediate states or steps. We visualized the analogy of a chain of states necessary for land to fulfill its conversion to residential land use. We even included, in the analogy, the idea of a chain being no stronger than its weakest link. A change of state in the conversion process requires a corresponding decision or combination of decisions. Hence, we now visualize a chain of decisions corresponding to the chain of intermediate states through which land must pass in order to acquire residential use.

In our judgement there are at least three really key decision agents involved directly in the process—the predevelopment landowner, the developer, and the consumer household. The financial intermediaries and public officials also play important roles, but in a much less direct manner. Figure 2 summarizes our conceptuali-

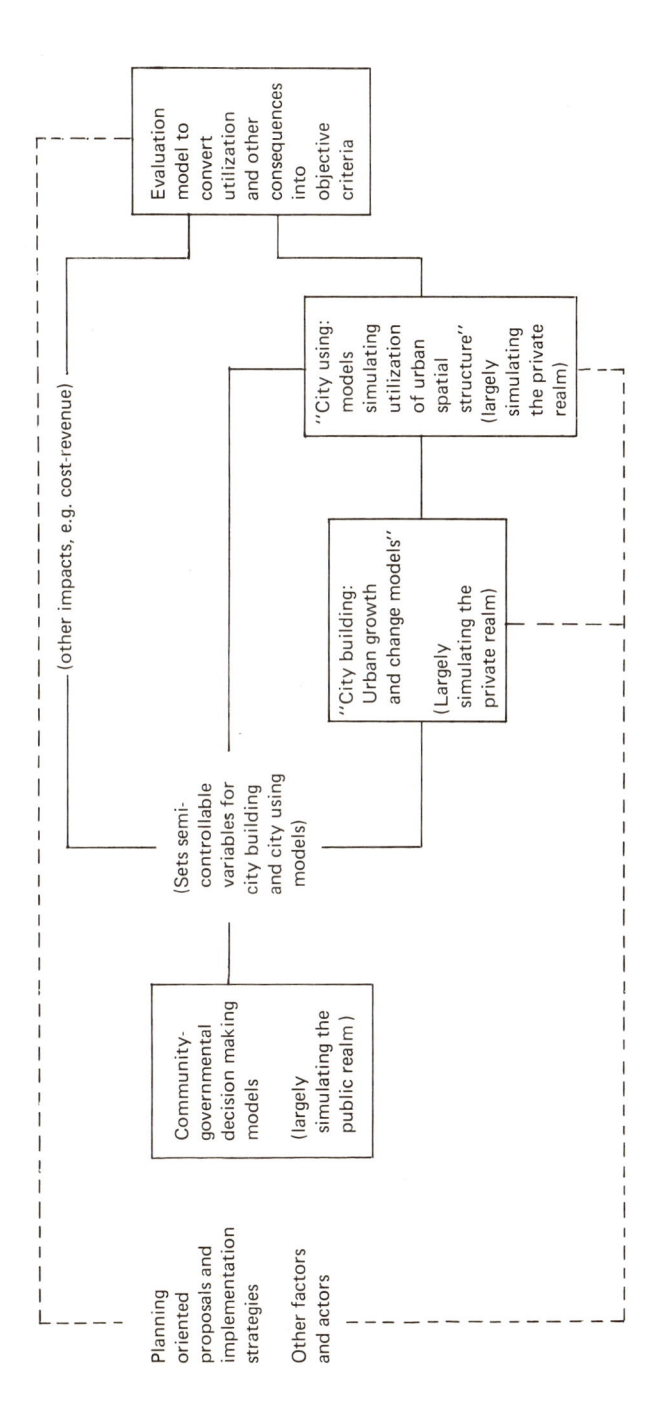

Figure 1. **Planning Models**

SEQUENCE OF STATES	1 URBAN INTEREST	2 ACTIVE CONSIDERATION FOR DEVELOPMENT	3 PROGRAMED FOR DEVELOPMENT	4 ACTIVE DEVELOPMENT	5 RESIDENCE
DESCRIPTION OF STATE	A DECISION AGENT CONSIDERS THE LAND AS HAVING DEVELOPMENT POTENTIAL WITHIN A GIVEN TIME PERIOD	A DECISION AGENT HAS CONTACTED ANOTHER AGENT REGARDING THE POSSIBLE SALE OR PURCHASE OF THE LAND	A DECISION AGENT HAS A DEFINITE IDEA OF THE TIMING AND CHARACTER OF DEVELOPMENT	A DECISON AGENT HAS BEGUN PHYSICAL DEVELOPMENT OF THE LAND	A DECISION AGENT HAS PURCHASED THE RESIDENTIAL PACKAGE OF HOUSE AND LOT
SEQUENCE OF KEY DECISIONS		DECISION TO CONSIDER LAND	DECISION TO PURCHASE LAND	DECISION TO DEVELOP LAND	DECISION TO PURCHASE HOME
DECISION AGENTS KEY:		LANDOWNER DEVELOPER	LANDOWNER DEVELOPER	DEVELOPER	CONSUMER
SUPPORTING:		REALTORS FINANCIERS PUBLIC OFFICIALS	REALTORS FINANCIERS PUBLIC OFFICIALS	REALTORS FINANCIERS PUBLIC OFFICIALS	REALTORS FINANCIERS PUBLIC OFFICIALS

DECISION FACTORS: CONTEXTUAL CHARACTERISTICS — DECISION AGENT CHARACTERISTICS — PROPERTY CHARACTERISTICS → DECISION PROCESS

LOCAL PUBLIC POLICIES: POLICY GUIDES AND IMPLEMENTATION INSTRUMENTS

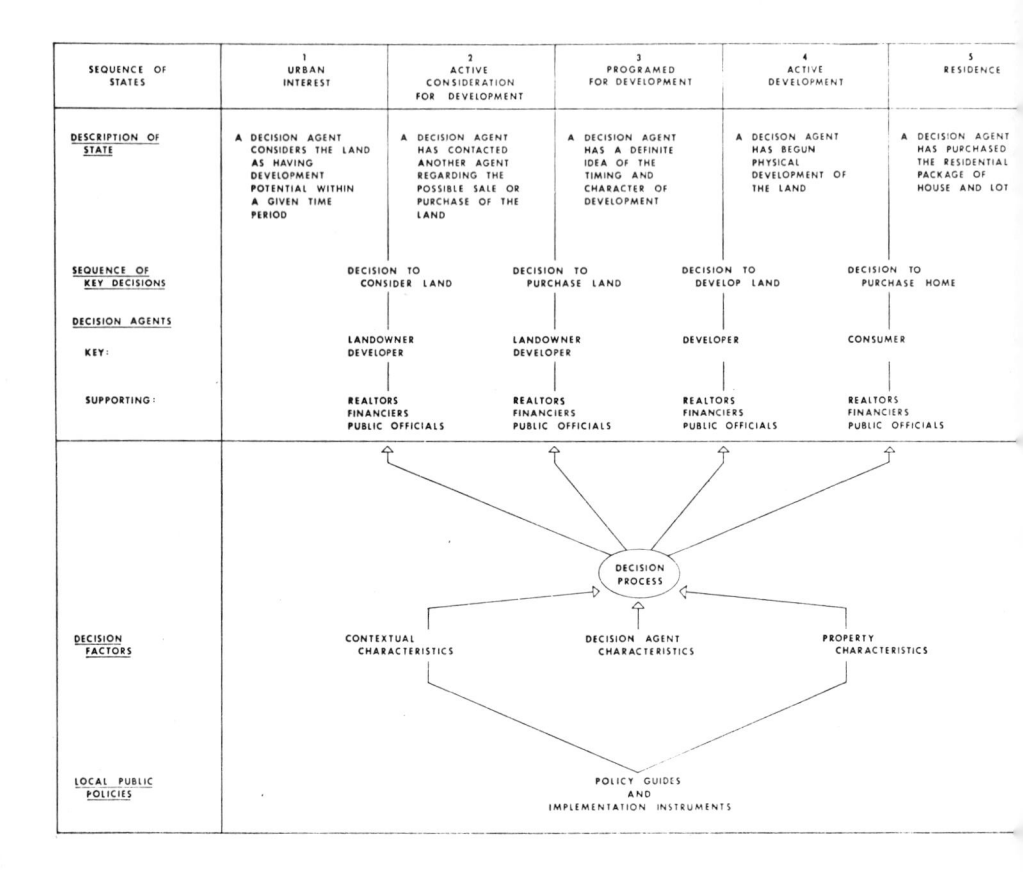

Figure 2. The Residential Land Development Process: Sequence of States, Key Decisions, Decision Factors, and Local Public Policies

(Adapted from Kaiser and Weiss, 1967)

283

Figure 3. **Elements in the Residential Development Process: An Analytic Framework**
(Adapted from Kaiser and Weiss, 1967)

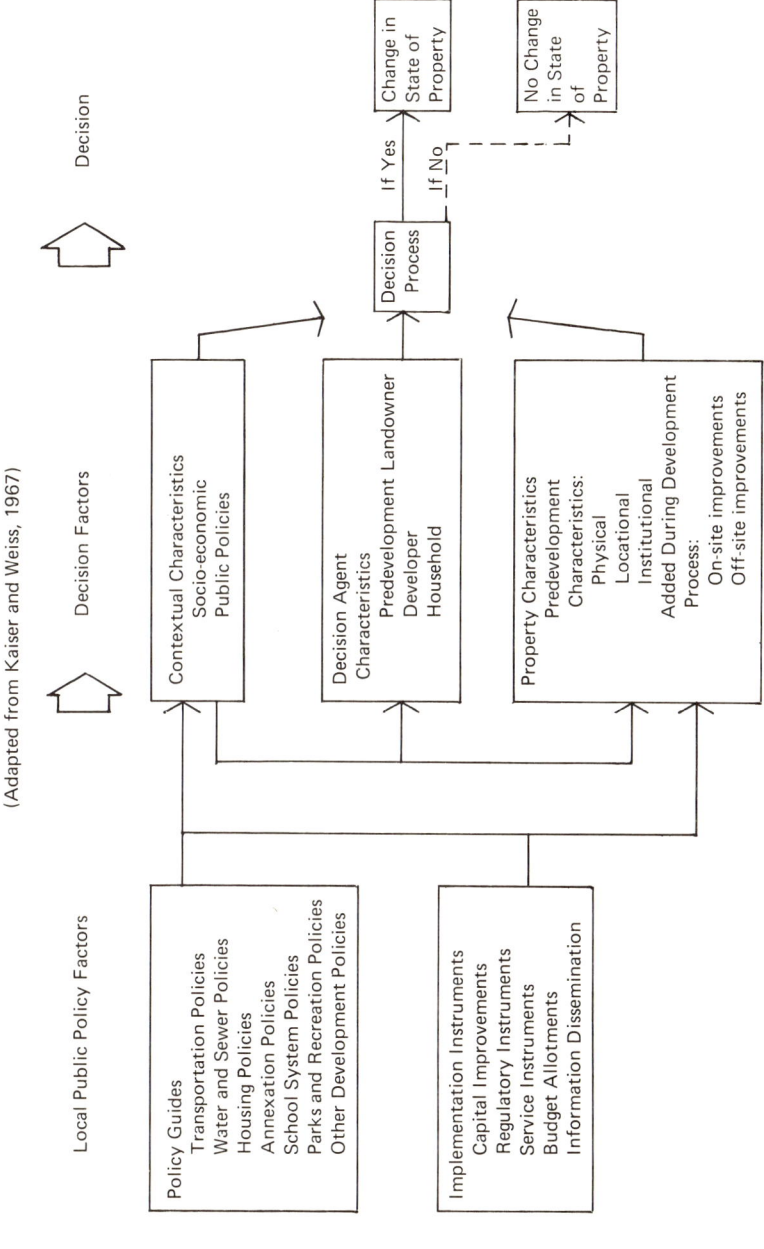

zation to this point—a chain of decisions made by three principal decision makers underlying a chain of intermediate changes of state by which land is converted from non-urban residential use. (See figure 2).

We can proceed one step deeper into the development process by examining Figure 3. In that figure, we try to introduce three types of factors which have consistently emerged from our research as influencing each of the key decisions required through the intermediate transition stages in order for land to change from its existing state to the following state. They are contextual factors, property characteristics, and decision agent characteristics. Each of these three sets of factors influences each of the key decisions in the land conversion process in a unique manner. The contextual factors provide the macro-environment for the decision, namely, the considerations which limit and determine the overall rate and type of change in the community and the general distribution of decision and property characteristics. The property and decision agent characteristics, on the other hand, describe the micro situation surrounding each decision. The property characteristics describe the property about which decisions are made. And the decision agent characteristics help explain the variation in decisional behavior among decision makers of the same type in the face of similar contextual and property characteristics affecting the decision. (See figure 3).

Again referring to Figure 3, we see local public policy as an attempt to influence the residential evolution of land by affecting the contextual and/or property characteristic factors in the key development decisions. Although decision agent characteristics are not directly influenced by public policy, an understanding of the role played by the decision agent in determining the response of decisions to any land use policy proposal is crucial to the design of such policy. You can relate the ideas diagrammed in Figure 3 to those presented in Figure 2 by examining the bottom part of Figure 2.

Linked Decision Agent Models

A possible configuration of decision agent models based on the decisions of the three key decision makers in the conceptualization above is illustrated in Figure 4. A distribution of the supply of new single family subdivisions and vacated housing. These supply estimating models and demand estimation models would be linked together by a residential choice model which is based on the household's residential choice decision and would allocate households to the supply of available housing. The output is visualized as joint distribution inclinations of housing characteristics and household characteristics spatially distributed in the urban area. (See figure 4).

Although this modeling idea is still in the first stage of construction, the results of the work done so far on several separate parts are fairly encouraging. I would like to discuss the developer model as a key example of one of these secision agent models.

The Developer Model Link in the Linked System of Models

The task of the developer-model link is to map the spatial pattern of the likelihood of developers locating single-family residential subdivisions based on certain suppositions about public policy represented by contextural and property characteristics in the model. The model is based on the concept of a developer's site selection being equivalent to the selection of an array of site characteristics as inputs into his residential production process (Weiss, Smith, Kaiser, 1966; Kaiser, 1968a; Kaiser, 1968b). Analysis of subdivision plat data from two North Carolina cities, Greensboro and Winston-Salem, provided the basis for calibrating a pilot version of a developer model based on discriminant functions. We believe this work supports the view that the developer is an important agent in the residential growth process, that property characteristics appear to be a fruitful means to explain his location decision, and that a model based on his locational behavior is feasible.

The findings thus far suggest that such characteristics as socioeconomic prestige level of the location, zoning, and availability of public water and sewerage are the most significant, while the amount of contiguous residential development, accessibility to downtown, employment, and elementary school and the major street

Figure 4. Linking Together Several Decision Agent Models

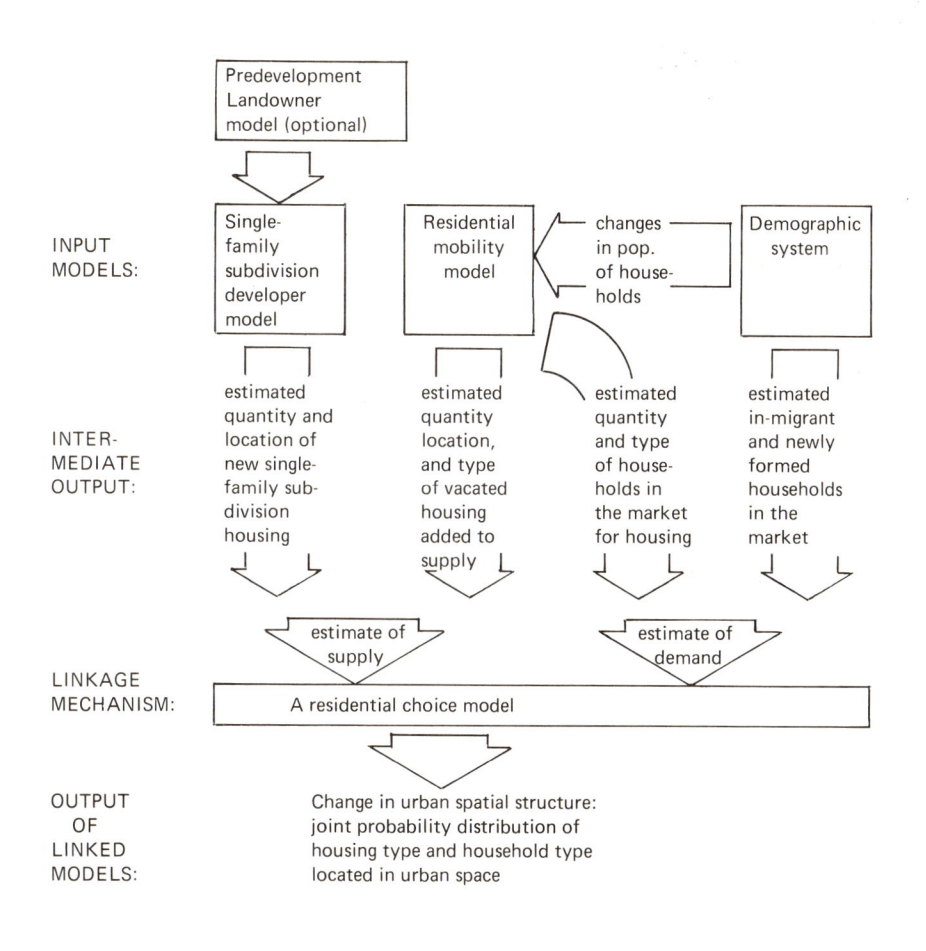

		Predevelopment Landowner model (optional)		
INPUT MODELS:	Single-family subdivision developer model	Residential mobility model	changes in pop. of households	Demographic system
INTER-MEDIATE OUTPUT:	estimated quantity and location of new single-family sub-division housing	estimated quantity location, and type of vacated housing added to supply	estimated quantity and type of house-holds in the market for housing	estimated in-migrant and newly formed households in the market
LINKAGE MECHANISM:	estimate of supply	A residential choice model	estimate of demand	
OUTPUT OF LINKED MODELS:		Change in urban spatial structure: joint probability distribution of housing type and household type located in urban space		

systems are also significantly related to developers' subdivision location decisions and should be utilized in the model. The physical characteristics of the land that were tested—proportion of marginal land due to excessive slope, subjectivity to flooding, or poor soil—do not appear to be important in the developers' locational decisions (Kaiser, 1968a, 27-31).

Our findings also suggest that at least two parallel model components be utilized in the single-family residential subdivision developer model—one would produce the spatial pattern of likelihood of large developers (those averaging over 100 lots per year in our study) locating subdivisions and another would do the same for small developers (under 100 lots per year in our study). While both are influenced by site characteristics, large developers are much more responsive to them than small developers. The pilot version of the model was able to classify up to 81 percent of the large-developer decisions correctly in our sample as opposed to 66 percent of small-developer decisions. (Kaiser, 1968a, 44). Further, the two appear to be looking for very different combinations of site characteristics. For example, small developers tend to choose sites farther from downtown, farther from an elementary school, farther from employment centers, and having fewer public utilities than sites not receiving subdivisions. Large developers tend to select the opposite kind of site (Kaiser, 1968a, 27-31).

The price market aimed for by the developer also influences his selection of property characteristics and hence the location of his subdivisions. Tests relating property characteristics to the price range of houses indicate that higher-priced subdivisions are more sensitive to socioeconomic prestige level of the site than lower-priced subdivisions, while middle and lower-priced subdivisions are more sensitive to zoning, availability of public utilities, and amount of nearby development. The pilot version of the model was able to obtain a better record of classification for higher-priced subdivisions than for lower-priced subdivisions—83 percent versus 67 percent (Kaiser, 1968a, 44). These findings suggest that the model be calibrated separately for each of several price ranges.

There was statistical evidence that differences in location between developer types is greater than differences between price ranges in our studies. Combining price and developer distinctions enabled the discriminant model to achieve as high as 95 percent correctly classified (Kaiser, 1968a, 44). The differences between developer types hold even while controlling for price range of the subdivision. This lends further support to the idea of developing several parallel models—one to estimate larger developers-low price subdivision location, one for large developer-high price, one for small developer-low price, and one for small developer-high price.

It may be useful to add another input variable to the developer model—the likelihood of selling on the part of the landowners for any site being evaluated. Such an input could be generated by a separate predevelopment landowner model with which we have done some work (Kaiser, Massie, Weiss, Smith, 1968). However, in preliminary tests with a small sample, we have not yet been able to improve the performance of the developer model significantly by adding a landowner prefix model which supplies a likelihood of selling as an additional input to accompany the site characteristics described above. Nevertheless, the tests are not considered conclusive and further research effort in this direction is deemed desirable.

Figure 5 illustrates the kind of output map the model would produce when combined with a University of North Carolina version of the Harvard Computer Graphic Laboratory's SYMAP program. This illustration indicates the variation in probability of subdivision occurring based on the property characteristics of sample sites in the urban area. The probability of all the sites not in the sample is calculated by interpolation from probability values of sample sites in the vicinity. Although this particular example is based on the developer model, the principle would be the same for the other decisions involved in the linked system of models—the predevelopment landowner's decision to sell, the household's decision to move, and the household's selection of a place of residence. The variables and parameters in the functions would change, but the principle remains the same and similar probability statements of these other decisions could be made as output. (See figure 5).

This type of model, it should be noted, cannot be interpreted very literally as to predictions. It does not estimate the amount of growth, nor does it allocate acres or other units of growth. It is what could be called a policy impact assessment model. It should be able to trace the impact on property characteristics

287

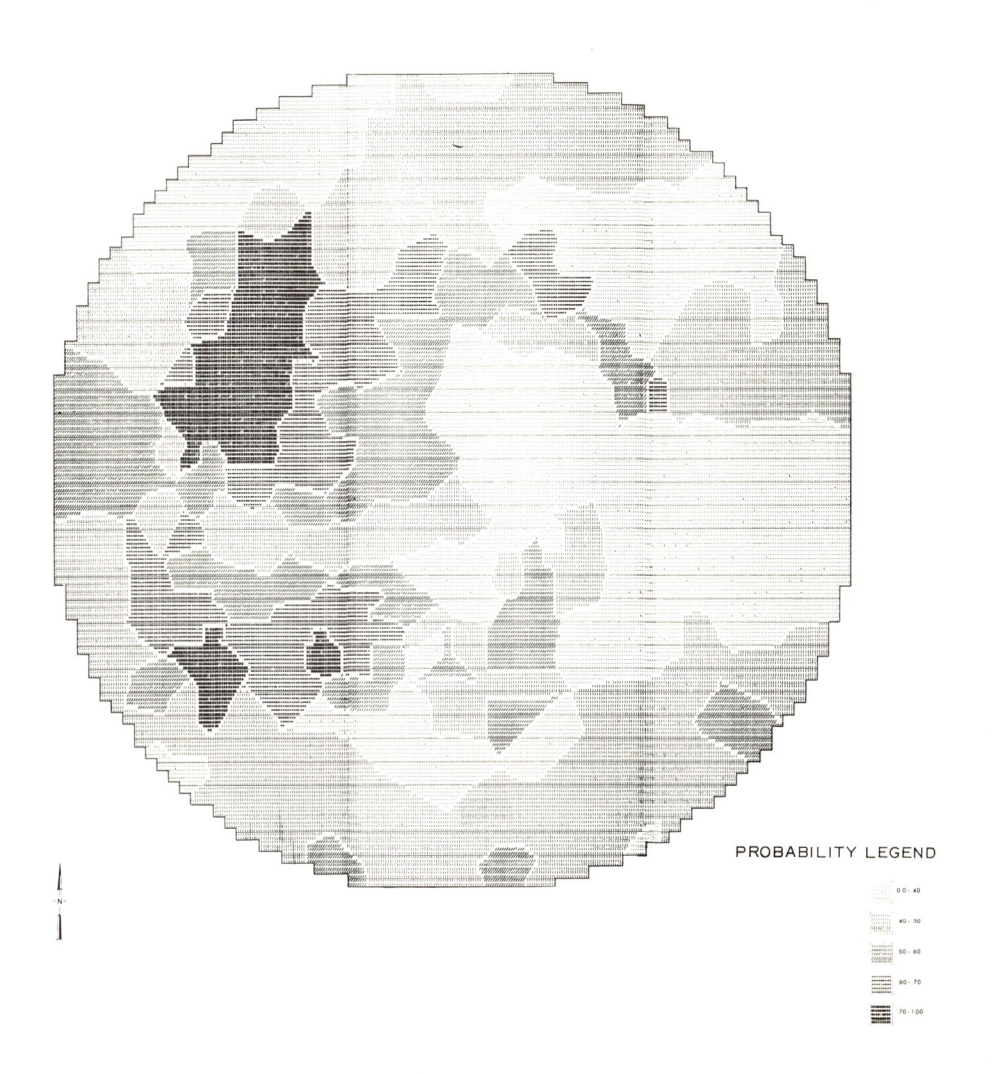

Figure 5. Map Showing the Spatial Distribution of the Probability of Subdivision Occurring in the 1961-63 Time Period as Forecast by the Model

such as zoning availability of water and sewerage due to changes of policy. It traces this impact as a change in the probabilities of changes of state occuring—be it predevelopment land sales, developers' subdivision location decisions, a households' decisions to move, or a households' selection of places of residence. Or looking at its output in the form of a map, the model should be able to trace the impact of policy suppositions on the conceptual topography of likelihood of a particular change of state of land—in the case of Figure 5, the topography of the likelihood of subdivision plats being recorded with the intention to develop as residential subdivisions.

Concluding Remarks

Many of the limitations of this research should be obvious, but it may be useful to review some of them in order to place our discussion in perspective. First of all, this is only a partial approach to planning models as opposed to a comprehensive one. It is partial in the sense that even the linked system focuses mainly on new single-family subdivision houses and does not include rental units, public units, or redevelopment of already built up areas of the city. It is partial in the sense that it focuses on spatial distribution and casts no insights as to the quality of residential growth.

Another area of qualification concerns the analysis that led to the model. There are some yet unexplained inconsistencies between observed associations and those expected on the basis of the literature and the earlier interviews. Also, the sample was limited in geographic area to two North Carolina cities, (for much of the "harder" data) thereby restricting generalization of the model and limiting our understanding of the effect of contextual factors which were insufficiently varied. Even though each urban area would have to calibrate a model for itself, the problem of a change in context still exists. If the planner wants to make any but marginal changes in the spatial distribution of the policy variables or if he wants to change the content of the policy, then he may actually be changing the contextual factors under which the model was originally calibrated, thereby invalidating the calibration. More study of the contextual factors is involved to surmount this problem.

Another limitation of the presently programmed computerized model to the planner is its cumbersomeness in tracing the implications of planning suppositions. Tracing the impact of a change in the spatial distribution of an input variable, say zoning, while holding others constant would require tedious coding and key punching to change values of the zoning variable for a number of sites. This difficulty may be alleviated by a "graphic processing system" (GPS) that several colleagues—Dale Saville working with George Hemmens and Terry Lathrop—are developing at UNC. They are attempting to provide a capability to reproduce maps of data on the cathode ray tube in such a way that the planner who would like to experiment with the model can interact with it and with the data in a very immediate and dynamic manner. He can call the input data onto the cathode ray tube and utilize a light pen and key punch attached to the tube to readily change the input. He can then send the new input to the model to produce an alternative output map on the cathode ray tube. This capability would make it possible for the planner to change the inputs in a much less tedious manner and almost immediately observe the consequences on the conceptual surface of development likelihood. Figure 6 attempts to diagram this interaction between man and machine. (See figure 6).

In spite of these limitations, there are sound reasons for continuing the application of this decision agent approach to the study of residential growth. For one thing, it provides an alternative to the more common modeling approach of aggregating the many decisions involved in the residential process into a simpler relationship. This is not meant to be a criticism of such models; it only suggests that an alternative approach may lead to insights that can supplement insights already gained through aggravative models. For example, contrary to assumptions and conclusions of many research and modeling efforts, decision agent modeling tentatively indicates that accessibility is of lesser importance than many other factors in explaining decisions to move, selection of new dwelling units, and location of subdivisions (Butler, Chapin, et. al., 1968; Butler and Kaiser, 1969), and that property characteristics seem to have little impact on the predevelopment landowners tendency to sell even though they are primary in the developer's locational decision.

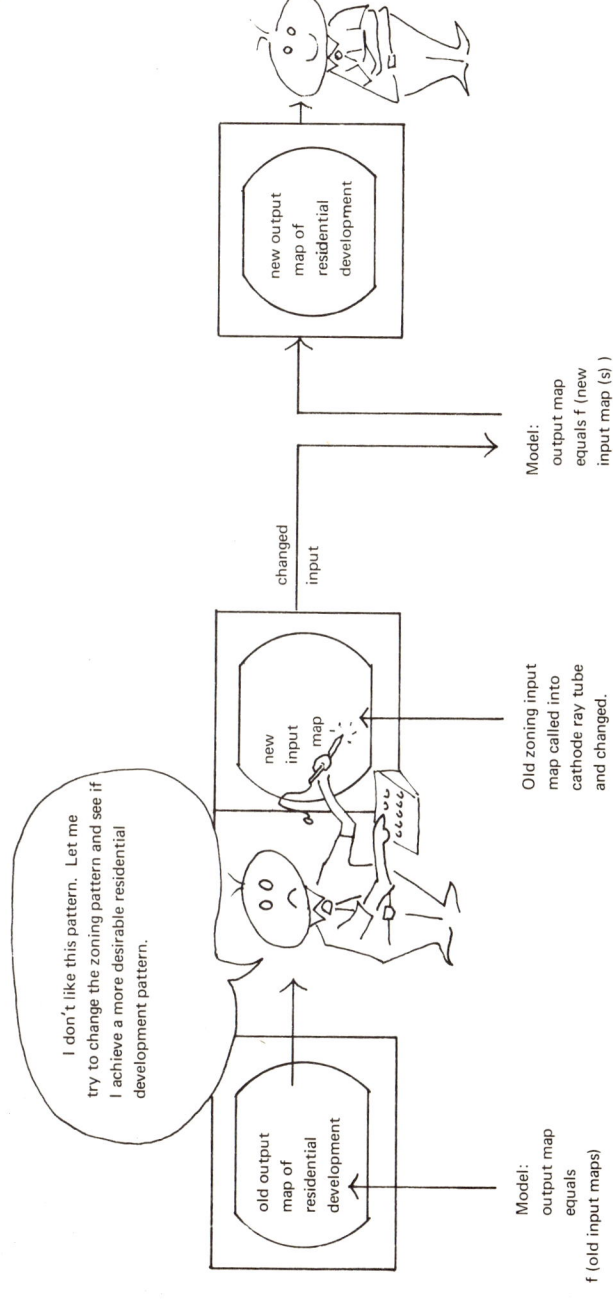

Figure 6. Illustration of Use of Graphic Processing System

We also feel that the insights provided in an emphasis of decisions, process, and decision makers are especially suited to problem solving situations involving public policy making because, in the end, it is this processual-complex of decisions that must be influenced in order to affect urban spatial structure. This approach can begin to suggest the ways in which different decision makers are affected by an array of policies. For example, some of our work suggests that tax policy could be an effective influence on predevelopment landowners but that its effect upon a site depends on the rate of value appreciation in raw land real estate and on the wealth of the owner. As a result, even if the tax policy encouraged a sale, it would not necessarily make the property more attractive to the developer—the next link in the chain (Kaiser and Weiss, 1967). The research suggests that mixes of policies, rather than single policies, are required, some of which are aimed at landowners, some at developers, and some at consumer households and that the arsenal may even have to be more specialized to account for differences among developers, among landowners, and among consumers.

References

1   By and large, we have used a research team approach, and I am essentially reporting the work of a number of faculty and student colleagues. Of these, Shirley F. Weiss has been the most consistent and it is under her research grants that much of the work on developer decisions, to be used as the prototype decision agent model in this paper, has been done. Some of the others have been F. Stuart Chapin, Jr., George Hemmen s, Edgar Butler, Michael Stegman, Ronald Massie, and John E. Smith. The research has been supported in part by the Environmental Engineering Policies and Urban Development Project, Public Health Service Research Grant UI 00128-07 from the Environmental Control Administration; in part by the Residential Choice and Moving Behavior Project, National Sciences Foundation Research Grant, GS 2427.

2   Butler, Edgar W., Chapin, F. Stuart, Jr., Hemmens, George C., Kaiser, Edward J., Stegman, Michael A., and Weiss, Shirley F., *Moving Behavior and Residential Choice: A National Survey* (Chapel Hill: Center for Urban and Regional Studies, Institute for Research in Social Science, University of North Carolina, 1968a).

3   Butler, Edgar W. and Kaiser, Edward J., "Prediction of Residential Movement and Spatial Allocation," a paper presented at the Annual Meeting of the Population Association of America, Atlantic City, New Jersey, April 11, 1969.

4   Kaiser, Edward J., *A Producer Model for Residential Growth: Analyzing and Predicting the Location of Residential Subdivisions.* An Urban Studies Research Monograph (Chapel Hill: Center for Urban and Regional Studies, Institute for Research in Social Science, University of North Carolina, 1968a).

5   Kaiser, Edward J., "Locational Decision Factors in a Producer Model of Residential Development," *Land Economics,* Vol. 44 (1968b), pp. 351-362.

6   Kaiser, Edward J., Massie, Ronald W., Weiss, Shirley F., and Smith, John E., "Predicting the Behavior of Predevelopment Landowners on the Urban Fringe," *Journal of the American Institute of Planners,* Vol. 34, (1968), pp. 328-333.

7   Kaiser, Edward J., and Weiss, Shirley F., "Local Public Policy and the Residential Development Process," *Law and Contemporary Problems,* Vol. 32 (1967), pp. 232-249.

8   Weiss, Shirley F., Smith, John E., Kaiser, Edward J., and Kenney, Kenneth B., *Residential Developer Decisions: A Focused View of the Urban Growth Process.* An Urban Studies Research Monograph (Chapel Hill: Center for Urban and Regional Studies, Institute for Research in Social Science, University of North Carolina, 1966).

George C. Hemmons
Department of City and Regional Planning
University of North Carolina

WORKSHOP ON MODELS IN PLANNING

Three papers were presented at the session on Models in Planning. The one thing common to all the papers is that they deal with location decisions in urban areas. Holton, Kramer, New and Marzot discuss factors involved in the location decisions on mental health clinics in Massachusetts. Kaiser discusses the decision processes of residential developers. Ray discusses an approach to urban spatial design. All three papers can also be characterized as process oriented. They are concerned with understanding change processes in urban areas rather than with evaluating static or equilibrium situations. Spatial patterns are viewed as the dependent variable to be explained by metrics dealing with the interaction among the components of urban form or decision rules used by actors in the development process. And the approach toward urban spatial design in all three papers appears to be a strategy that would give first priority to identifying the actual metrics and decision rules operating in the urban area, and then would examine ways of modifying these to affect spatial patterns. With the focus on location and the orientation toward process, the commonality in the papers and the discussion ended.

Different opinions or different ideas were presented in two areas. One is the question of whether the urban location problem should be approached from the bottom with a focus on the individual decision maker or individual urban form element, or whether it should be approached from the top, from a system point of view. The approaches taken by the authors span the range of this question. Ray takes a system viewpoint. Kaiser stakes out the middle ground. Holton, *et. al.* take a micro view of the problem of understanding urban spatial organization. The methodology employed varies quite appropriately with the definition of the problem. Ray is attempting to find a heuristic equivalent to a mathematical programming solution to implement his systems view. Kaiser relies on statistical inference, and Kramer develops his materials through case studies.

The other way the papers differ is in the amount of faith the authors appear to have in the idea that a "rational planning model" can be used as a framework to understand what is going on in the world. Not accidentally, I think, the paper that took the most individualistic view on the question of how do you understand location decisions and how do location outcomes occur took a very anti-rational planning approach and presented a strong critique of the rational planning model. He particularly stressed the inadequacies of problem definition and the lack of clear objectives in most urban location decisions. The paper that took a system-wide approach toward working out location decisions seemed willing to accept the idea that a rational planning model was good framework. There wasn't a lot of discussion about this. Nothing was resolved as to whether one or the other approach was good or bad, but I think the session was a very good one in that it did cover the full range of different approaches that are being taken in current research.

Turning to the conference as a whole, let me first say that I have found it very rewarding. Not being a designer, I have found some of the discussions and ideas to be quite novel. But also I fear I have also observed a continuing problem in urban and environmental analysis. This is what might be called the "old wine in new bottles" syndrome, and I will return to it shortly.

First, let me comment on the multiplicity of ideas and topics presented. As I noted we had a range of viewpoints in the session on Models in Planning, but little comparative discussion of them. I found a similar situation in many of the sessions. And I was particularly struck by the way reinforcing and contradicting ideas come up in different sessions. Perhaps all that this indicates is the great difficulty in organizing information in a developing, multi-disciplinary field. Of course many other fields and disciplines are in similar disarray. But I think that in view of the communications problems across discipline boundaries and between methodologies, and considering the lost opportunities for creative interaction of people and ideas resulting from this disarray, it might be well worthwhile to expend some additional energy on the classification problem in this field. Trying to sort things out in the right boxes and finding appropriate labels for them may be an unrewarding task, but it is necessary. To the extent it forces a focus of attention on delimiting the principle ideas in the field and their interrelation, it might pay great dividends—regardless of how often the boxes are relabelled and rearranged.

Now, back to the "old wine in new bottles" syndrome. It occurred to me as I went from session to session that I was observing a rather remarkable reinvention of a number of concepts—mostly from the social sciences. One author actually concluded his presentation by saying as much. The context was often new, at least new to me, and the "language" was often different so that it took some time before the strange sense of familiarity I experienced turned into recognition. This perfectly healthy, and in some ways reassuring situation simply reinforces my comment on the classification problem. And it leads to thought that focusing on the basic ideas in design research problems—primarily ideas about human behavior—may be the best way to order the bewildering variety of method and context.

Finally let me note my strong appreciation of the adventuresome spirit and open conduct of this conference. As what is meant to be the first in an annual series of multi-disciplinary meetings on design, this initial experience offers the hope of opening a valuable dialogue.

Robert Ray's paper entitled "Towards An Approach to Computer Assisted Design" was not available at the time this volume was ready for publication.

Charles Hamilton Burnette, Ph. D.
Research Associate
Institute for Environmental Studies
University of Pennsylvania

TOWARD A THEORY OF TECHNICAL DESCRIPTION FOR ARCHITECTURE

Introduction

    Despite the fundamental role of abstract representation in the making of architecture, there exists no theoretically coherent, technically explicit, and practical basis for organizing architectural description. Working drawings and specifications, major forms of technical communication between the architect and the contractor, are organized according to descriptive conventions whose main justification is that they are traditional and familiar. These conventions do not facilitate the various uses of the information which they organize nor do they efficiently represent the thing described.

    While specialized concepts, theories, notational schemes, and computationally oriented articulations dealing with the subject matter of architecture have been offered in recent work by Theil[1], Norberg-Schultz[2], Rose[3], Teague[4], and others, no comprehensive system for articulating, organizing, and modeling architectural information during technical communication has emerged. Fundamental problems of ambiguity, form, scope, efficiency, and overall coordination remain to be solved.

    Although communication within the field of architecture is hampered by information that lacks precision and that is difficult to generate and awkward to use, a better basis for structuring the substantive information of architecture for technical communication does not seem to be a prominent concern within the discipline. Many of the major difficulties in the practice of architecture today may be directly traced to faulty description, poor communication practice, and archaic media. A poorly organized body of knowledge, professional isolation and the inefficient use of human resources in the drafting room are among the more critical problems caused by these factors. Nevertheless, architects seem to be occupied with questions of professional scope, management,, and style to the exclusion of an appropriate concern for their instruments and habits of expression.

    This paper presents a theoretical system for structuring information of the scope and character traditionally conveyed by architectural working drawings and specifications. Fundamental concepts are identified and structured to provide a reference framework for the categorization of such information. The rationale by which this model is deemed to be valid, universal, comprehensive, and practical basis for structuring description is also presented. Finally, forms for representing this description in the computer are suggested and the system's potential for unambiguously representing descriptive concepts at the level of computational form is illustrated.

    A more extensively developed exposition of the theory and its practical application may be found in a recent Ph.D. dissertation[5]. In that work a series of eight formats implement the theory for computation and representative examples from the full range of description contained in a reference set of working drawings and specifications are couched in the formats to illustrate their use.

    The proposed system for structuring information (or one that is equally comprehensive, flexible, and language-like) is a pre-requisite for the effective application of the electronic digital computer during the solution of architectural problems. The use of the computer in architecture is likely to continue to be specialized, uncoordinated and generally unsuited to the conversational communication characteristic of the architectural process until a coherent system for assuring complete, unambiguous, and useful information is at hand.

A theoretically coherent system for ordering information is essential to the efficient retrieval, handling and communication of the large bodies of highly fragmented information used. This information varies in character and scope, and changes in relevance and worth according to the circumstances which it informs. Indeed as architectural problems become at once more diverse, complex, pressing and pluralistic, the computer has become the only appropriate medium for handling and communicating information. Its requirement for precision must be met.

The answers to questions such as, "When is a reveal a soffit?", "When is an element of a building a component?", or, "When is a space vague or well defined?" could all remain matters of individual judgment as long as interpretation was subjective, redundancy unregulated, and the circumstances in which a message was formulated reasonably deduced. Today, when the computer shares the task of interpretation, and when messages originate in complex and ambiguous circumstances, explicit conventions by which to assure correct interpretation and handling must be developed and implemented.

The Role of Theory

A theory of architectural description is necessarlly distinct from a theory of what a work of architecture ought to be or mean. It is normative regarding the process of definition (apprehension) rather than evaluation (comprehansion). It may be used to guide verbal behavior but does not explain or predict it. It offers a system by which description may be established to assure a causal effect rather than one which explains or predicts the effect. It is essentially linguistic in nature.

Much of the theorizing about architecture has confounded this distinction and largely for that reason has failed to be of practical value in technical communication aimed at the synthesis and realization of the built environment. For example, Ruskin[6] and Viollet-le-Duc[7] confounded substantive description with poetic and rational valuations, respectively. More recently, Sullivan[8] and Kahn[9] chose anthropomorphic evocations and philosophical positions, respectively, as the significant reference systems for communicating about architecture. Today, a more fragmented and explicit identification of meaning with architecture is in vogue. This is exemplified by Venturi's[10] association of meanings from popular culture with formal motifs, Alexander's[11] empirical attempts to isolate the meanings of formal patterns, and the various efforts to use the semahtic differential technique to measure the affective meaning of selected images[12]. Such correlations of different vestiges of meaning with more or less abstract images of architecture are the means by which its significance is established. The descriptive articulations stemming from such correlations, however, are not reliable foundations for the technical communication by which architecture is realized. Such articulations change with the emphasis of the times and with the individual. To be useful for technical communication a theory must distinguish between affective evaluations and substantive description.

Perhaps the most significant attempt to get beyond didactic articulation to a universal, comprehensive theory having implications for practical communication is that of Christian Norberg-Schultz[13]. Within a philosophical context of phenomenology, Norberg-Schultz attempts to correlate semiotic distinctions (the semantic, syntactic, and pragmatic theory of signs) with knowledge of perception and substantive aspects of architecture. Although just such a phenomenological-structuralist approach seems appropriate to the complexity of architectural experience, Norberg-Schultz falls short of offering a coherent theory of practical value to the technical description of architecture. He neither comprehensively relates his concepts nor provides them with useful forms.

Stimulated by Norberg-Schultz, the present work presumes that an adequate theory of architectural description should have cultural, perceptual, and empirical validity. It must assimilate the understandings of the past, clearly model theoretical distinctions, and assure that the information it organizes may be readily applied.

A premise of universality is also shared with Norberg-Schultz. Descriptive concepts should be capable of application to all possible architecturel structures[14]. It is likely that such concepts are recurrent and therefore culturally established.

A further premise is that a theory must be comprehensive if it is to serve the variety of needs in the architeectural process and accommodate all levels of generality. This is perhaps the most difficult premise to fulfill. The wholeness necessary to the notion of comprehensiveness is implied when an entity to be described is identified or labeled. A name implies something specific and integral. Any description of this nominal whole, however, is biased by the purpose for which the description is generated. For example, the articulation of architecture (assume all knowledge of architecture) into its "mass", "space" and "surface" aspects is, on examination, biased by the intention to appreciate architecture as an exclusively sculptural object. This intention or purposeful point of view, defines and limits the extent of description as denoted by the three terms, and through them, indicates the context (sculptural experience) necessary for the correct interpretation of the information which the terms represent.

While an integral referent and a limiting focus of concern are required to establish the scope of description— its practical limit of comprehensiveness—people regularly judge the adequacy of information within the context which results by referring to some empirically acquired or rationally established categorical structure[15]. Thus, a comprehensive theory of description requires an empirically valid, normative reference system by which it is possible to judge completeness. Scientists refer to such a normative model to judge the adequacy of experimenta description. The categories of their model are: a priori theory and goals; elements considered; conditions of observation; instruments used; methodology; observed data; a posteriori conclusions and theory. In this model, "both elements considered" and "conditions of observation" are statically conceived in that they define the objects and circumstances of action while "methodology" and "observed data" define actions and consequences and are thus operational.

A model of rational comprehensiveness for architecture may refer to this empirically established model for scientific disclosure but its purpose, referent, and use are different. In addition to the need for replication and comparison as in science, architectural information is constantly being changed and reinterpreted. It is not necessarily causal. While the seven categories of the scientist remain applicable for the architect, "conditions of observation" may be more generally represented as relational constraints, "observed data" as procedural restraints, and "instruments" as formgiving mediators. With these changes, the scientific model becomes: background and goals; elements; relational constraints; mediators; procedures; procedural restraints; evaluations. This may be more clearly stated in terms of the types of information which each distinction identifies: directive; designative; relational; representational; procedural; regulatory; evaluative. All types must be included in a comprehensive description.

To summarize:

1. Description is theoretically valid if it has long standing cultural, psychological and empirical reference.
2. Any object of description is by implication wholistic.
3. The purpose for which something is described determines the extent of the description.
4. The purpose of description must be explicitly stated to establish context and guide interpretation.
5. The concepts chosen as the basis for categorizing information must be universal and applicable at all levels of description.
6. Description is theoretically comprehensive if it contains: definitive description in which statically conceived elements of the object are enumerated and related; mediating description by which the object is represented and known; operative description in which actions with respect to the object are programmed and regulated. (Figure 1)
7. The distinctions of the theory must have a practically useful representation.

Fundamental Concepts

What concepts universally applicable to architectural description recur historically and what is the extent of description which they denote? Architecture. like other physical three-dimensional objects, is often described in terms of its "mass", "space" and "surface"[16]. Similarly, from Vitruvius (firmitas, utilitas, venustas) through the Victorians (firmness, commodity, delight) to the moderns (technology, function, form) a single sense of articulation has prevailed regarding the general aspects of architecture that have operational significance for human beings. (Technology implies concrete operations and involves the capacity to realize or effect. "Function" manifests behavioral events and implies the capacity to utilize or experience. "Form" denotes abstractions or the result of synthesis and required a capacity to recognize and represent. All are significant with respect to what people do regarding architecture.)

Given these universally applicable, culturally validated concepts, it must be shown that they are rationally comprehensive and logically related if they are to be acceptable as a basis for the description of architecture.

We have learned from earliest experience that behind a sensable surface is a physical substance (mass) and before it there is distance (space). (Figure 2) Both architectural mass and space exist as mental constructs built from the direct sensation of surface. This logical empirically valid dependency, in which surface is the perceptual referent, articulates a continuum that is comprehensive is the extent of description is limited to a physical, three-dimensional object that is statically appreciated (as the object of the action of perceiving). Surface is the critical aspect in the definition, communication and experience of architecture-as-object in that it mediates the extent of space and the content of mass by functioning as the common boundary which explicitly delimits them. This dependency is acknowledged in practical communication by the use of plans, elevations and sections to represent the form, fit and relative location of assembled elements. Nevertheless, architectural theorists have missed the significance of surface almost completely, conceiving of it as merely a tactile element rather than as an essential mediator. Today, when many of the problems of architecture relate to communication, this dependency is more apparent and useful. In particular, mass may be appropriately described by enumerating individual properties or elements and identifying them with the appropriate surface while space may be unambiguously defined by the relationship of an observer to surfaces surrounding him.

Given this rationale based on the perception of an object, we must determine if it satisfies the theoretical requirements for comprehensiveness. As "mass" defines the object, "surface" represents it, and definition of "space" depends on the operational location of the observer, the three concepts are theoretically comprehensive within the context they manifest.

The comprehensiveness and logical relationship between the operative concepts "technology", "function", and "form" must also be established.

Norberg-Schultz has argued for the comprehensiveness of these distinctions without specifying the operative bais which they reflect. He writes;

> *We can describe a work of architecture functionally, technically and formally and the aspects of the environment have to be described under these headings.*[17]
>
> *The form as well as the building task (function) and the technical aspect are abstractions from the work of architecture which we designate as an architectural totality. The totality is a building task realized technically within a style*[18].

A rationale to support these claims may be based on the requirements for comprehensiveness of the story. "Technology" is directed at realizing the object or event and is therefore definitive. "Function" is concerned with natural or human behavior and is operational, per se. "Form" is the means by which technical organization and functionally dictated events (phenomena) are given images or represented. (Figure 3) Functional intentions must be fomally represented (planned) before they may be realized.

It has beeen argued that both of the three-term concept sets: (1) are universally applicable: (2) separately model static and operative aspects of the architectural experience: (3) are comprehensive within their apparent limitations: and (4) are culturally valid and may be referred to experience.

The static concept of "mass" may be related to the operative concept "technology" in that the physical substances or elements which constitute "mass" are organized or fit together by "technology" "Mass" is thus defined by substantive properties or elements while "technology" is distinguished by questions of fit, order or organization (Figure 4).

We also note that "surface" and "form" are "mediators" within the limited context of their respective concept sets. "Surface" mediates by being a concrete boundary for "mass" and "space" while "form abstractly manifests a functional shape or realizable pattern. Thus the defining attribute of "surface" is boundary while that of "form" is shape (image) Figure 5).

Finally, space and function are operatively related in that space limits functional behavior or is defined by it. The description of events is categorized under the label "function" while "space" distinguishes distances and functional limits. (Figure 6).

A logical relationship which reflects these correlations and defining attributes is indicated in Figure 7. It may be seen that the elements or properties of mass are physically related (ordered) through technololgy. The surface through which this organized object is realized provides the objective basis for the perception of form. This perception results from a spatial relationship (distance) from the surface and from functional behavior regarding it. The phenomenological character of this model (and the theory it manifests) is evident in that a mediated object (what appears as form) is composed of extensional aspects (properties inherent to the object whose surface is sensed) mediated by intentional perceptions (attributes projected on the object by its perceiver). The perceiver sees what he intends, subject to what exists. Perceived forms are selected from surface maps which inventory properties and relationships.

This conceptional scheme which is theoretically comprehensive may be used as the basis for categorizing all information related to architecture. In particular, elements or properties are identified with the label "mass" and constitute the substantive elements to be considered. Circumstantial relationships which organize or determ these elements are categorized as technological description. Particular representations by which mass entities and/or their relationships are constituted as a whole are identified as surface mappings. Form is the label applie to the set of this mapped information identified in spatial or functional specification. "Space" designates surface entities identified in terms of their distance from some reference while "function" indexes surface entities in terms of the actions linking them. All are related by specifications of purpose and operationally controlled b procedural regulation.

Computational Form

In order to present the forms which make so fragmented an articulation useful, it is desirable to assure a minimum level of familiarity with computational elements and structures. Basically, distinct sequences of coded descriptions of varying lengths may be established in particular addressable locations in computer memory. These distinct statements may be collected into lists of varying size which are called "blocks". These blocks then occupy a series of contiguous filing locations in computer memory and may be treated as filing units with names and formats. But references called links may be specified to relate blocks, formated sub-blocks or statements in blocks more or less as required. The linked lists which result may connect various fragments of information to model a complex decription. Several such data structures may reference the same information fragment. The chain of information constituting a complex description is selected according to the code identifying the description and its parts by the program which is controlling the computer's progress along the links. Therefore, we must examine the explication of cues by which the computer and/or the programmer could quickly identify particular information both within the context of a specified purpose and with respect to conventional understandings and word use. For example, a six part format for a block may provide a basis for coding the information identified with each of the six concepts of the proposed schema. If any information appeared in a certain formatted location, it would be interpreted according to the conventions identified with that location. But the distinctions may be modeled at yet a more basic level, that of computational form itself.

The limited vocabulary of computational form, labels, lists and links between lists may be used to model the information identified with the six concepts of our categorical reference framework. The description of mass, being restricted to individual elements and properties, is easily represented as a list. Distances (space) are attributes and as such may also be listed. Distinct events in the description of function (behavior or process may be itemized as well. "Form" and "surface", however, must be represented at the level of data structures (linked lists) as it is necessary to relate operative and definitive information that is separately listed to arrive at a surface map, formal image or perceived event. The way in which these links are made may be used to distingu whether items listed belong to mass, space function or technology.

The phenomenalogical appreciation of the theory suggests that one way links which specify a coordinate relation may signal a specific theoretically understood and computationally recognizable interpretation. As a surface is a two dimensional map of a sensible whole and as all points are equally necessary to its definition it must be described by a coordinate data structure. Since mass is an extension of surface into three dimensions its description also requires a coordinate structure. Each property or construction task contributes to the existence of a wall; or, both the operation of baking and the property of clay are necessary to the existence of baked clay brick. Hierarchies may exist in the description of mass as a more or less specific description of a property or element, but these are functionally purposeful distinctions.

Since space is operationally defined, the data structures representing it would be hierarchical. If one attends to a surface in the near distance and to one farther on, neither can be described as coordinate to the other. This is true for both perception and function, the two aspects of space that are theoretically meaningful. In the first instance, visual focus is different. In the second, the physical distance impinging on a function is different. Thus hierarchical structures may be used to model operative description and coordinate structures may be used to model operative description Special interpretive codes would identify this particular mode of interpreting linkages. A change in a hierarchical structure would introduce a computation while a change in a coordinate structure would not. For example, the movement of the vantage point in the data structure modeling a perspective view of a space would cause the view to be recomputed. Conversely, the description of a unit brick may be changed to a unit block without affect on the other elements in the coordinate data structure. When some functional consideration of information in a coordinate description is required, dimensioning for example, a hierarchical, functional structure would be imposed.

An additional use may be made of the phenomenal model and the concepts of static and operational description. The model suggests that coordinate, extensional information describing mass may be referred to the gestalt principle of proximity (collective wholism). This suggests that the fragments of such description be listed contiguous in computer memory. (Figure 8) Similarly, the principle of sequential wholism (continuity) applies to hierarchical models of operational information and suggests that spatial and functional description be formulated as computational subroutines. (Figure 9) The combination of these two types of information to produce a representation of surface or form may be identified with the principle of closure (configuring wholism). (Figure 10) Technical information which is statically treated yet operational in nature is ambivalent and may be formulated as a contiguous part of mass or surface description and/or as a subroutine.

While the concepts "mass", "surface" and "space" manifest the statically definitive aspects of the general theory and "technology", "form" and "function" its operative active aspect, the computational form, phenomenal interpretation, and operative principles manifest its representative aspect and complete the cultural, perceptual and empirical references required.

Application of the Theory and Forms

There are many practical benefits from categorization of information according to the pattern as identified by the terms: entity, purpose, mass, technology, surface-form, function, space and evaluation.

In the description of mass the individually enumerated properties and elements produces information organized to facilitate analysis and accounting operations. Sorting and editing is facilitated by the list form and elements or properties may be added as they arise.

The hierarchically ordered links describing operational relationships provide a means to establish sequences, references and patterns to facilitate the coordination of information, the synthesis of fit and the programming of events in space.

The hierarchical and coordinate description of surface and form provides a means for modeling complex wholes while facilitating information retrieval, visual communication and the comprehension of information.

Thus, various needs of the architectural process are appropriately served by articulating information according to the system proposed.

A Clarification of Terms

With the theoretical relationships proposed many descriptive conepts may be clarified and given unambiguous form. For example, the concepts of "primary" and "secondary" spaces, "screens", "surfaces", "objects", "ports", "merges", "ends", suggested by Philip Thiel,[19] may all be distinctly modeled by data structures. "Vagues", "suggests" and "volumes" being affective evaluations can only be entered as a posteriori subjective information.

For example, "paths" are specified as hierarchical subroutines linking the locations of particular function-events coordinately related to a map (surface) (Figure 11).

A "view" is functionally constrained space and thus computed on the basis of direction and line of sight. (Figure 12).

"Primary space" may be difined as surface description at the first level of hierarchy. A "secondary space" then is one which is dependent on the network description of a primary space for elements of its description. (Figure 13).

"Openings" are represented as mass elements (doors, windows, holes) in a coordinate structure describing surface. There can be no such thing as an opening in the spatial description just as there is no such thing as an opening in space. A "screened view" may be represented as a functional perception by the application of a viewing operation to the description of surfaces containing openings so as to select description of surfaces in the secondary space according to the view through the openings. Screens, per se, may be represented as that part of a data structure that is not influenced by links between primary and secondary spaces given that such links exist.

Merging spaces have several definitions. "Portals" may be described as edges common to the surface descriptions of two or more spaces. (Figure 14). More simply, merging spaces may be recognized by common points shared by the surface descriptions of more than one space. (Figure 15).

Ends of spaces may be perceptually or functionally defined by simply ending the specification of the network description.

"Vague spaces", "suggestive spaces" and "volumes" are entirely a function of the generality of description. A vague space may have the same spatial organization as a volume, yet, because its description is incomplete, it is considered vaguely defined. To Theil a "vague" is a weakly organized space. Such an interpretation depends on the existence of rules by which the quality of a spatial organization can be determined. Since, in the present system, space may be described only in terms of function or perceptual intent it may be evaluated by comparing the functional program of a space with its physical map or by an aesthetic response to a selected image of the space. The behavior which follows these functional or perceptual events (a change in the surface of the space, perhaps) is the only measure of spatial evaluation which this system facilitates.

Conclusion

The present paper offers, in a cursory form, a theory based on both definitive and operative description. I offers a rational basis, founded on cultural, perceptual and empirical reference, for the assignment of meaning t computational forms and suggest the descriptive power of such a conceptual scheme.

Previous papers have explored the system of distinctions from the point of view of classification in library science[20], problem solving[21] and as the basis for an abstract machine with which to model adaptive behavior.[2] The feasibility of the system for structuring all inforamtion contained in a set of typical working drawings and specifications is thought ot be established.[23] It now remains to render the theory operational in the particular formats developed to implement it and to test it in practice.

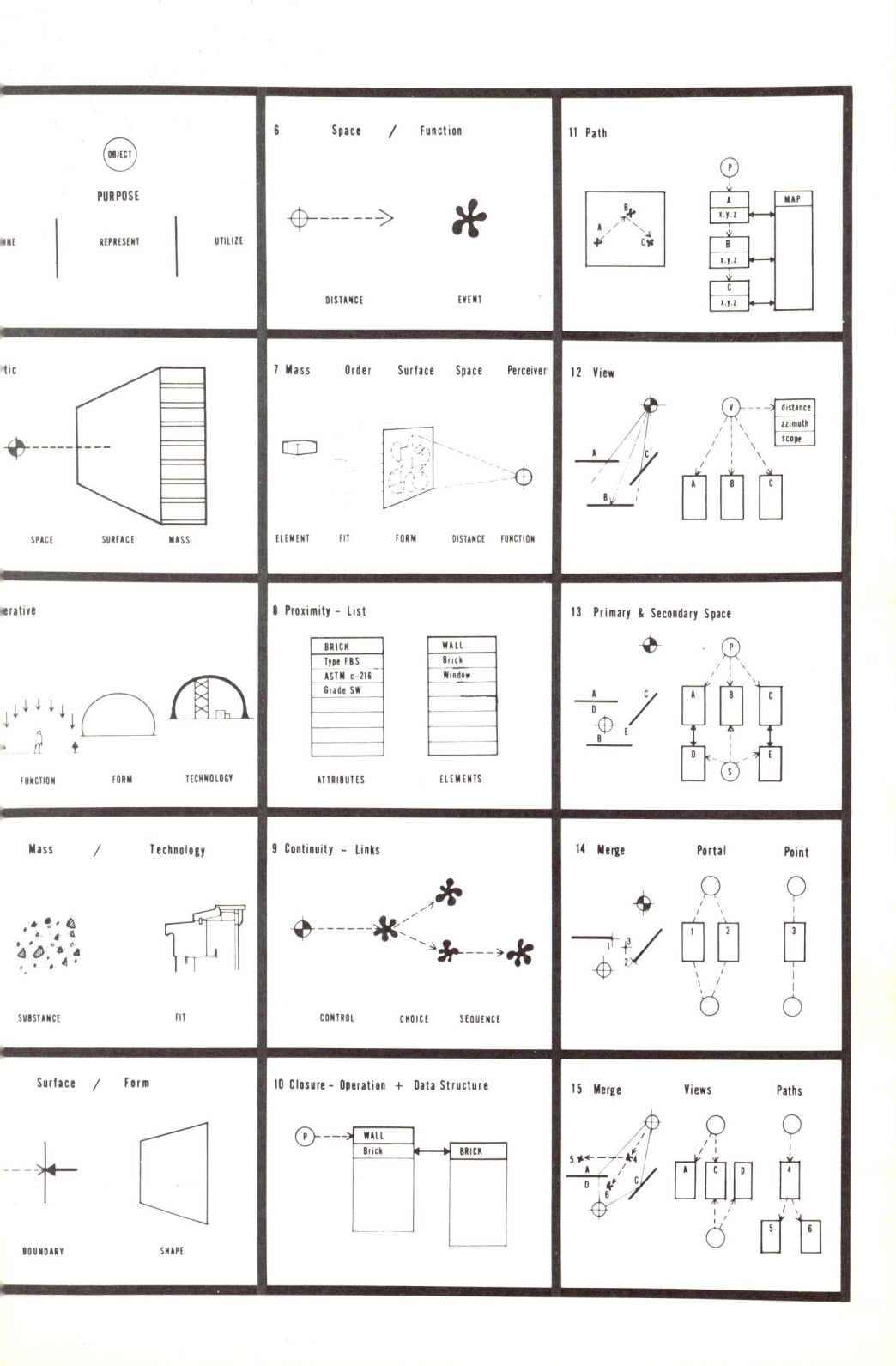

	6 Space / Function	11 Path
OBJECT PURPOSE REPRESENT UTILIZE	DISTANCE EVENT	
tic SPACE SURFACE MASS	7 Mass Order Surface Space Perceiver ELEMENT FIT FORM DISTANCE FUNCTION	12 View
erative FUNCTION FORM TECHNOLOGY	8 Proximity – List BRICK / Type FBS / ASTM c-216 / Grade SW WALL / Brick / Window ATTRIBUTES ELEMENTS	13 Primary & Secondary Space
Mass / Technology SUBSTANCE FIT	9 Continuity – Links CONTROL CHOICE SEQUENCE	14 Merge Portal Point
Surface / Form BOUNDARY SHAPE	10 Closure – Operation + Data Structure P --- WALL / Brick --- BRICK	15 Merge Views Paths

References

1. Philip Theil, "An Experiment in Space Notation", The *Architectural Review,* May, 1963.

2. Christian Norberg-Schultz, *Intentions in Architecture,* Universitelsforlaget, Oslo, Norway, 1963.

3. Stuart Rose, "A Notation/Simulation Process for Composers of Space", College of Education, Michigan State University, East Lansing, 1968.

4. Lavette C. Teague, Jr., "Research in Computer Applications to Architecture" in *Computer Applications in Architecture and Engineering,* Harper ed., McGraw-Hill, New York, 1968.

5. Charles H. Burnette, "An Organization of Information for Computer-Aided Communication in Architecture University of Pennsylvania, May 1969 (University Microfilms, Ann Arbor, Michigan)

6. John Ruskin, *The Seven Lamps of Architecture,* Noonday, New York, 1961.

7. John Summerson, "Viollet-Le-Duc and the Rational Point of View" in *Heavenly Mansions,* Norton, New York, 1963.

8. Louis Sullivan, *Kindergarten Chats,* Wittenborn, New York, 1947.

9. Louis Kahn, in "The Philadelphia School", *Progressive Architecture,* 1962.

10. Robert Venturi, *Complexity and Contradiction in Architecture,* Museum of Modern Art, New York, 1966.

11. Christopher Alexander et al, "A Pattern Language Which Generates Multi-Service Centers", Center for Environmental Structure, Berkely, California, 1968.

12. Henry Sanoff, *Techniques of Evaluation for Designers,* Design Research Laboratory, N. C. State University Raleigh, May, 1968.

13. Christian Norberg-Schultz, Op. Cit.

14. Ibid, p. 132.

15. G. A. Miller, E. Galenter, and K. H. Prebram, *Plans and the Structure of Behavior,* New York, Holt Reinhart and Winston, 1960.

16. Ibid, p. 86.

17. Ibid, p. 103.

18. Ibid, p. 105.

19. Philip Theil, "Notes on Environmental Space and Elementary Space Notation", College of Architecture and Urban Planning, University of Washington, mimeo.

20 Charles Burnette, "An Heuristically Oriented Approach to Classification", Institute for Environmental Studies, University of Pennsylvania, mimeo, 1967.

21 _____, "A Linguistic Structure of Architectonic Communication", in *Response to Environment,* The Student Publication of the School of Design, N. C. State University, Raleigh, June 1969.

22 _____, "Toward A Theory of Interaction for Planning and Design", mimeo, May 1967.

23 _____, op. cit. (5).

William R. Miller
Arendt/ Mosher/Grant, Architects
Santa Barbara, California

Vahe Khachooni
Head of Computer Services, Daniel, Mann, Johnson and Mendenhall
Los Angeles, California

James Olsten
School of Architecture, California State Polytechnic College
San Luis Obispo, California

MATRIX METHOD FOR GROUPING AN INTERRELATED SET OF ELEMENTS

This paper describes a method for grouping an interrelated set of elements into subsets such that there is a maximum amount of communication within a subset and a minimum amount of communication between subsets. Some work has already been done in this area. In the early 1960's, Christopher Alexander was working on his HIDECS programs[2] and in 1964 his book, *Notes on the Synthesis of Form,* was published.[3] In the mid 1960's Allen Bernholtz and Edward Bierstone applied Alexander's studies to a specific architectural design problem.[4] Today, Herman Miller, Inc., is using relational problem solving techniques in planning interior spaces.[5] The value of the method is in the simplicity of its underlying concept and that it can be operationally employed by architects.

Relational Problems

Relational problems can be described as those problems which involve the manipulation of an interrelated set of elements where each element and each element relation has associated with it a specific set of attributes. The manipulation is such that the final arrangement satisfies some predetermined objective. A hierarchical structure of this description is shown in Figure-1.

G = combined set of all elements and relations
E = set of all elements
R = set of all relations
e_i = the ith element
r_{i-j} = the relation from element e_i to element e_j
$e_{i,\ell}$ = the ℓth attribute of the ith element
$r_{i-j,\ell}$ = the ℓth attribute of the relation r_{i-j}
$v_{e_{i,\ell}}$ = the assigned value of the ℓth dimension of the relation r_{i-j}

Figure-2 shows a graphical representation of a portion of this structure where the elements are represented by nodes and the relational links between elements by arrows. The method described below operates on the following type of relational problem:

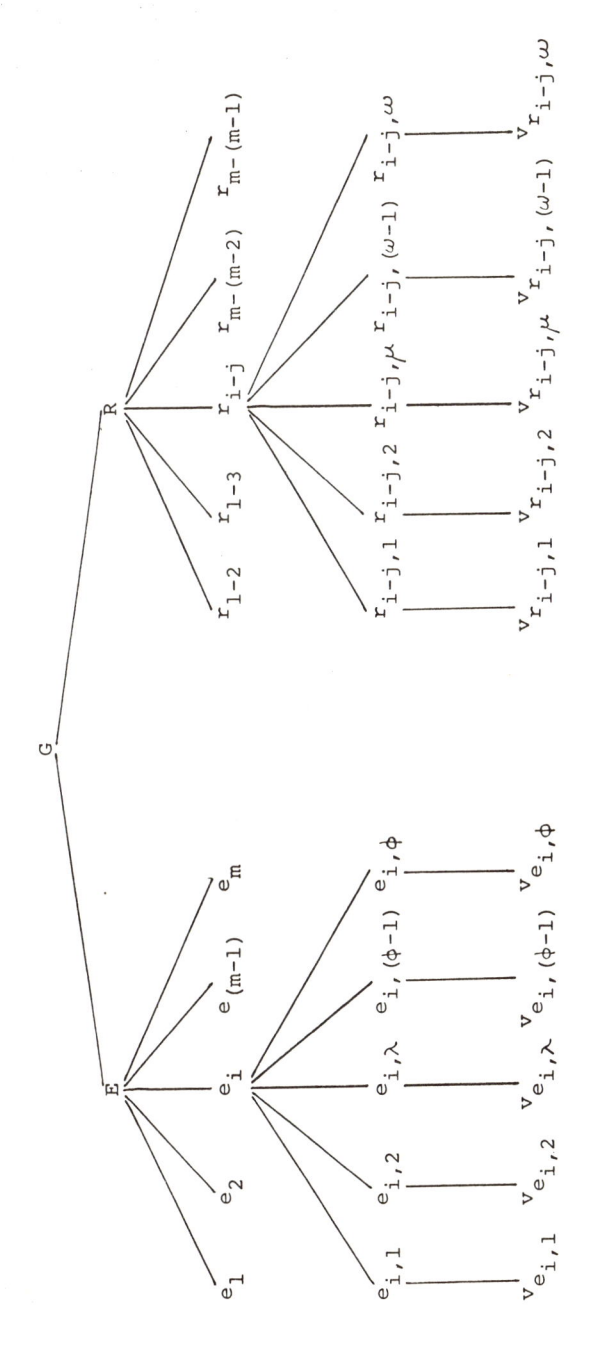

Figure 1. Hierarchical Structure of Relational Problem

(1) where the only attribute associated with a given element or relation between element is that of existence; (2) where the assigned values are all of equal value, that is, there are no "weighted" elements or relations; (3) where all relations are bi-directional; and (4) where no more than one relational link can occur between any two given elements.

Matrix Method

This class of relational problems can be described using graph theory. A graph consists of a set of junction points called "nodes" with certain pairs of the nodes being joined by "links." Figure-3 shows an example of a graph. Such a graph can also be described by merely listing the nodes (elements) and those to which they are linked (related). Figure-4 shows a listing for the particular graph in question. This listing can be tabulated in a vertex (or adjacency matrix (Figure-6). The rows and columns of the matrix are labeled with the element identification numbers. Note that the numbering sequence must be the same for both the rows and the columns. The presence of a one (unity) in the interior of the matrix represents the existence of a relation between the addressing elements. If the square in the matrix is blank, then no relation between the addressing elements exists. In actuality, a zero is used to indicate a non-relation; however, for clarity, the zeros have not been shown. For example, element 3 relates to elements 5 and 6. Again, for graphical clarity, it is assumed that all elements relate to themselves; thus there will asways be a diagonal of one's running from the upper lefthand to the lower righthand corner of the matrix. This diagonal is referred to as the "base diagonal." The matrix is symmetrical about this diagonal because we have assumed that all relations are bi-directional.

The existence or nonexistence of an element or a relation between elements can be based on any criteria. It should be noted, however, that in order to maintain a clear understanding of the physical situation being represented, this freedom will be restricted to a certain degree.

A simularity of pattern exists between the graph of Figure-3 and the matrix tabulation of Figure-5; there are two clusters of elements in the graph and there are two clusters of one's in the matrix. That is, elements 1, 2, and 3 are clustered in the graph with a corresponding cluster of ones lying in the intersecting rows and columns, representing the same elements in the matrix. This relationship is made clear in Figure-6. The unit value in the matrix which is not part of these two clusters, that is, the 1 in squares 3, 4 and 4, 3 represents the relational link between clusters.

This visual relationship will be examined as the interpretation may appear arbitrary. For example, the relational link between elements 5 and 6 will be eliminated while all other relationships will remain constant. Now what would be our visual interpretation of this new condition? Figure-7 shows three possible interpretations each equally valid. Figure-7a interprets elements 4, 5, and 6 as being in the same cluster. Figure-7b interprets 4 and 6 as forming a cluster and element 5 as forming a cluster. In order to show this latter interpretation in the matrix, we have interchanged rows 5 and 6 and, correspondingly, columns 5 and 6. This change leads to a very important conclusion: *The visual relationship between a graph and a matrix tabulation of the graph depends entirely upon the order in which the elements are numbered.* If the nodes of the graph were numbered as in Figure-8a, the matrix tabulation of this graph, shown in Figure-8b, would be nonsensical. However, if the corresponding rows and columns of the matrix in Figure-8b were rearranged to that of Figure-8c, the visual relationship between the graph and the matrix would again be achieved.

This visual relationship holds ture when the one's in the matrix are clustered near the base diagonal. This fact allows us to find the clusters for any set of interrelated elements, assuming:

1. The only attribute ascribed to an element or relation is that of existence.
2. All relations are bi-directional.
3. All elements relate to themselves.

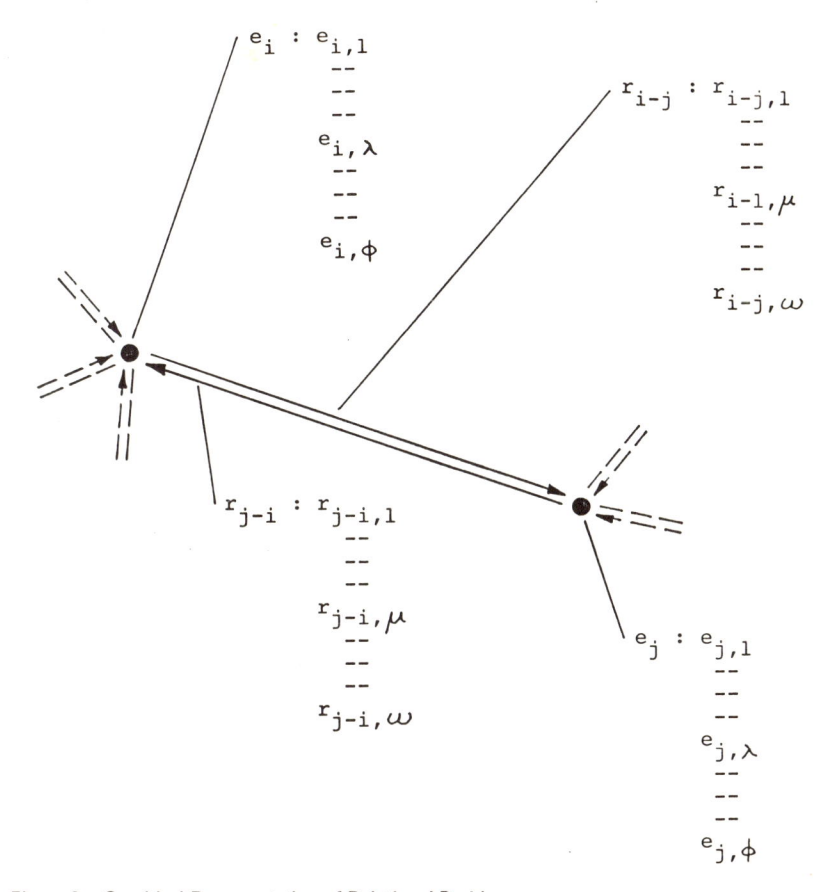

Figure 2. Graphical Representation of Relational Problem

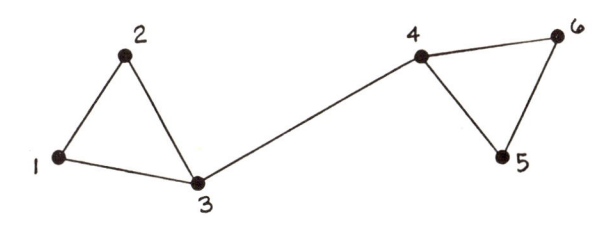

Figure 3. Graph

Element	Relates to Elements
1	2, 3
2	1, 3
3	1, 2, 4
4	3, 5, 6
5	4, 6
6	4, 5

Figure 4. Listing

	1	2	3	4	5	6
1	1	1	1			
2	1	1	1			
3	1	1	1	1		
4			1	1	1	1
5				1	1	1
6				1	1	1

Figure 5. Vertex Matrix

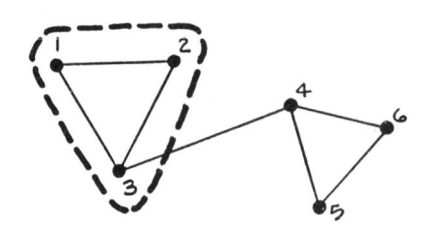

Figure 6. Visual Relationship

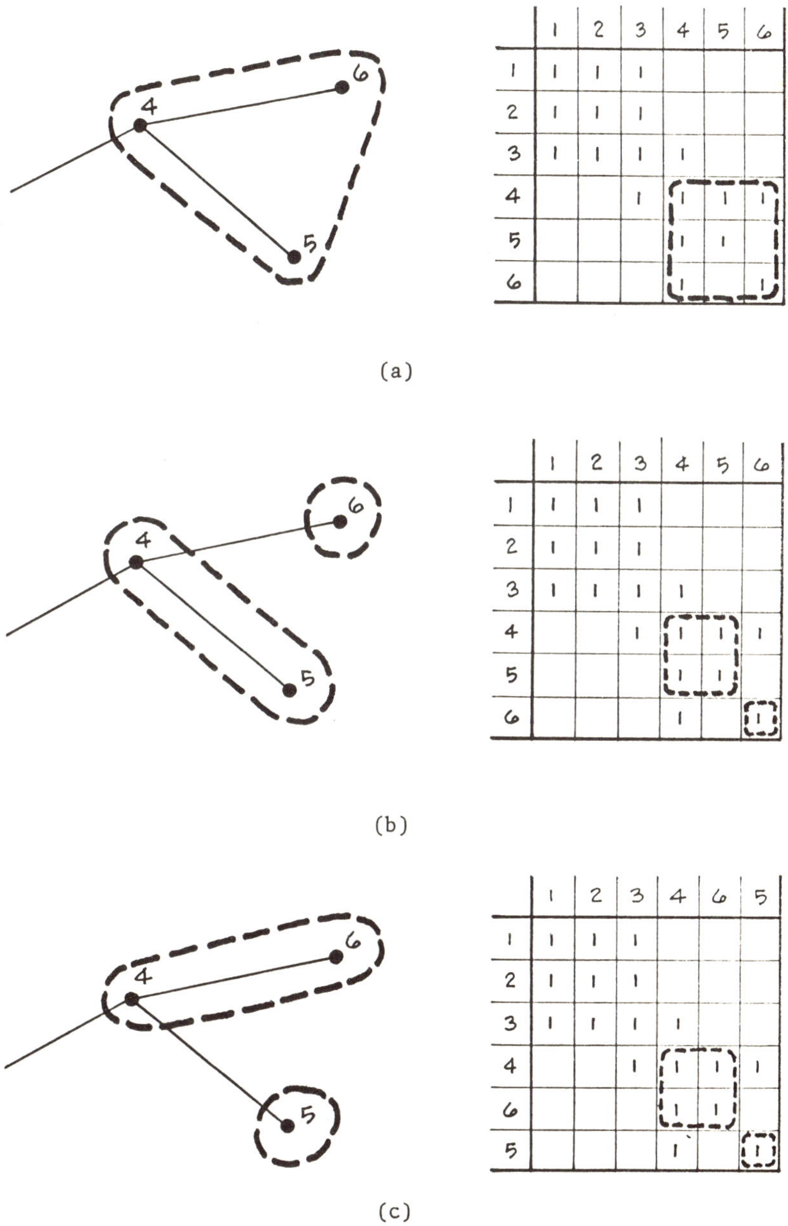

(a)

(b)

(c)

Figure 7. **Interpretation**

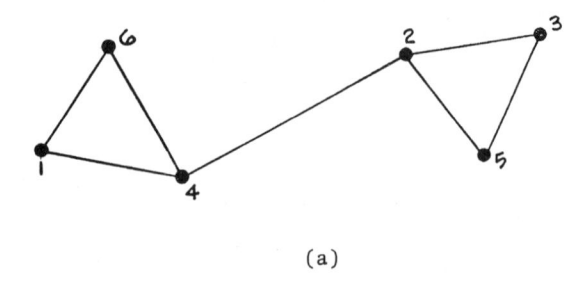

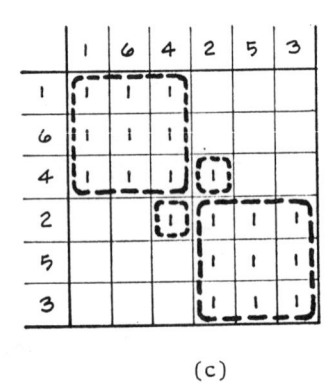

Figure 8. Ordering the Matrix

The property of a cluster is such that there is a maximum amount of communication (number of relations) within a cluster and a minimum amount of communication (number of relations) between clusters. Example:

Suppose a designer has a problem composed of a set of interrelated elements as described in the list shown in Figure-9. Each element represents the existence of a space within a building and each relation represents the existence of a geographical proximity between spaces. The designer wants to know how to arrange the spaces such that those that are strongly related geographically will group together, thus keeping the distance between related spaces to a minimum. The solution procedure is as follows:

Step 1) The information listed in Figure-9 is tabulated as shown in Figure-10. This matrix shall be referred to as the "data matrix." The data matrix is then examined for symmetry about the base diagonal. If it is found that the matrix is not symmetrical, it must be made so prior to proceeding.

Step 2) The corresponding columns and rows of the data matrix are rearranged such that the unit values within the matrix are as close to the base diagonal as possible. This is no easy task, for given N elements, there are N! possible arrangements of the columns and rows. (Consider: if $N = 10$, then $N! = 3,628,800$). The results of this process shall be referred to as the "solution matrix." Figure-11 shows the solution matrix for the problem in question.

Step 3) The solution matrix is then partitioned such that its clusters are isolated within a partition as shown in Figure-12. This partitioning will not only isolate the clusters but will also isolate the relational links between clusters. Since the matrix is symmetrical about the base diagonal, consider the upper righthand half of the matrix. The relational links between clusters are found to lie in that area which represents the intersection of the horizontal projection of one of the clusters with the vertical projection of the other cluster. For example, the unit value in the upper lefthand corner and the cluster in the lower righthand corner.

Step 4) Once the solution matrix has been partitioned as described above, a generalized graph is drawn as in Figure-13. This graph shows the major cluster, the elements in each major cluster, and the number of relational links between each major cluster.

Step 5) The final step is to draw a detailed graph depicting the elements and all of their relations as shown in Figure-14.

Conclusion

The critical step of this process involves manipulating the columns and rows within the matrix such that the unit values are forced to cluster along the base diagonal. If one were to attempt this task manually, he would be forever exchanging columns and rows. Indeed, only the most trivial problems could be solved by hand. With the aid of a computer, however, it is possible to write a program that will perform this manipulation automatically. Such a program, MATRAN-I, has already been written.[6] This portion of the paper concludes with a brief description of the MATRAN-I program.

MATRAN-I is composed of the following subroutines: PACKER, PRINTER, INPUTR and MATRXT. MATRXT is the subroutine that block diagonalizes the input matrix. All of the other subroutines are supportive to this subroutine. The manner in which MATRXT block diagonalizes the matrix is briefly described in the following paragraph.

MATRXT causes the computer to interchange various rows and corresponding columns within the matrix

Element	Relates to Element(s)
1	5,7,8,9,10
2	4,9,12
3	7,11
4	2,6,8,9,12
5	1,8,10
6	4,9,11,12
7	1,3,11
8	1,4,5,10
9	1,2,4,6
10	1,5,8
11	3,6,7
12	2,4,6

Figure 9. **Input Listing**

	1	2	3	4	5	6	7	8	9	10	11	12
1	1				1		1	1	1	1		
2		1		1					1			1
3			1				1				1	
4		1		1		1		1	1			1
5	1				1			1		1		
6				1		1			1		1	1
7	1		1				1				1	
8	1			1	1			1		1		
9	1	1		1		1			1			
10	1				1			1		1		
11			1			1	1				1	
12		1		1		1						1

Figure 10. **Data Matrix**

	12	2	9	4	6	8	5	10	1	7	3	11
12	1	1		1	1							
2	1	1	1	1								
9		1	1	1	1				1			
4	1	1	1	1	1	1						
6	1		1	1	1							1
8				1		1	1	1	1			
5						1	1	1	1			
10						1	1	1	1			
1			1			1	1	1	1	1		
7									1	1	1	1
3										1	1	1
11					1					1	1	1

Figure 11. **Solution Matrix**

	12	2	9	4	6	8	5	10	1	7	3	11
12	1	1		1	1							
2	1	1	1	1								
9		1	1	1	1				1			
4	1	1	1	1	1	1						
6	1		1	1	1							1
8				1		1	1	1	1			
5						1	1	1	1			
10						1	1	1	1			
1			1			1	1	1	1	1		
7									1	1	1	1
3										1	1	1
11					1					1	1	1

Figure 12. **Partitioned Solution Matrix**

until the unit values within the matrix are brought as close as possible to the base diagonal. The computer first identifies two rows for a possible exchange. The value of the matrix is then calculated. (The value of the matrix at any given point in time is equal to the sum of the products of the value within a cell, either 1 or 0, multiplied by its distance, number of cells, from the base diagonal.) This value is placed in temporary memory. The computer then exchanges the two rows and their corresponding columns and calculates the new value of the matrix. If the value of the matrix prior to the exchange is greater than the value of the matrix after the exchange then an imporved condition has been identified and the computer allows the exchange to remain. If the value prior to the exchange is less than or equal to the value after the exchange then no improvement has taken place and the computer proceeds to exchange the rows and their positions. The computer then identifies two more rows and corresponding columns for another possible exchange. This process continues until no further improvement can be made. The computer then prints out this final "block diagonalized" or "solution" matrix.

Application Description

The following involves the analysis of a major courthouse complex in Southern California.

Though it is possible to define each functional space within the courthouse as an element, broad areas have been treated as the elements of the matrix at this stage. Fourteen of these broad areas such as the Superior Court, Municipal Court, District Attorney's Offices, etc., were identified. The geographical relationships between these areas were then determined. No physical areas or priorities were assigned at this time. A listing of the defined elements and their relationships appears in Figure-15.

The duplicate relationships are not listed or inputed to the computer; rather, the program automatically assigns these relationships based upon previous entries. From this list the input matrix was generated. (See Figure-16). The bold graphics permits the architect user to better identify with this type of pattern symbolism than do numerical listings.

MATRAN-I then rearranged, or optimized, the data matrix and the results are shown in Figure-17. Interpretation of the solution matrix involved some judgment at this point. For example, the location of the partition between element 14 and 4 presented a problem. It was decided to partition the matrix as shown in Figure-18, thus isolating element one (1) as a center of communication. From this partitioning, a generalized graph was drawn (See Figure-19) where a definite structure is apparent.

A detailed graph was then drawn such that a clear representation of the structure of the graph was obtained as shown in Figure-20.

Analysis of Figure-20 indicated that the graph could be greatly simplified if the relation between elements 10 (Garage) and 14 (District Attorney) could be deleted from the problem. Although this deletion was not judged viable, element 10 was dropped from the problem by assuming that the garage would be placed below the other areas.

Based on this assumption, a second detailed graph was drawn as shown in Figure-21. At this point the areas could be renumbered and a second matrix analysis performed. However, since the problem was now fairly clear in the minds of the architects, this second analysis was not performed. Note that element 13, Veterans Service Organization, is now isolated. This indicates, based on the assigned relationships, that the Veterans' area can be located almost anywhere in the total complex. Although other factors apply and actually determine its final location, this early analysis was helpful.

Further study of Figure-21 led to the intuitive grouping of elements 1, 5, and 2 as ahown in Figure-22. Elements a, 5 and 2 comprised what was then referred to as "the courts complex," that is, the Municipal Court, the Superior Court and the Defender. The graph shown in Figure-22 was the final graph presented to the architect and was used as a basis for further design and study.

Application of MATRAN-I to this very elementary example pointed out some capabilities and shortcomings of the techniques.

314

Figure 13. Generalized Graph

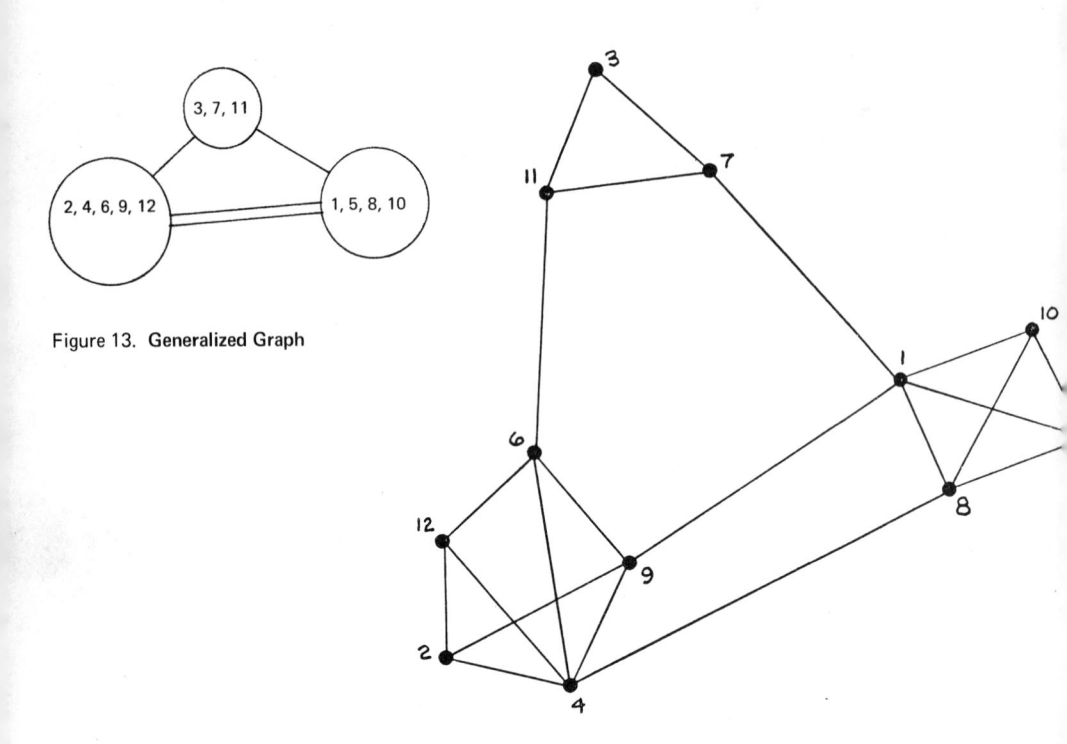

Figure 14. Detailed Graph

Area	Number	Relates to Area(s)
Municipal Court	1	2,3,4,6,9,11,14
Defender Inc.	2	5
Law Library	3	5,14
Marshall	4	5,10
Superior Court	5	6,8,9,11,14
County Clerk	6	7,10
Recorder	7	
Grand Jury	8	14
Probation Dept.	9	10
D.P.W. Garage	10	12,13,14
Jail	11	12
Sheriff	12	14
Vets Serv. Org	13	
Dist. Attorney	14	

Figure 15. Input Listing

Figure 16. Data Matrix

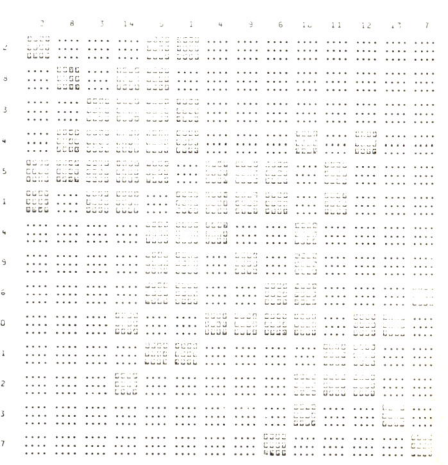

Figure 17. Solution Matrix

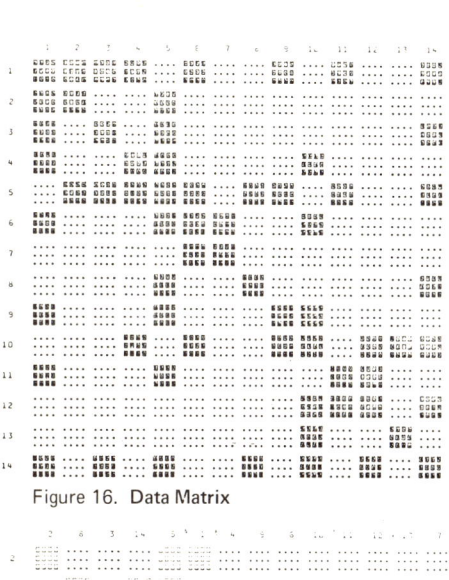

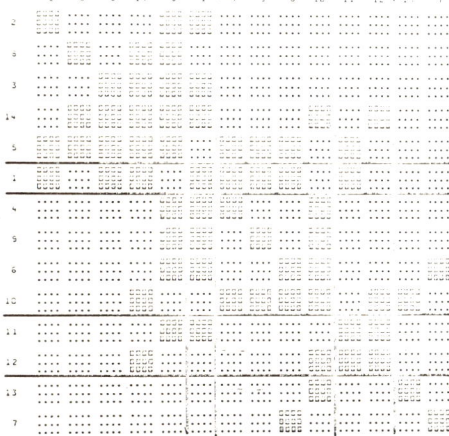

Figure 18. Partitioned Solution Matrix

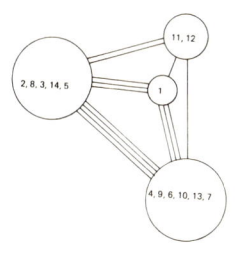

Figure 19. Generalized Graph

316

Figure 20. Detailed Graph

Figure 21. Detailed Graph (Revision No.

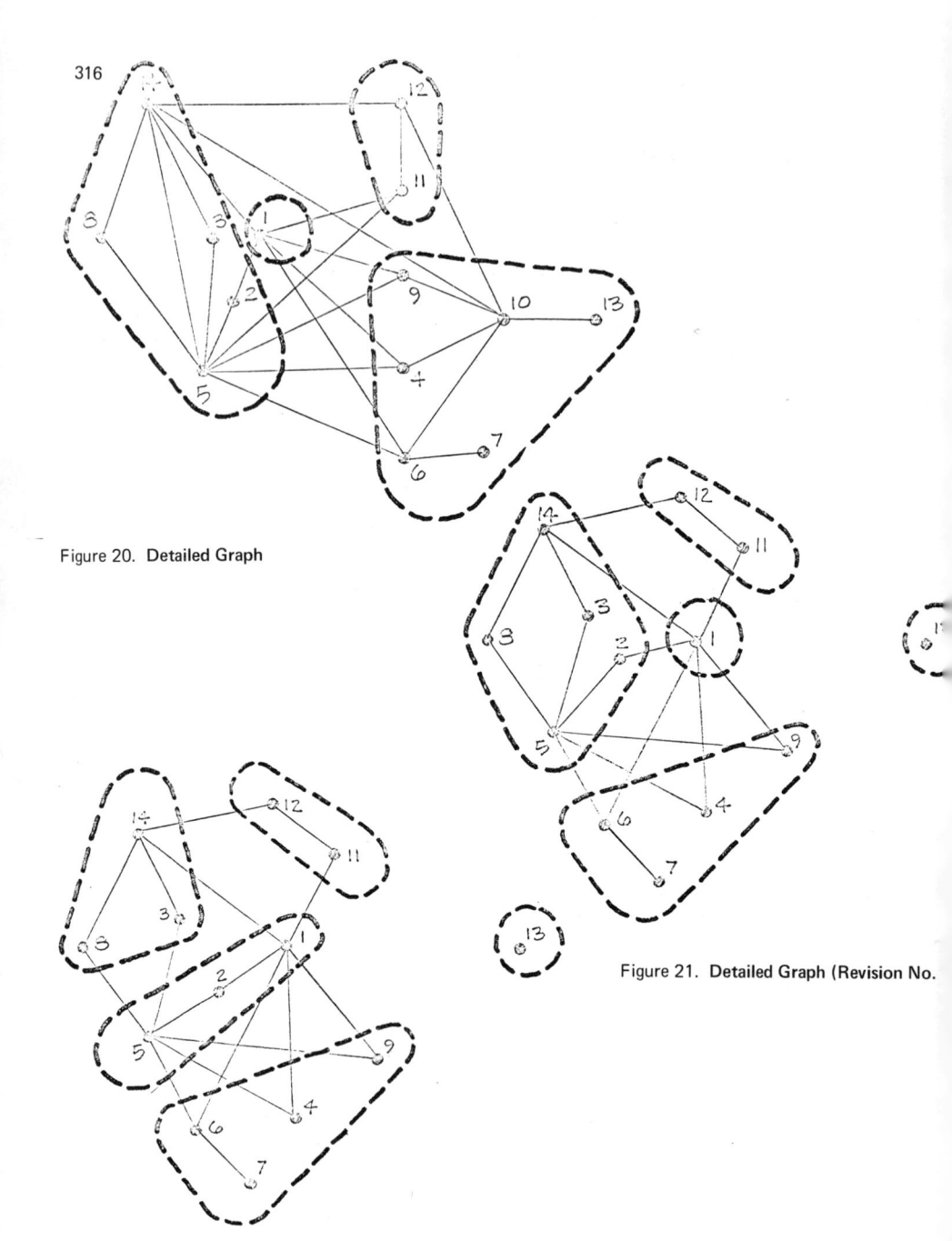

Figure 22. Detailed Graph (Revision No. 2)

The capabilities of this approach are:

1) Element groups and relationships are isolated as a basis for further study.
2) The architect, or user, is forced to be specific in the definition of the problem.
3) The ability to study alternate concepts very early in the design analysis phase is a definite advantage

The shortcomings are:

1) MATRAN-I is limited to YES and NO relationships only. This will be improved with completion of MATRAN-II, currently under development.
2) MATRAN-I cannot describe negative relationships. This, too will be implemented in MATRAN-II.

In spite of these shortcomings (above), MATRAN-I as applied to the courthouse problem pointed out the feasibility of the approach, and the need for further research and development in this area.

References

1 The authors of this paper wish to acknowledge the assistance provided by various groups and individuals. The firm of Arendt/Mosher/Grant, Architects in Santa Barbara has not only provided comment and encouragement to the project but has also contributed to its financial support. Daniel, Mann, Johnson and Mendenhall of Los Angeles has provided free access to a UNIVAC 1108 as well as their own in-house system.

Mr. Miller in particular acknowledges appreciation to Lacy Johnson and Peter Janca of the General Electric Company, Information Service Department in Santa Barbara for the services of the Mark-II time sharing system provided during the early phases of the project. Also, to the early comments of Gary Shelton and Ted Harsham. Final appreciation is given to Mrs. Christine Stanley for the final preparation of the paper.

2 Alexander, Christopher and Marvin Manheim, *HIDECS 2: A Computer Program for the Hierarchical Decomposition of a Set with an Associated Graph*, MIT Civil Engineering Systems Laboratory Publication No. 160, 1962; and Christopher Alexander, *HIDECS-3: Four Computer Programs for the Hierarchical Decomposition of Systems which have an Associated Linear Graph*, MIT Civil Engineering Systems Laboratory Research Report R63-27, 1963.

3 Alexander, Christopher, *Notes on the Sysnthsis of Form*, Harvard University Press, 1964.

4 Bernholtz, Allen and Edward Bierstone, "Computer-Augmented Design," *Design Quarterly*, 1966-1967.

5 Propst, Robert, *The Office: A Facility Based on Change*, The Business Press, Elmhurst, Illinois, 1968.

6 This program was originally devised by William Miller. The same program has been recoded in FORTRAN-IV and serves as the matrix manipulation subroutine in MATRAN-I. Vahe Khachooni has been responsible for adding the many embellishments now enjoyed by MATRAN-I users.

Donald A. Watson
Architect
Guilford, Connecticut

MODELING THE ACTIVITY SYSTEM

Of all the types of information on which an architectural designer depends, the category of activity is often the least developed. It is no wonder that many contemporary buildings are unresponsive to behavioral requirements. To develop knowledge about the use of space, environmental researchers will ultimately depend upon an operational theory of human activity, as well as a language by which designers can understand and communicate the concepts of the activity program. These needs form the background to the case study offered in this paper.

This paper presents a case study in activity analysis wherein patterns of use are observed and plotted as they interrelate in a building conceptualized as an "activity system." A three-dimensional model is used to correlate and display social interaction and physical location in terms of frequency and intensity of use. A taxonomy for activity analysis as well as design principles for accomodating dynamic activity programs follows directly from these considerations.

Analysis of an Activity System: A Case Study

Activity can best be identified and defined through direct observation of existing situations. Social psychologists have provided a variety of analytic frameworks for this task, such as time and motion, [1] social interaction,[2] and human communication.[3] The criticism of such approaches can be anticipated. Communication, interaction, and movement are inseparable manifestations of human activity. Furthermore, a particular activity cannot be fully understood without considering its role in the context of a larger activity network.

The following case study describes the activity system in an existing building, the Art and Architecture Building at Yale University. The data gathering method is one of "structured observation." In-depth interviews, questionnaires, and logs were not included here, but were employed by a separate team in order to compare various research methods.

In spite of the intent of the design of the building, the individual work stations in the building, that is, the faculty and student desks are easily identified and plotted. These points, plus the locations used for classes and informal gatherings, establish the recurrent trip patterns within the building. (See Figure 1). Only the activity places which actually are used are plotted. Such areas can be said to physically structure social interaction, even if they are as simple as a few chairs in a corridor outside an office. In addition, indirect communication is effected in buildings through the use of bulletin boards, display walls, and exhibitions and their locations are also noted.

Figure 2 indicates the trip patterns that are generated by these activities and identifies the separate orbits of the students of each department. Figure 3 shows the percentage of use of each circulation path and activity area of the building (the number of users from the total population in a twenty-four hour period).

Three different types of areas can be distinguished: (1) low-use, long-term areas, or "private zones" which include the circulation spaces and areas taken for stand-up functions (information desks, displays, coat racks); and (3) the mediating or "semi-public zones." The latter are of particular interest. It is in the semi-public zones that one observes the interactions and interface activities crucial to community relationships. All semi-

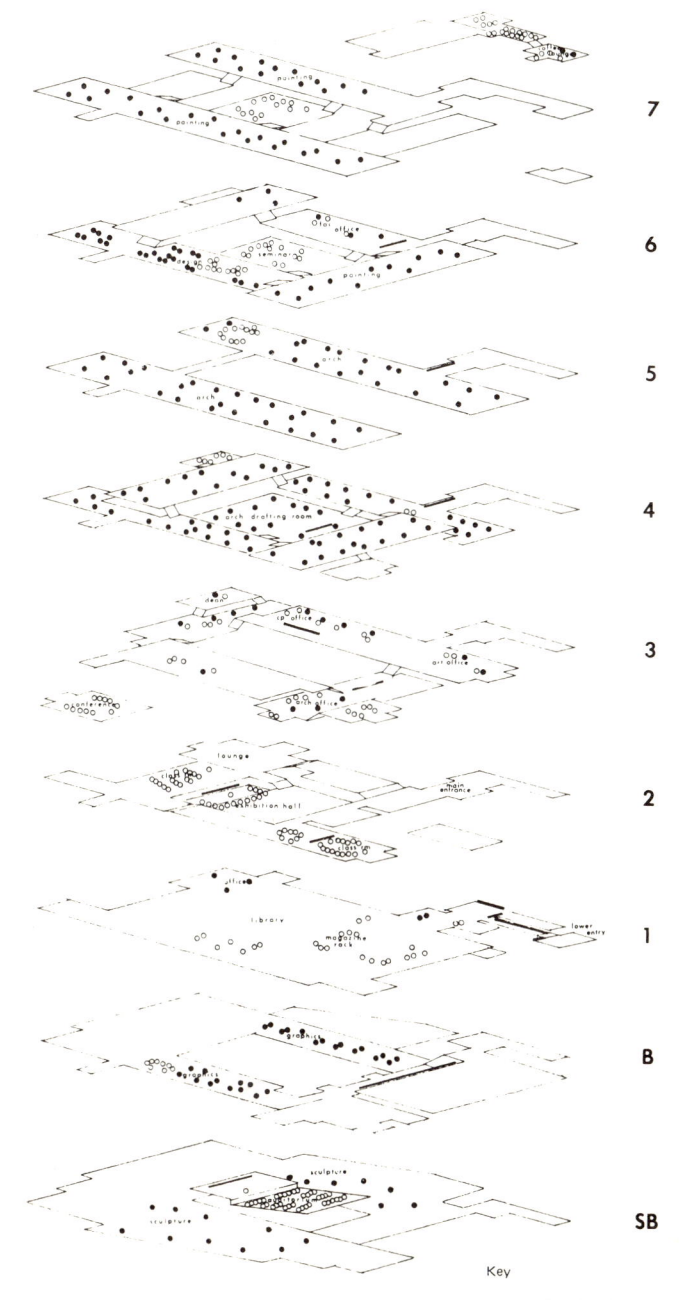

Figure 1. Interaction Patterns Sch. of Art & Arch. Yale University

Key
● assigned work station
○ social interaction (physically structured)
---- *indirect communication device

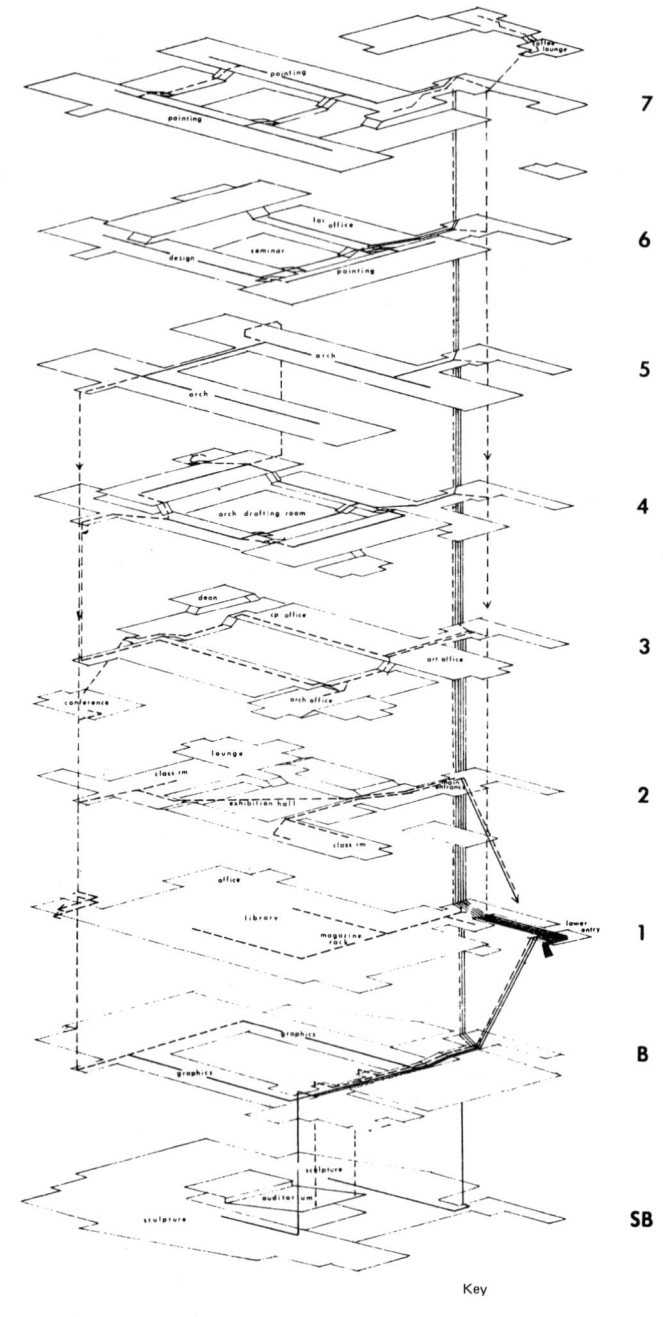

Figure 2. Movement Patterns Sch. of Art & Arch. Yale University

Key
——— trip to assigned work station (by dept)
- - - trip to semi public zone

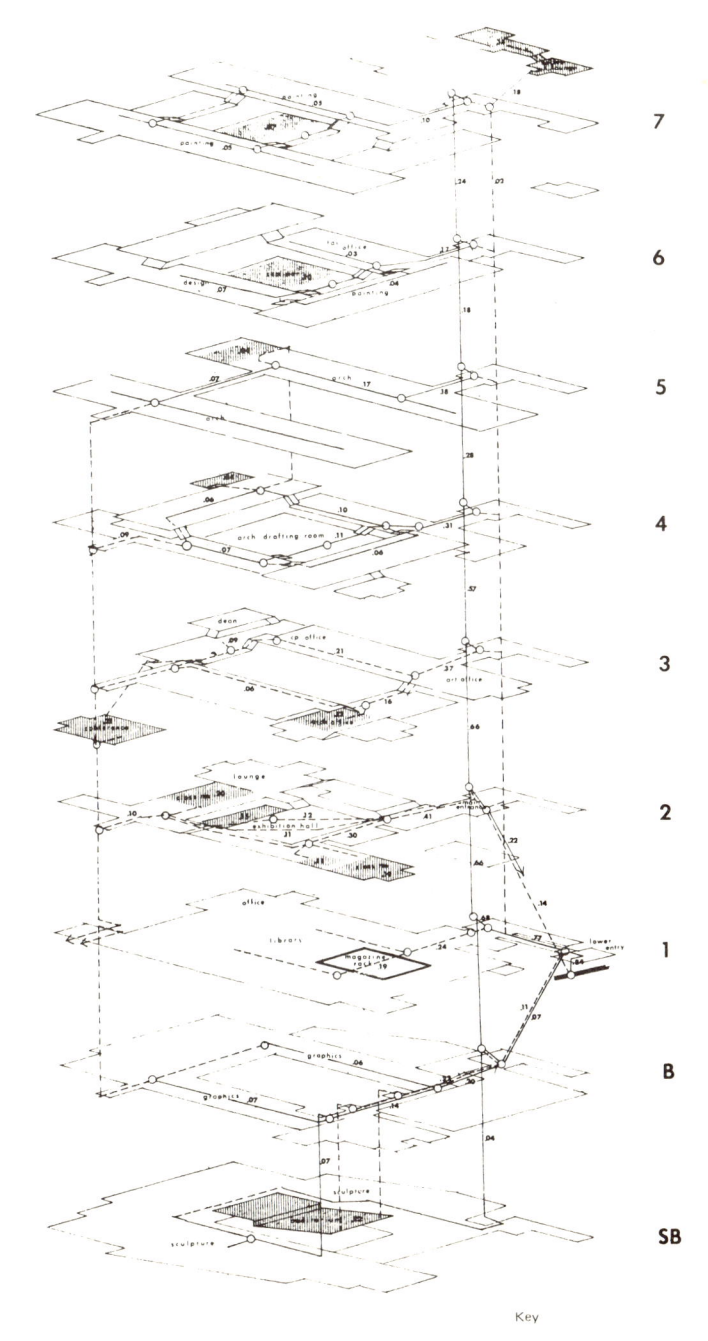

Figure 3. Intensity of Use Sch. of Art & Arch. Yale University

Key

semi public zone activity dependent
semi public zone location dependent
% of total user participation

public zones are usefully distinguished. Location dependent areas are those where a successful pattern of social interaction depends upon its immediate and accessible location on existing circulation paths (for example, the "magazine rack" in the Yale Building). Activity dependent areas are those which draw the user because of the assigned activity and is independent of its location in the circulation network (for instance, the "coffee lounge"). Such an area then generates its own trip patterns.

When this information is displayed in a three dimensional location model such as Figures 1 through 3, interrelationships of activities and areas are easily grasped. For example, a change in an assigned area, activity, or circulation link will affect the total system. Operational conflicts are readily identified as well. Interaction among students of different departments is hindered by the assignment of departments to work areas on separate floors, the lack of interdepartmental interface zones, and the failure of the main Exhibition area to serve its intended purpose as a place of informal social interaction (data from November 1967).

Time-lapse photography was used experimentally as a communication device. Its success in this function deserves comment. One of the changes considered by the administrators of the Yale Building was the closing of the lower entry door. Time-lapse films were made concurrently of the several interconnected spaces between the lower entry and the library. Various users, including library personnel, administrators, and students were then able to view the film and to comment on the value of the activity pattern in the library, then seen to be quite successful as an informal meeting and relaxing area. The research team could not hope to elicit the same response using the more abstract diagrams shown above. The time-lapse film, a form of descriptive model, may be worth the expense as a communication device in critical problem areas. The diagrams offer more as analytic models.[4] The activity system could also be described in mathematical form, and with the introduction of probability functions, attempts at a predictive model could be made.[5]

Concepts of Activity System Analysis

With the above analysis in mind, the interdependent elements of activity systems can be set forth. They are: linkages, the connecting circulation paths; activity zones, areas that have various use potentials due to their locations or assigned activities; and the activity cells, units of activity that develop spontaneously or as programmed.

Linkages and Activity Zones. The actual circulation patterns that result from the use of activity zones, as distinct from available but unused physical circulation areas are the linkages. In some cases, principal linkages develop despite the original physical design—for instance, the narrow corridor of the lower entry of the Yale Building.

The linkages between activities create what, in the physical areas of the building, can be conceived as a magnetic field which obtains varying degrees of intensity and frequency of use. These fields of influence are the activity zones which are, as defined above, public, semi-public, or private.

Activity Cells. Architectural programmers and theorists have sought to define a "least common denominator" or basic unit of analysis of activity systems from which to construct a structuralist conception of human activity.[6] As seen in actual environment settings, however, activity subsets are so dependent upon their systemic relationship to both larger and smaller elements that they can be taken as separate entities only as a provisional operational convenience. As such, there are a number of levels at which a pattern of activity can be partitioned into analytic elements: goals and purposes, activity groupings, separate activity units or cells, individual transactions or interactions, the messages contained in each transaction, and ultimately, the cues perceived by the user from the physical and social environment that trigger such messages. Each of these elements can be seen as a subset of the next largest element, as depicted in Figure 4. A given activity program is a semi-lattice of such elements. A given activity cell "A" is a unit common to several activity groupings. For instance, an "interdepartmental seminar" is part of the curriculum of two departments. In turn, the activity cell is composed of interaction sequences that are also found in other activities, the cell "interdepartmental seminar" containing three sorts of interactions: faculty delivery, student discussion, slide presentations, and so on.

It may be possible to completely chart the cues and the interactions that compose an established social program, as for example, to analyze the contribution of light and sound to the meaning of a religous service. The point however, is that this sort of completeness and stability is hardly applicable to the programs that are presently challenging environmental analysts. Activity cells and groupings in modern life are subject to constant change due to learning on the part of the participants, new participants, technological developments, and other modifications in the physical, social, and communications context. Activity groupings that appear to demand constant adjustment may be by nature transient, mobile and dynamic in ways that require an indeterminant physical context. After all, users are more often able to organize themselves and to make effective changes in their environment beyond the most sophisticated prediction and control of the building designers and sponsors.

The architectural program, however, does constitute the initial set of decisions to which the subsequent users respond. The conception of the interdependence of elements in an activity system suggests that constant adjustments be made by relocating activities and changing linkages. Thus, in making decisions during the design and use of the building, one should distinguish the extent to which an activity is a generator, and its operation independent of location, or a generated cell, in which case the operation develops as a result of its initial location.[7]

It is worth digressing for a moment to consider an hypothesis that emerges when explicating changes in space use over time. This involves the notion of the primacy of the patterns of communication. The participants in a given set of activities are in effect members of a community and establish some means of controlled communication in relating to one another. If there are different types in the community, the means for each sub-community are as different as their varying purposes and may also be in a private language that is undetected by the outsider. In some cases, the physical spaces provide a principal means of communication and control (examples 1 and 2 below). In other cases, however, the spaces are unimportant or are used in a manner not intended in the original design, ostensibly ineffectively, (examples 3 and 4 below).

1. The Coffee lounge in the Yale Building was set up three years after the building was put in use in an area originally used as a Guest Apartment. Before the Coffee Lounge was established, weekly "coffee hours" were organized to bring together members of different departments. Once the Coffee Lounge was open, the original "coffee hours" were unable to attract participants and were discontinued. It would appear that the on-going social-interaction made possible by the Lounge more than met the need for informal interdepartmental contacts.

2. Although the Exhibition Hall originally contained a "student lounge," it was very little used for informal social interaction. It would appear that the Hall was too far removed from the main circulation to attract incidental use, and instead the Library served this purpose. However in May, 1969, the Exhibition Hall was used for the all-school Assemblies that characterized the Spring Term in many schools. Bulletin boards, tables, chairs, and a coffee table were set up, mostly by students. As a result, students would make the special trip up to the Hall in order to "check out what's going on."

3. The Lower Entry of the Building is the predominant entry and one of the few spaces used by students from all departments of the School. At some points, the corridor is 3'-6" wide. Both personal and "community" notices cover the walls and doors.

4. The Faculty Office Level also serves as a circulation route to a corner Class Room which was not originally intended for student use. To have students walk through secretarial and faculty areas may interrupt the office work, but it also provides the context for faculty-student encounters. Secretaries have adapted to the situation to the extent that departmental communication is effected "on sight." So many faculty and students walk through that no other effort is needed to contact a person for incidental business.

Strategies for Design

To complete the set of ideas that follow from the activity analysis undertaken in the case study, I propose to mention briefly the design principles by which physical systems can meet the dynamics of activity systems over time. They are applicable to the architectural programs for hospitals, schools, and research facilities which are particularly subject to the internal and external changes mentioned earlier. The following are four general principles or strategies for making adjustments in an activity system together with the prerequisites of the corresponding physical plant.[8]

1. Relocate Activity Cells. In every activity system, there are a number of principal generator activities that establish the main patterns of the activity system. Their physical prerequisites include relocatable furniture, walls, and so on, and the use of short term and re-useable materials. These would allow adjustment in the system if a generator activity becomes obsolete or discontinued, a particular activity changes or is regrouped, or the original space is required for another activity.

2. Create New Linkages. Pedestrians will seek out the path that appears to require the least effort in going from one point to another. New linkages can be effected if it is possible to add or eliminate stairs, close off or open elevator stops, or open or close horizontal connections such as halls and doors. Conditions that can be corrected are traffic overloads in existing activity zones due to undesired traffic, and previously unrelated or separated activities that are seen to require more direct connection.

3. Prepare a Master-Plan that allows expansion of each activity zone. This option is particularly important for an activity system that will be enlarged as, for example, a multi-building complex or city. Here, transportation technology such as elevators, shuttles, and the like, can be used to link similar activity zones that are physically distant, for example as in the Yale Building where the elevator extends the "public zone" vertically to the 7th floor Coffee Lounge. This strategy avoids the overload inevitable with "single point" circulation networks and allows new activity cells to be added in a chain of activities that require time-distance proximity.

4. Allow for separate service routes. This principle, familiar to architects and planners, is simply to provide separate routes for service needs in order to avoid disrupting an existing activity system by service installations and repairs.

These design considerations result from the investigation of the dynamic nature of activity systems in space. Obviously, they are limited in this case by the interests of the analyst who is an architect. Others with different training and backgrounds hopefully will see different possibilities for the form of activity analysis described in this paper.

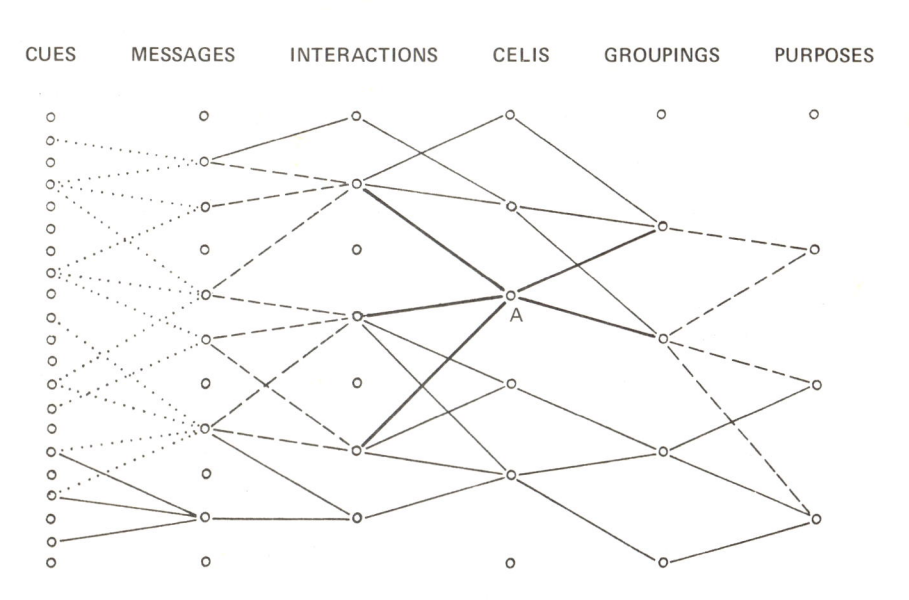

Figure 4. Analytic Terms of an Activity Program

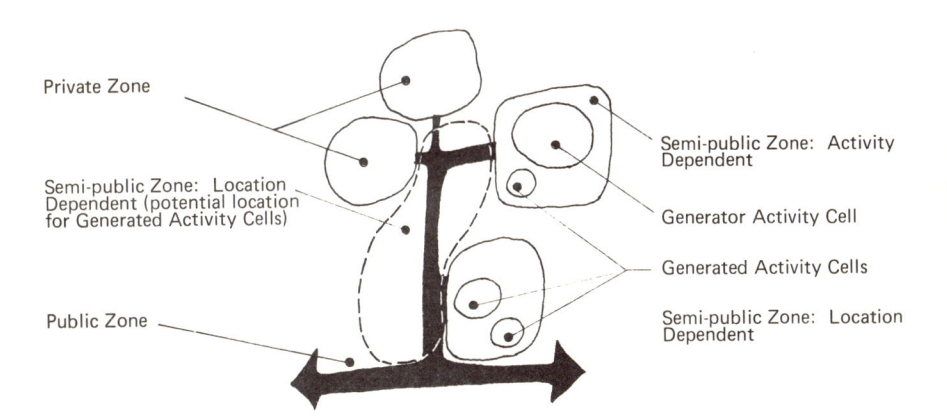

Figure 5. Illustration of Terms

References

1   A. G. Shaw, *Processes and Practices of Motion Study* (1926); R. S. Weiss and S. Boutourline, Jr., *Fairs, Exhibits, Pavilions, and Their Audiences.*

2   J. Collier, Jr., *Visual Anthropology;* R. S. Weiss, *Processes of Organization.*

3   P. Watzlawick, Beavin, and Jackson, *Pragmatics of Communication;* T. J. Allen, "Information Channels in R & D Organizations" (in preparation—Sloan School of Management, M. I. T.)

4   A. Froshaug, "Visual Methodology" in *Ulm 4* April, 1959; "Circulation Graph" in *Ulm 19/20,* August, 1967; D. Watson, "conceptual Models in Design" in the Proceedings, Interdisciplinary Conference on Decision Making Aids (forthcoming, Ohio State University Press).

5   V. Bazjanak, "A Study of Movement in Educational Buildings", paper delivered at "Campus Planning Conference", Washington University, St. Louis, Mo. April 1969.

6   D. Haviland, "Activity/Space", AIA Architect-Researchers Conference 1967; R. Lindheim, "Putting Research to Work" in *AIA Journal* Feb. 1966. C. Alexander and B. Poyner, "Atoms of Environmental Structure"; F. Duffy, "Office Buildings" in *DMG Newsletter* Vol. 3, No. 2 (Feb. 1969).

7   Cf. S Chermayeff and A. Tzonis *Advanced Studies in Urban Environments,* (1967) (limited edition: Yale University Press).

8   D. Watson, "Working Papers: The Study of The Environment" in *Connection,* Spring Issue, 1969 (Dept. of Architecture, Harvard University).

Murray Milne
School of Architecture and Urban Planning
University of California at Los Angeles

WORKSHOP ON DECISION MODELS

There is something beautifully fresh and optimistic in naming this conference "The First Annual..." The organizers of this gathering felt that at least a few people were ready to come together to consider the possibility that the physical environment might be designed in terms of a fundamentally new set of values and concerns. The response here shows that they are probably correct, but it will be many years before we know for sure if we have, in fact attended the birth of a healthy new discipline. We can only hope that in the interim this infant idea is not smothered to death by rhetoric without substance.

Karl Deutsch and Thomas Kuhn have observed that the birth and rebirth process of any discipline cycles through at least four distinct layers or stages. As a reaction to a state of confusion and disarray, the first stage is a period of naming and list-making. In this phase all the conceivable entities and their attributes are identified. A certain amount of argument about questions of relevance and value also goes on at this time.

Second comes the taxonomic or classification phase, which attempts to answer questions about the relationship between these fundamental entities, their order and priority. This usually requires a certain amount of quantitative measurement involving everything from counting and weighing to the design of experiments and the statistical analysis of data. This encompasses the classic kind of normative research.

In the third phase, models, simulations, and conceptual constructs are proposed and tested. A great deal of consideration is given to the kind of questions that may legitimately be asked and the kind of techniques that may be employed in their solution. Specialized shorthand notation systems may also appear, because by this stage enough people usually agree on the definition of terms and the relationships between elements. About this time, ingroup jargon often develops because the practitioners are attempting to legitimize their activity and to protect it from outside control. The farther this activity is from the general public, the more incomprehensible will be its jargon.

The final phase is one in which universal paradigms are constructed or are modified to account for anomalies in the evolving discipline. Perhaps most important, this becomes the rebirth phase whenever a brand new theory is porposed which throws the discipline into complete disarray and confusion, forcing the cycle to begin again.

Every instant in the evolution of a new discipline involves one or more of these phases, although not necessarily in this particular sequence. Interestingly enough, fewer and fewer people can productively contribute at each successive level. If a new viable theory or paradigm emerges, it is usually put forward by only one or two individuals. But after this breakthrough, a great many people at the bottom of the pyramid may become actively engaged in identifying and classifying activities.

It seems to me that many of the people at this conference are beginning to grope for the entities that are fundamental to this approach. They are also beginning to sort out and classify the order and relationship of these basic elements. Among the people who gave papers here, there seems to be some commonly-held though vaguely inexplicate agreement on the kinds of problems and questions that are legitimate and the procedures or methods that can be used to arrive at solutions. But as far as I can see, there is as yet no new discipline of environmental design.

In fact, the three papers presented in the session on Decision Models illustrate many of these points. One proposes a new taxonomy, the second is a method for establishing physical relationships, and the third is a descriptive model of activities.

The first paper, "Towards a Theory of Architectural Description," proposes a taxonomy for naming and classifying relevant attributes of the designed environment. Two closely related influences seem to be evident here; one is the associative storage and retrieval aspects of list processing computer languages and the other is Noam Chomsky's theory of transformational grammar. Papers like this are absolutely necessary to establish the cornerstones of any new emerging discipline.

The paper entitled "A Method for Grouping an Interrelated Set of Elements" describes MATRAN, a computerized procedure for architectural planning. Although this particular program was written by a team of architects, it follows in the tradition of CRAFT, FLP, ALDEP, and many other spatial allocation programs which were originally written by Industrial Engineers to automatically solve plant layout problems. Spatial allocation models make up a significant percentage of the more than 450 CLUSTER-finding computer programs that Geoffry Ball of Stanford Research Institute reported on in 1966. By now this number will have grown considerably. This is an important area of research as environmental designers begin to make increasing use of these specialized techniques for finding the relational structure of the elements in a system.

The third paper, "Modeling the Activity System," offers a description of the organizational structure and physical relationship of the various systems of circulation and activity that can be observed in a building. Here we have an excellent example of the contention of metalinguists like B. L. Whorf that the way we perceive our world is determined by the pattern and structure of our language. By carefully observing the occupants of a building and then isolating and naming generic activities, it is possible for an environmental designer to see, for the first time, a total system or pattern of activity where (for him at least) something closer to randomness existed before.

Ralph Brill - Eric Castro
Node Four Associates, Incorporated, New York

A. J. Pennington
Drexel Institute of Technology, Philadelphia

THE COMMUNITY DEVELOPMENT WORKSHOP

    There are encouraging trends in our society toward expanded individual participation in decision making processes. The decentralization of business operations, the complexity and diversity of governmental structures, and more recently the demands for student participation in shaping educational programs are all symptoms. Perhaps the most significant of all is the assumption of responsibility by residents of urban communities for decisions affecting their own environment. These developments represent a revitalization of democratic principles and hence are greatly to be welcomed. Two great dangers exist, however, one is that the complexity and specialization of tasks in a technological society will prevent a true participatory community from emerging. Democratic forms flourished primarily in agrarian societies. Many people have taken the pessimistic position that democracy and freedom are inconsistent with an industrial civilization, and have proposed either accepting such a situation and learning to live with it, establishing small communities employing primitive technology (Skinner's Walden Two, for example). or violently destroying the present economic base in hopes that somehow a more humane political structure will emerge from the ashes. We reject all of these.
    Another obstacle to the development of a more participatory society is the great difficulty of communication among various sub-groups. This is particularly true with regard to emotional polarizations concerning race, class, and lifestyle, and also over the traditional issues of economic and political power. Again, the pessimistic position asserts that communication is fundamentally impossible on these questions, and hence that some form of passive or active despair is necessary.
    We also reject this position, but fully recognize that the pessimists may be right. There are certain problems for which no solution exists. The disadvantage of pessimism, however, is that it has a tendency to be self-fulfilling, i.e. the predicted and feared catastrophic outcome tends to come true through the prediction and fear itself. A realistic optimism, on the other hand, leaves two alternative futures open: failure, in which case we are no worse off than before, and success, in which case we are much better off. Hence, we have chosen to take the optimistic position. This paper describes some experiments designed to help overcome the two obstacles to realization of a humane, technologically based society which were stated above:
    (1) The obstacle of technical complexity
    (2) Emotionally based obstacles to communication.

The Community Development Workshop

    The Community Development Workshop (CDW), now in its formative stages, is the name we have given to a collection of techniques designed to implement participation in the planning process. It is an eclectic approach, making use of current work in the psychology of groups, mathematical modeling and systems analysis, simulation gaming and other techniques. Hence, the CDW is more an attitude and orientation than a specific technical method. The following outline for one session indicates some of the psychological techniques employed, i.e. Confrontation, Synectics, and Encounter Micro-Labs.

Outline for February 22-23, 1969
Tarrytown House, Tarrytown, New York

1. Format: Weekend marathon, 30 hours, noon Saturday to 6 P. M. Sunday.
2. Participants: up to 24 people drawn from Node Four Associates, Drexel Institute of Technology, and the future Marcus Garvey Gardens community.
3. Purposes:
    (1) For the above to establish a working relationship.
    (2) To develop new mechanisms for goal-directed group effort.
    (3) To develop a proposal for financial support of the same.
    (4) To develop specific task assignments in planning, technology, management, etc. for the community.
    (5) To have fun.
4. Schedule (approximate):
    1½ hours. Lunch and socializing.
    2½ hours. Confrontation.
    1 hour. Coffee break and party.
    2½ hours. Encounter Micro-Labs.
    1½ hours. Dinner and socializing.
    1½ hours. Synectics Groups.
    ½ hour. Music break.
    1 hour. Art break.
    ½ hour. Snack and socializing.
    2½ hours. Confrontation.
    3 hours. Sleep break.
    1 hour. Breakfast and socializing.
    1½ hours. Synectics groups.
    ½ hour. Music break.
    1½ hours. Proposal and assignment writing.
    2½ hours. Encounter Micro-Labs.
    1½ hours. Lunch and socializing.
    2½ hours. Confrontation.
    2 hours. Review and evaluation, written and oral. Party.

Confrontation

Rule 1: No physical violence or threat of physical violence.

Rule 1 above is superficially very simple, yet when explicitly stated becomes a powerful psychological mechanism for inducing "confrontation," i.e. a high degree of honest communication. The emotions involved may be hostile or cooperative. By explicitly recognizing the latent potential for violence in even the most "civilized" setting, and then overtly and emphatically ruling out such behavior small group interaction takes on many new dimensions. For this reason it is desirable that such a group contain several members with prior experience in Confrontation.

It has been found that 12-15 participants is about optimum, and that 2½ hours is a reasonable length of time for a session.

Synectics

Synectics is a creative problem solving technique developed by William J. J. Gordon, President of Synectics, Inc. of Cambridge, Mass. The word "Synectics" was coined from "synthesis" and "eclectic" which refers to the creative technique involved, i.e. the synthesis of eclectic ideas. The concept is that creation implies the combination of diverse (and apparently improbable) ideas. The "operational mechanisms" above are designed to stimulate creative activity on the part of task-oriented small groups.

A Synectics Group could range in size from three to about ten or twelve, depending upon the nature of the problem. Two and a half hours is a reasonable time for one session, but this can be quite flexible.

The Synectic process:
    (1) Making the strange familiar (fact gathering and analysis).
    (2) Making the familiar strange (creation).

Operational mechanisms:
    (1) Direct analogy.
    (2) Personal analogy.
    (3) Symbolic analogy.
    (4) Fantasy analogy.

Encounter Micro-Labs

Representative Experiments: Physical unlocking, milling, blind walk, falling, lifting, stretching, pushing, breathing, association, fantasy, doubling, role playing, dyads, introspection, essence game, wordless meeting, inclusion/exclusion, verbal encounter, physical encounter.

Encounter micro-lab exercises emerged out of the work of a number of groups including Esalen Institute, Big Sur, California; Orion, Tarrytown, New York; National Training Laboratories, Washington, D. C. and Bethel, Maine and others. They have been applied in a number of contexts and for a variety of purposes ranging from pyschotherapeutic to management development. The central theme is that of exploring alternate mechanisms for thinking, feeling, and relating.

There is no specific group size or time period indicated. The emphasis will be on trying a variety of exercises under carefully controlled conditions.

Another major ingredient of the Community Development Workshop concept is technical gaming - the use of computer-based simulation games to enable planners, architects, engineers, government officials, and members of the community to work together. Although a game situation is necessarily "unrealistic" to some degree it does provide a very useful vehicle for communication on a variety of questions ranging from the concept and philosophy of technical projects to detailed resource tradeoffs. Preliminary experience of this type had been gained with the game "City I" developed by Peter House, Phillip Patterson and their associates at the Washington Center for Metropolitan Studies (now established as Envirometrics, Inc., Washington, D. C.). A description of City I follows.

City I

City I is played by nine teams with three to five members per team who act as entrepreneurs in a partially urbanized county divided into four political jurisdictions. The playing board is divided into 625 square miles most of which are unowned by the teams at the beginning of play. These land parcels may be purchased and developed by the teams during the course of the game. There are nine types of private land use which the teams can develop on a parcel of land: heavy industry, light industry, business goods, business services, personal goods, personal services, high-income residences, middle-income residences, and low-income residences.

Each of the nine teams is elected or appointed by elected officials to assume the duties of one of nine governmental roles, which are played simultaneously with the entrepreneurial functions common to all teams. The elected officials (the County Chairman and the Central City Councilman) must satisfy the electorate (the other teams) in order to stay in office each round. The Chairman team appoints other teams to play the roles of the School, Public, Works and Safety, Highway, Planning and Zoning, and Finance departments. The two residual teams play the Mass Media and Citizen's Organizations. The governmental departments build schools, provide utilities, build and upgrade roads and terminals, maintain roads, buy parkland, zone land, and estimated revenues.

Teams set their own objectives for both the public and private actions they undertake. Team decisions are recorded each round (approximately two hours in length) by a computer, which acts as an accountant and indicates the effects of the teams' actions on one another and on the county itself. The interaction of public and private decisions and their influence over time is illustrated by regularly provided computer print-outs. Even though conflicts may develop between urban and suburban interests, among businesses, and among governmental departments, teams often find that cooperation is equally as important as competition in fulfilling their objectives.

Participants of a play of City I receive a comprehensive view of central city and suburban growth and development. Teams are free to try alternative solutions to problems created within the model by their own actions in previous rounds. The governmental, economic, and social systems of the model are defined broadly enough so that they may be altered by a team majority vote. Through their own actions players become aware of the interrelation of public and private decisions, the interdisciplinary scope of urban problems, and the effect over time of public and private decisions.

Another game, called "BUILD", specifically oriented toward local community development within the urban context is now under development at Drexel Institute of Technology. The prospectus for BUILD follows.

BUILD

BUILD will be both a mathematical model and a role-playing computer game designed to assist in advocacy planning of new communities within the city. The model will be designed to represent the typical situation of extreme deterioration of housing, services, and economic activity in an urban area designated for rapid physical transformation, but with a major emphasis on preservation of community values. It is intended that the game itself will provide a communication medium among community members and outside professional planners.

The structure will be simple and yet will provide the minimal framework which still typifies the political/social/economic interactive nature of the "ghetto". The model roles are broadly divided into three classes - business, government, and people (residents). The residents are further divided into the roles or working force, agitators, and parents. Business roles include both national and local business interests, and the builders, developers, or planners. Government roles include the local Mayor's office, Police Force, Board of Education, Health & Welfare, Zoning & City Planning, and the Social Planners' Office. This list identifies twelve distinct roles which broadly covers the community structure. It is easy to identify many more roles, however, this list represents a compromise between accuracy and size.

The model will include the detailed functions and interactions, both dollar flow and communication, of each of these roles.

As indicated above the Community Development Workshop is a set of techniques designed to enhance community participation in the planning process. It is an experimental and formative state. As time goes on we visualize the establishment of additional activities of the general type described above. These would be made available to the community on an essentially continuous basis. Preliminary experience with these methods has strengthened our original optimism. We hope that by this time next year it will be possible to report definitive positive results.

Henry Burgwyn
Department of Architecture
North Carolina State University, Raleigh

IDENTIFYING COMMUNITY LEADERS

The town of Zebulon, North Carolina, is unique in that it was one of few small towns to attempt an application for a Model Cities planning grant. Sharing with many other rural areas, a lack of planning and design expertise and realizing that its resources were insufficient to retain professional services, Zebulon sought assistance from the North Carolina State University, Department of Architecture.

The Town Attorney provided the architecture students a completed Model Cities application which was based on studies previously executed by various State agencies and industrial concerns. This application provided a general quantitative description of the symptoms of rural poverty. It was, however, inadequate as a foundation for any program which would substantially affect the causes of poverty in the area.

It was essential, then, that the students conduct their own investigation of the factors influencing life and growth patterns in Zebulon and its surrounding region. After a meeting with the Mayor's Advisory Council, the class was divided into five Task Forces to study the Region, Public and Private Sectors, the Physical Fabric, and the Social Fabric. Together these five Task Forces formed the Community Development Group which opened its Zebulon office December 3, 1968.

The group placed its emphasis on the planning of resources with special awareness of sociological implications. In the early phases there was a tendency to advocate the formulation of criteria and solutions which dealt with previously neglected segments of the community. In subsequent investigation, however, it became apparent that the decision-making structure of small rural communities required total citizen representation, since an urban neighborhood is comparable in scale to a rural community. The basic concept, then, became one of involvement of the entire community in the necessities and activities of community planning.

The group did not have the time or resources to personally contact each citizen in the community. Hence, it was necessary to identify the leaders in the community and persuade them to promote the activities of the Community Development Group. Since the Group was primarily concerned with the existing Model Cities program, an investigation was initiated to identify the community leaders of that program, which is the focus of this paper.

The most important research concerning power structures can be divided into two categories: reputational and plural analysis. The first, reputational, is characterized by the work of Floyed Hunter.[1] In general, his procedure is to determine community-power structures on the basis of judgments by community members who are considered "knowledgeable" about community life. These "judges" select names from lists of potential candidates based on imputed degrees of influence. Those persons most frequently selected according to the given criteria are said to constitute the power structure. One of the basic assumptions of the reputational method seems to be that power is exercised behind the scenes; it is next to impossible to obtain an accurate picture of the structure of power by attending to overt behavior. Thus, the researcher must rely on the inside information supplied to him by a panel of knowledgeable community members. For example, Hunter describes men of power in Regional City enforcing their decisions by persuasion, intimidation, coercion, and, if necessary, force.

This method of research has been severely criticized because of its arbitrary sleection of formal judges and the wide variety of issues encompassed. The reputational method is often said to only identify those persons who have the reputation for being influential. Nelson W. Polsby, for example, charges that the assumption of covert leadership often leads researchers "to disbelieve their senses and to substitute unfounded speculation for plain fact. Another danger underlying the assumption that power is exercised covertly is what Robert Dahl has referred to as the "fallacy of infinite regression." If observable behavior is not to be regarded as a reliable index of power, then one must search behind the scenes for centers of power found behind the actors who carry out the drama on stage. Once these covert power-holders are identified, the question arises as to whether there may not be another group of power-holders behind these, and others yet behind them, and so on.

While the reputational method has its faults, it is useful in describing the perceived distribution of power in the local community. If, for example, it can be shown that the way in which the power structure is perceived helps determine the way in which people react to it, reputations for power will definitely provide a useful variable in the study of power. It is for this reason that a "reputational stage" was included in the research.

The second category of research, pluralism, is defined by the work of Robert A. Dahl[2] and Nelson W. Polsby.[3] These men argue that no one person or group will make all the decisions on all issues either covertly or overtly. Rather, different leaders exert power in different issue-areas such as education, economic development, or urban renewal. Often there is little or no communication between these leaders. Polsby and Dahl advocate studying the actual decisions made concerning a specific issue. The degree of control exerted by various actors can be determined and then compared to the researcher's operational definition of power. The pluralists completely reject the idea of ruling "elites" and often take the notion that nobody dominates in a town. The third stage, issue-analysis, of my research draws heavily on the work of the pluralists. However, I have combined information concerning potential, perceived, and actual behavior in my operational definition of power.

Research Design

The research design consists of three stages. The first stage of the research consists of a positional analysis. This technique serves to define the potential power offices in the community's institutionalized economic, political, and civic structures. The "positional" analysis deals only with the formal, overt positions, not the covert ones which are many times more important.

In the second stage the "reputational" technique was utilized to determine perceptions of power and influence. This stage can be divided into five subsections, which are as follows:

a. images of powerful groups
b. images of influentials in different areas of life
c. perceived relative influence by local leaders
d. images of the unnamed power source
e. varying conceptions on how to stimulate public interest

The third stage of the research employed issue analysis as a means of determining the power structure. The Model Cities Program was used as the primary issue and analysis to render a true picture of the actual behavior of the men involved.

Once information has been gathered concerning the potential, perceptual, and actual aspects of power, an operational definition of power was established to evaluate the various actors. There are several issues concerning the community power structure in Zebulon.

a. nobody would dominate the Model Cities Program
b. the issue of the Model Cities Program is not important enough to merit the attention of the local citizens

c. several individuals specialize in the issue of the Model Cities program and thus dominate
d. one organization or minority group dominates the Model Cities program
e. the elected officials dominate the Model Cities program
f. there is a covert group which controls the Model Cities program behind the scenes
g. the ruling elite that the town follows on all issues dominates the Model Cities program

The most justifiable hypothesis was one consistent with Dahl's statement, "typically, a community is run by many different people, in many different ways, at many different times." I hypothesized that in the town of Zebulon, North Carolina, there are several affluent men who have taken a special interest in the Model Cities program. These men hold formal positions in the political structure and are responsible for making many technical or legal decisions. Through their administrative positions, these men control the Model Cities program policy. These men are not elected officials or particularly influential in any other issue-area. Also, these men feel that local communities should have complete control of Model Cities program.

In addition to these men who dominate the Model Cities program positively, there is a homogeneous group of men who oppose it. These men are not respected by the leaders of the Model Cities program. These "enemies of change" put the interests of the entire community, and hence the status quo, before the interests of the underpriviledged. These men have vested interests which conflict with the Model Cities program.

Testing of the hypotheses were divided into three analytic stages. The positional analysis, stage one, consists of identifying all formal positions in the economic, political, and civic structures. The information needed was obtained from documents and interviews with the formal leaders. Once this list of potential leaders was prepared, the second stage could begin. The second stage, the reputational analysis, consists of gathering data concerning how both formal leaders and ordinary citizens perceive power in the community. This information was obtained from a questionnaire administered in the town. The third stage, issue-analysis, involved a questionnaire and reconstructed all the actual decisions made concerning one specific issue, the Model Cities program. The decisions were reconstructed from documents, newspaper articles, and interviews. The questionnaire was used to determine how knowledgeable the citizenry was concerning the issue of the Model Cities program. Once data was collected pertaining to potential, perceived, and actual behavior, an operational definition of power could be defined. Decision-makers could then be evaluated in terms of this definition and the different hypotheses could be tested to determine the composition of the power structure.

The first stage, the positional analysis, identified 49 formal leaders in the community. The names of the most important leaders were included in a questionnaire to determine the strength of their reputations concerning the Model Cities program. The second stage, reputational analysis, was initiated to determine who were the perceived leaders of the Model Cities program. Both formal leaders and ordinary citizens were questioned. Questionnaires completed by formal leaders identified in the positional analysis were kept separate from those completed by ordinary citizens.

There are approximately 850 people over twenty years old in Zebulon. Six percent (6%) of these people completed the questionnaire; 55 people. The number of questionnaires completed by formal leaders was 21 or 45%. The number completed by ordinary citizens was 32 or 4%. An attempt was made to administer a questionnaire to all the formal leaders. This however, proved to be impossible. A random sample was used in defining the number of and location of the ordinary citizens that were to be questioned. The town was divided into sectors with an equal number of houses in each sector. Each house was given a number and the numbers were randomly selected.

The respondents were first asked to "please name the persons in Zebulon whose opinions would likely persuade you to accept their point of view."

Responses of formal leaders	voting in elections	borrowing money	model cities program
Ferd Davis	4	0	1
Dr. L. M. Massey	3	0	0
W. B. Hopkins	2	0	5

Responses of formal leaders	voting in elections	borrowing money	model cities program
John Mangum	2	1	2
Leary Davis	2	0	7
Aaron Lowery	1	0	0
Wilbur T. Debnam	1	0	0
T. E. Hales	1	0	0
Jerry Niswonger	1	0	0
Avon Privette	1	1	0
Amos Estes	1	0	0
Lucille Pippin	1	0	0
Robert D. Massey	0	6	0
Horace Gay	0	1	0
Hal C. Perry	0	2	0
Vance Brown	0	1	0
F. D. Finch	0	1	0
H. C. Wade	0	1	0
George H. Temple	0	1	0
Nobody	9	5	2
No Answer	4	3	6
Total Response	29	21	17

Responses of ordinary citizens	voting in elections	borrowing money	model cities program
Ferd Davis	3	1	0
Dr. L. M. Massey	1	0	1
W. B. Hopkins	3	1	3
John Mangum	7	1	3
Leary Davis	1	1	8
Aaron Lowery	1	0	3
Wilbur T. Debnam	1	0	1
T. E. Hales	0	0	0
Horace Gay	0	4	0
Robert D Massey	0	6	0
Hal C. Perry	0	6	0
F. D. Finch	0	1	0
Gloria George	0	1	0
Elmo Harris	0	0	1
Odell Wright	1	1	1
Mavis Montaque	0	1	1
Frank Kannon	0	1	0
Filmore Dunn	1	0	0
J. T. Locke	2	0	0
Dr. Ben Thomas	1	0	0
Nobody	7	3	3
No Answer	8	9	12
Total Responses	29	28	25

The respondents were next asked to comment on the following statement.

It is sometimes said in Zebulon that "they control things", "they get things done", "they get things their way", and so on. Who are *they* in such statements?

Responses of Formal Leaders		Responses of Ordinary Citizens	
leaders in civic affairs	7	town government	6
town government	5	W. B. Hopkins	3
the citizens of the town	3	Dr. L. M. Massey	2
other	6	leaders in civic affairs	2
no opinion	0	no opinion	7
no answer	0	no answer	9
total responses	21	total responses	29

The third question in the reputational stage focused directly on the formal leaders identified in the positional analysis.

Here are the names of persons living in Zebulon who have been mentioned on the front pages of the *Zebulon Record* recently. Please tell how statements about the town's Model Cities program by these people would influence your personal opinion, by checking one of the spaces after each name.

Responses by formal leaders	favorably	no opinion	unfavorably	never heard of the person
Dr. M. P. Grogan	15	3	0	0
Patrick Farmer	15	5	0	0
C. K. Corbitt	14	5	1	0
Griffin Todd	6	9	0	1
Ken Wilson	10	7	0	0
Ferd Davis	19	1	0	0
J. T. Locke	3	8	6	1
Ralph Bunn	11	6	1	0
Frank Kannon	10	8	0	0
Gloria George	11	4	2	0
Fred Vick	7	6	1	1
Floyd Edwards	16	3	0	0
Dr. L. M. Massey	17	2	1	0
John Mangum	11	6	0	0
Barrie Davis	16	2	0	1
Aaron Lowery	17	2	0	0
J. M. Potter, Jr.	13	4	2	3

Responses by formal leaders	favorably	no opinion	unfavorably	never heard of the person
Garland Crews	10	4	2	3
Wilbur T. Debnam	19	1	0	0
John Alford	16	1	1	1
T. E. Hales	18	1	0	0
Lizzie Askew	6	11	0	1
Leary Davis	20	1	0	0
Robert D. Massey	16	2	1	0
George D. Morgan	9	5	3	0
John H. Hilliard	10	6	1	1
W. B. Hopkins	18	2	0	0
Hal C. Perry	15	1	0	1

Responses by ordinary citizens	favorably	no opinion	unfavorably	never heard of the person
Dr. M. P. Grogan	18	3	0	3
Patrick Farmer	12	6	1	6
C. K. Corbitt	14	4	0	8
Griffin Todd	14	3	0	7
Ken Wilson	17	4	0	5
Ferd Davis	24	3	0	0
J. T. Locke	9	6	2	5
Ralph Bunn	9	9	0	3
Frank Kannon	14	7	2	1
Gloria George	14	7	1	2
Fred Vick	9	7	0	3
Floyd Edwards	13	6	1	4
Dr. L. M. Massey	17	7	0	1
John Mangum	17	6	0	1
Barrie Davis	19	1	0	4
Aaron Lowery	23	3	0	2
J. M. Potter, Jr.	12	5	1	4
Garland Crews	9	5	1	3
Wilbur T. Debnam	23	2	1	1
John Alford	12	6	1	4
T. E. Hales	21	3	0	2
Lizzie Askew	11	6	1	6
Hal C. Perry	20	1	1	3
Leary Davis	22	3	0	0
Robert D. Massey	19	4	1	2
George D. Morgan	14	5	2	3
John Hilliard	11	4	2	6
W. B. Hopkins	21	5	2	0

The issue analysis, third stage, was designed to determine the actual decision-makers involved in Zebulon's Model Cities program. The first part of this analysis was designed to test the respondent's knowledge of the Model Cities program. They are as follows:

1. Are you familiar with the Town's Model Cities program?

Responses of formal leaders		Responses of ordinary citizens	
yes	20	yes	18
no	1	no	14
total responses	21	total responses	32

2. The Model Cities program can help the Town of Zebulon grow and prosper.

Responses of formal leaders		Responses of ordinary citizens	
yes	20	yes	29
no	0	no	0
no opinion	1	no opinion	3
total responses	21	total responses	32

3. The design students from State University can provide assistance to the citizens of Zebulon in planning and carrying out a Model Cities Program.

Responses of formal leaders		Responses of ordinary citizens	
yes	20	yes	25
no	1	no	3
no opinion	0	no opinion	4
total responses	21	total responses	32

4. The design students from State University should plan Model Cities projects which mainly help poor people.

Responses of formal leaders		Responses of ordinary citizens	
yes	7	yes	16
no	14	no	11
no opinion	0	no opinion	5
total responses	21	total responses	32

5. The Federal Government should say how federal funds are to be used in Model Cities Projects.

Responses of formal leaders		Responses of ordinary citizens	
yes	8	yes	8
no	12	no	22
no opinion	1	no opinion	2
total responses	21	total responses	32

6. What would you suggest doing to get people in Zebulon interested in Model Cities program?

Responses of formal leaders	Responses of formal leaders	Responses of ordinary citizens
town government should conduct meetings at which the program is explained and the opinions of the citizens sought	16	19
advertize the program in the news media	4	7
no answer	4	7
select one area; make a success, then publicize the success		
no opinion	0	2
total responses	24	35

The second part of the issue analysis phase consisted of an analysis of all the decisions made in Zebulon pertaining to the Model Cities program. Interviews, newspaper articles, and official town government documents were used to reconstruct the decisions.

A small group of concerned individuals met in the fall of 1967 to discuss the new anti-poverty program in Zebulon. From this initial meeting evolved an alliance of prominent citizens, city officials, and anti-poverty workers. The alliance, composed of W. B. Hopkins, Hal C. Perry, Aaron Lower, Ferd Davis, Gloria George, and Leary Davis, decided to investigate the Model Cities Program. Various federal and state officials were interviewed and asked to meet with the Black citizens to explain the program. John Mangum, minister of one of the Black churches, evolved from the meeting as the spokesman for the Blacks and was brought into the alliance. During the Spring of 1968, this group gathered information and prepared an application for a Model Cities Planning Grant. John Mangum convinced the Black citizens of the merit of the program, Hal C. Perry and Ferd Davis influenced the white citizens, and W. B. Hopkins and Aaron Lowery influenced the City Council. In every instance, Leary davis provided the technical information concerning the Model Cities program. Hence, the alliance convinced the citizenry that improving slum conditions was an important issue and should be dealt with immediately.

After examining the three stages of the research, an operational definition of community power was developed. First, a man must hold a position from which he can have the potential to wield power. Second, he must be perceived by other people as a leader in a certain issue-area. If not, he will be unable to mobilize popular support in a time of crisis. Third, he must have the support of other perceived leaders in his issue-area. fourth, he must be actually involved in the decision-making process.

In Zebulon, the Model Cities program produced six men who meet the requirements defined above. They are as follows:

Leary Davis	Town attorney, district solicitor, ranked first in reputational analysis, deeply involved in actual decisions
W. B. Hopkins	Town manager, ranked second in reputational analysis, deeply involved in actual decisions
Aaron Lowery	Mayor, ranked third in reputational analysis, deeply involved in actual decisions
John Mangum	Mayor's advisory committee, ranked eighth in reputational analysis, deeply involved in actual decisions
Hal C. Perry	Chairman of industrial development, ranked seventh in reputational analysis, involved in initial decisions

	Mayor	City Manager	Town Comission	Little River Action Center Advisory Committee	Mayor's Advisory Committee	Town Attorney	Recreation Comission	Minority Housing Sub-committee	High School Principal	Wake County Board of Education	Negro Community Leader	Director, Little River Action Center	Town Planning Board	Deputy Sheriff	Merchant	Chairman, Wake County Housing Authority	Former Mayor	President, Chamber of Commerce	President Zebulon Industrial Project	Editor of Local Newspaper	President Rotary Club	Banker
Aaron Lowery	●																					
Willie B. Hopkins		●		●	●																	
Ralph Bunn			●																			
M. P. Grogan			●																			
Patrick Farmer			●																			
C. K. Corbitt			●																			
Floyd Edwards			●		●			●														
Leary Davis						●																
Fred Corbitt																						
Ken Wilson					●															●		
Jerry Niswonger					●		●															
Frank Wall					●																	
Odell Wright					●																	
Thurman Helper								●														
Stuart Brodie								●														
Henry Montaque								●														
Ralph Lewis								●														
John Mangum					●																	
John Alford				●	●			●														
Garland L. Crews					●				●													
Haywood Jones					●				●													
Rubic Gill					●																	
Luther Massey				●	●																	
Heber Windley					●																	
Griffin Todd				●	●																	
Ferd Davis										●												
J. T. Locke											●											
Gloria George												●										
Fred Vick				●																		
J. M. Potter													●									
Lizzie Askew											●											
John H. Hilliard													●	●								
Mrs. Howard Massey				●																		
Filmore Dunn				●																		
Joe Richardson				●																		
Mrs. Ferd Davis				●																		
Harold Davis				●																		
W. O. Blackley				●																		
Mrs. J. T. Locke				●																		
Frank Kannon															●							
Barry Davis															●						●	
Wilbur T. Dednam															●		●					
T. E. Hales										●					●		●					
Hal C. Perry																			●			●
Robert D. Massey																●						●
George D. Morgan															●							
Amos Estes															●							
Max Williams															●			●				
Avon Privette															●							

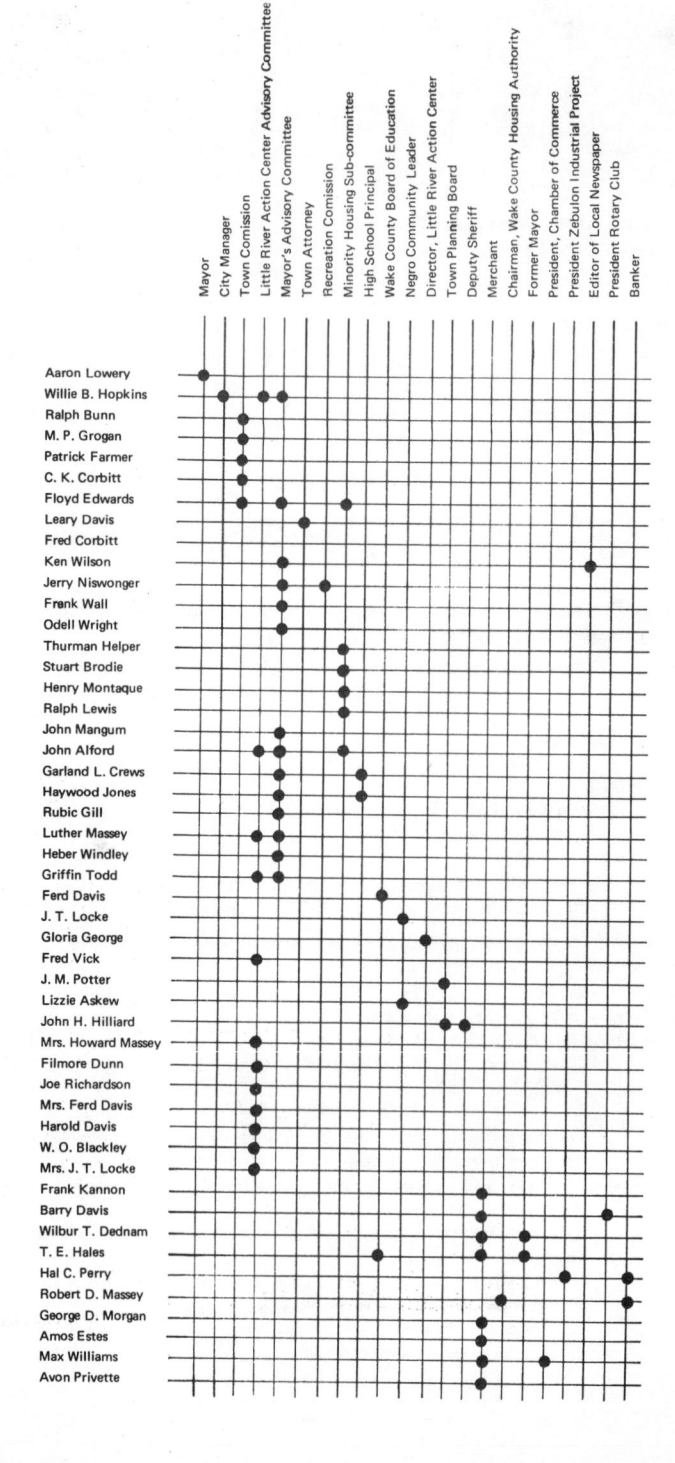

All of these men had the support of T. E. Hales, Wilbur T. Debnam, and Robert D. Massey. These men were perceived to be leaders in the Model Cities program; however, they were not involved in any of the actual decisions.

The initial hypotheses that certain non-elected officials who had become interested in the Model Cities program control the power is largely true. Both Leary Davis and W. B. Hopkins are appointed officials who, through their involvement in the legal and technical aspects of the Model Cities program, control the decision-making process. However, there are certain community leaders, Ferd Davis and John Mangum, who are influential in aspects of community life other than the Model Cities program.

It was very difficult to identify men opposed to the Model Cities program. However, many people felt that the program should mainly help the entire community rather than poor people. This was very evident in the responses of the formal leaders.

The Town commissioners were found to be relatively non-influential concerning the Model Cities program.

References

Henry Sanoff and The Community Development Group, This paper also appeared in *Planning for Rural Communities,* Design Research Laboratory, Carolina State University, June 1969.

1   Floyd Hunter, *Community Power Structure,* Doubleday and Company, Inc., New York, 1963.

2   Robert Dahl, *Who Governs.*

3   Nelson Polsby, *Political Theory.*

4   Robert Dahl, *op. cit.*

References

This paper also appeared in *Planning for Rural Communities,* Henry Sanoff and the Community Development Group, Design Research Laboratory, North Carolina State University, June, 1969.

1. Floyd Hunter, *Community Power Structure,* Doubleday and Company, Inc., New York, 1963.

2. Robert Dahl, *Who Governs Democracy and Power in an American City,* Yale University Press, 1961.

3. Nelson Polsby, *Community Power Structure and Political Theory,* Yale University Press, 1963.

4. Robert Dahl, *op. cit.*

David R. Godschalk
Department of City & Regional Planning
University of North Carolina, Chapel Hill

NEGOTIATE: AN EXPERIMENTAL PLANNING GAME

While often paying lip service to the involvement of citizens in community planning, in practice, professional planners have remained somewhat cool to this concept. The general values of democracy imply that citizens should participate in those decisions which affect their environment, especially those that involve spending public funds. Arrayed against this normative injunction have been at least two influential "common sense" ideas: the idea of efficiency through intra-agency plan preparation prior to public disclosure, and the idea of plan quality being forced down to the lowest common denominator when public group decision making is used. These ideas lie at the roots of such maxims as, "Too many cooks spoil the soup," and, "Who ever heard of a committee designing a creative plan?"

My purpose is to oppose the traditional common sense position and to argue that not only should there be more citizen participation in planning, but also that increased participation can lead to increased efficiency and creativeness in planning outcomes. I call my approach "collaborative planning", and define it generally as that process of community planning which attempts to incorporate the clients' activities and attitudes into decision making, both in order to educate planners and clients about the situational potential for planning, and to influence the behavior of planners and clients toward achieving more beneficial results for themselves and the community.[1] The key factor is the interaction between the citizen client, who supplies behavioral information, and the planner change agent, who supplies information on the possible alternative innovations.

To explore some of the relevant variables in collaborative planning theory, a game has been designed around a model of a general planning situation. The underlying conceptual hypothesis is that acceptance of innovative features in community plans varies with participation in the planning process, tending to increase as participation increases. A second hypothesis is that participation in planning for a controversial development project can cause citizens to redefine a conflict (threat) situation to one of competition or cooperation. At the operational level the combined hypotheses state that in a game situation players who have participated in preparation of a land use plan will tend to cast more affirmative votes for a low income housing project near a middle income subdivision than those who have not participated.

Participation in the planning process is the major independent variable. The group decision outcome, tallied by player votes, is the major dependent variable. Another set of important variables (furnished by the player and measured on a pre-game questionnaire) are existing attitudes toward risk, territory, cooperation, public versus private benefits, and racial and income integration in housing areas.

On the social-psychology learning curve, the game is a "starting mechanism", causing players to systematically reconsider their own attitudes toward various social processes and values. Thus, the game is a possible source of new attitudes, values, and social behavior.

A conscious decision was made to use a relatively realistic model of the contingencies operating in the situation. Hence the game outcomes cannot be considered completely unambiguous results, but they are expected to contain useful evidence bearing on the general hypothesis.

Game Situation and Roles

A nonprofit, low income, federally supported housing project has been proposed to be located on a tract of land near two developing middle income subdivisions. When the necessary request for rezoning the land for multiple family residences is submitted to the City Council, after having been recommended for approval by the Planning Board, middle income residents protest violently. Before voting on the request, the City Council appoints a committee of two representatives from each affected group to study the problem with the help of the City Planner. These representatives, two residents of the middle income subdivisions and two residents of the low income area which would supply the tenants for the housing project, are instructed to agree on a course of action, if possible, and to report back to the City Council.

The Council has been informed by the regional representative of the U.S. Department of Housing and Urban Development that, if the two groups can agree to support the project, the chances are good that a sizeable federal grant can be obtained. If only one party supports the project then the chances are less good, though still not impossible. If both reject the project, no grant application will be accepted.

In its general form this is similar to problems faced by many American communities in trying to locate housing areas or facilities for the poor.[2] In the ferocity with which they defend their residential areas against such "invasions", middle class Americans resemble in some ways various species of animals defending their territories. Their first actions are largely ritualistic shows of strength, designed to convince the invader that he cannot hope to succeed, but if he persists the result may be a fight to the death. Like sticklebacks furiously burrowing at the edge of their territories, the middle class residents throng to protest at public meetings, displaying petitions and verbal evidence of their aroused state. In both cases the participants are deadly serious.

The player roles are defined by the general body of information on value preferences and life styles of low income slum residents and of middle income suburban residents. The planner's role is that of a professional, who is both a technical expert and a social arbiter. To the dismay of many professional planners this technical concern for the public interest is often a weak role. As Norton Long has observed, "...the protagonists of things in general are few, vague, and weak,"[3] but at least they, the government and the newspapers, often tend to support each other.

Game Rules and Payoffs

At stake in the dispute if the project is built are the perceived possibility of a drop in property values for the middle income player, and the personal difficulties of being uprooted from a supportive social situation for the low income player. Rewards to the middle income player if the project is built come from the indirect effects of the federal investment in the community plus the public opinion approval of the newspaper, churches, city planner, and others supporting the project. He also faces the loss of opinion status from his neighbors who oppose the project, and whether the opinion outcome is a net profit or a net loss depends upon the point values which the player assigns to criticism from the newspaper as compared with his neighbors. The low income player's rewards if the project is built represent a direct return in the form of housing cost subsidies and improved environmental conditions, plus a minor return from opinion support sources. These values, taken from a pre-game questionnaire, can be displayed in a matrix form.

Game equipment includes a three dimensional scale model of the project site and surrounding area showing roads, buildings, and topographic contours at a horizontal scale of 1" - 100' and a contour interval of 10'. In addition, a zoning ordinance and map and a set of drawings showing the project site plan, building plans, and elevations, are used.

It is intended to play the game a number of times with players of different backgrounds. In each trial there will be control group, whose participation in planning is nil, and an experimental group who are invited to participate in designing the land use plan for the project site and surrounding area. Working on a model, the experimental group may try out various arrangements of land uses, houses, drives, parking areas, and the like. For a more complete test of the participation effect, it may be desirable to have a second experimental group who are given more than one planning session. The control group will simply be given a synopsis of the situation and an "official" land use plan.

Prior to getting together as a group, each player will be asked to fill in a brief confidential questionnaire indicating some factual data about his age, education, and occupation, along with some Likert scale questions on his attitudes toward the relative values of his neighbor's opinions versus newspaper opinions, the value of community benefits versus personal benefits, the value of innovation versus tradition in housing and neighborhoods, the value of taking risks versus security, the value of maintaining segregated housing areas versus opening up the community to all racial groups, and the value of solving problems cooperatively versus competing with others for rewards. Each player will also record his initial opinion about the desirability of accepting or rejecting the project.

During the game players comments will be tape recorded. At the close of discussion the players will vote to accept or reject the project as a group. Then each player individually will be asked to complete a short confidential questionnaire indicating his personal opinion about the project, and his attitudes about the game itself.

The rules are simple. The two teams are asked to discuss their differences concerning the project, and to arrive at a recommendation for the City Council as to whether it should be built, and if so, whether there should be any changes in the plan or conditions attached to its approval. No other explicit directions are provided; however, the planner (experimenter) does use a prepared list of technical rebuttals to the most common objections to the project:

1) land use arrangement,
2) traffic congestion,
3) property value decline,
4) school overcrowding,
5) separation of low income residents from their friends and community facilities.

Evaluation

Tapes of the game runs reveal a sequence of moves akin to chess plays. The major low income attacking moves consist of demands for their opponents to explain their objections to the project and of references to the community purposes to be served by it, while their defensive moves were justifications of the project-location and small size in terms of the choice provided and the symbolic importance of positive action. The middle income players, while largely on the defensive and using such time-honored tactics as attempts to broaden the issue and to delay its resolution, also attacked the project on various technical points, such as lack of public transportation, and suggested alternative locations.

The two major hypotheses proposed concern the positive role of collaborative planning in community innovation diffusion and community conflict resolution. [4] The experience from the pretests of the experimental game devised to test these hypotheses and the related propositions suggests that the game situation is much more related to the community conflict hypothesis, and that a separate game might be designed to explore the important phases of the innovation diffusion process prior to conflict. However, many attempts to introduce community development innovations do cause conflict at some stage, and therefore this game can illuminate certain aspects of both hypotheses.

In evaluating the preliminary experience, it seems obvious that many of the most frustrating problems which occur in this game situation are directly related to:
1) Entering the planning process after most of the major decisions have been made, and possible compromise alternatives are extremely limited;
2) Working with a disputed project which is essentially paternalistic in nature, and whose initial planning did not include participation by the low income groups it was designed to serve;
3) Attempting to resolve technical issues and to build an atmosphere of trust between people of different social class in a single short session;
4) Trying to convince middle class subjects, who have already taken a public stand on an issue, to change their position in exchange for rather slim incentives such as minor planning revisions or approval of the planner and other supporters of the public welfare.

In playing the game it soon becomes obvious that the project planning is already completed, and because of the non-availability of alternate sites and the restrictions of the zoning ordinance the players have very little room for compromise. The number of units, a basic matter of contention, is fixed by federal minimums and land costs. A zoning buffer has already been proposed by the Planning Board (which one player characterized as a "bribe"). There were some project exchanges proposed, such as another zoning classification or a lowered density, but essentially the middle class offer was to help in solving the overall housing problem in return for locating this project somewhere else.

It is also obvious that the project is quite paternalistic - conceived and executed by middle income white liberals for, rather than with, a low income, black and white, population. The lack of public transportation and the separation from other low income people and their social institutions could prove to be difficult for project residents to cope with, especially in view of open hostility from many surrounding middle income residents.

Although the pretest subjects exhaustively explored the various factors bearing on the problem, the "one-shot" aspect of the game situation left no opportunities for longer term reciprocal behavior sequences. The interdependence of the players terminated after the one session of discussion and the final vote, which was simple too short to find an effective technical solution and to develop trust between the participants at the same time. This suggests the usefulness of a continuing discussion group, perhaps using an ABAB... research design in which A is simply a discussion session and B is an exchange situation. [5]

Finally, the class and race nature of the long standing cultural conflict underlying the dispute over the project is represented in deep seated attitudes which are very resistant to debate and persuasion, especially after a public position has been taken. Reversing these attitudes is simply too ambitious an undertaking for a planning discussion. Even though the planner sides with the low income players, using the persuasive influence of his technical arguments on behalf of the project, only some weak incentives are available to get the middle income players to accept the project. The planner can exchange the approval of himself and other supporters of the public welfare. The poor can add to the guilt costs of the middle income players, but in the final analysis can only offer threats to disrupt public order through violence and riots if the project is not approved. Most of the decision power is held by the middle income players, subject to their ability to withstand the negative opinion of them held by the planner and the low income players. In order to make the exchange process more viable, as a central element of the game, it should be possible to give specific rewards for cooperative behavior. Players could be given chips representing their stakes and the planner could distribute additional chips representing opinion points and investment rewards. Point losses could be levied for project disapproval, riots, or other negative acts. At the end of the experiment these chips could be exchanged for real rewards, similar to an ongoing poker game. While it would make the game more artificial (some might say "Mickey Mouse") this type of exchange mechansim would offer more opportunities to measure and analyze outcomes than a simple voting measure. It would also expand the scope of communications between players, making their exchanges contingent.

In conclusion, we feel that this research project does offer good possibilities for further research. Certain modifications in the questionnaires and in the exchange mechanism of the game should provide a sharper research focus. In its present version, the pretest runs provided a good small group simulation of the polite confrontation from behind fixed positions which characterizes much of the conflict over urban development problems today, and which so painfully documents the "official channel" powerlessness of the poor and their advocates.

References

1   For a further discussion of the collaborative planning approach see David R. Godschalk and William E. Mills, "A Collaborative Approach to Planning Through Urban Activities," *Journal of the American Institute of Planners,* XXXII (March, 1966), pp. 86-95. See also Godschalk and Robert M. Griffin, Jr., *The Titusville Downtown Development Project: A Report on Collaborative Planning in Action* (Tallahassee: Institute for Social Research, Florida State University, 1967); and Godschalk, "The Circle of Urban Participation," in H. Wentworth Eldredge (ed.), *Taming Megalopolis,* Vol. II, (Garden City, N.Y.: Anchor Books, 1967), pp. 971-979.

For a similar rationale, which includes a discussion of the conditions for an optimum collaborative relationship, see Bennis, Warren G. "Theory and Method in Applying Behavioural Science to Planned Organizational Change," in J. R. Lawrence (ed.), *Operational Research and the Social Sciences* (London: Tavistock Publications, 1966), pp. 33-76.

2   The classic study, concerning the location of public housing in Chicago, is Martin Meyerson and Edward C. Banfield, *Politics, Planning and the Public Interest* (Glencoe: The Free Press, 1955).

3   Norton E. Long, "The Local Community as an Ecology of Games," *The American Journal of Sociology,* Vol. 64 (November, 1958), p. 255.

4   For a summary of innovation studies, see Rogers, Everett M. *Diffusion of Innovations* (New York: The Free Press of Glencoe, 1962).

Some studies of conflict are:

Ackoff, Russell L. "Structural Conflicts Within Organizations," in Lawrence (ed.) *Operational Research and the Social Sciences,* pp. 427-438.

Marek, Julius, "Conflict, a Battle of Strategies," in Lawrence (ed.), *Operational Research and the Social Sciences,* pp. 483-498.

Rapoport, Anatol, "Laboratory Studies of Conflict and Co-operation," in Lawrence (ed.), *Operational Research and the Social Sciences,* pp. 369-397.

Sisson, Roger L., and Ackoff, Russell L. "Toward a Theory of the Dynamics of Conflict," in Stewart Mudd (ed.), *Conflict Resolution and World Education* (The Hague: Dr. W. Junk Publishers, 1966), pp. 125-142.

5   For an applied example of the ABAB...research design, see Hamblin, Robert L. and Associates "Changing the Game from 'Get the Teacher' to 'Learn'," *Transaction* (January, 1969), pp. 20-31.

Barry Jackson
Architect
New York City

WORKSHOP ON PARTICIPATORY PLANNING AND URBAN LOCATION MODELS

Participatory or Advocacy Planning is a political issue and in some way might not seem to fit into this conference, however the Node-Four group, discussed the use of systems or a systematic approach to participating in planning which I can contrast to my own work. We don't use a systems model when working with the community, but approach that aspect on an ad hoc basis. We use our systems experience for the problem-solving so these two elements merge. What concerns me is that design methods in the real world can get between the problem and the solution. And somehow I think this is something we have to guard against.

I had a feeling that the work I saw in the other workshops was regressing toward very simplistic things. I couldn't see how we could apply some of these studies to the solution of real world problems. The researchers seem to be trying to redefine and segment problems, and seem to be sliding backward from the research levels of previous years. That was my feeling except that one doesn't see all the sessions, so it is hard to tell whether this is the general tone of the whole conference.

I also had the feeling that things were getting somewhat academically oriented and that the people who had the potential of designing building systems weren't really here at the converence; that industry has the potential and the money and the capacity, and that the people here with the ideas are frustrated because they don't have the number of man hours or the resources to build the systems that they have in their minds; that somebody else is either on the verge of building or implementing these systems.

I also felt that there wasn't really a cohesive direction toward all the work that is being done (although I guess this is true in every field), and that there is no particular problem that anybody is focusing on; everybody is focusing on something else. There ought to be a collective effort if we are going to be put in the squeeze between the large systems companies. We have to organize ourselves in a more collective fashion to effectively utilize our resources. Perhaps I am too pragmatic because I keep having to work with solutions to real problems, but I keep remembering Marvin Manheim's remark about the economics of problem-solving: "We keep having to throw methodology out the window every time we have a small problem because we haven't got time to be systematic about it, and we don't learn enough about the irrational problems we deal with to try to superimpose rational methods."

John M. Peterson
Department of Architecture, University of Cincinnati

Leonard M. Lansky
Department of Psychology, University of Cincinnati

AN EXPERIMENTAL APPROACH TO THE STUDY OF CRITICAL JUDGMENT IN DESIGN [1]

At the outset, we must comment on the team of architect and psychologist. Besides the joys and frustrations of educating one another, we believe that such teams, artist and behavioral scientist, are necessary to encourage critical judgment in research in the design fields. We have discovered that each of us as an expert in one field is woefully ignorant and naive in the other. If we are to develop solid data, to design productive research, and to develop comprehensive theories about critical judgment, we believe we have to talk together and struggle to communicate with one another. Our "two language" problem can be solved if we want to do so.

This research program began quite accidentally. It resulted from some erroneous predictions and a casual conversation. The freshmen architectural design class had been given the carving of a wood cube as a problem: in three weeks, using only jacknife, gouge, sandpaper, and linseed oil for the finish, complete a handcarving, working directly (no preliminary design), in a six-inch wood cube. For a number of reasons, the staff had randomly assigned different groups of students different limitations in the number of edges to be left intact on the cube: 0, 3, 6, 9, or 12.

From experience with the task and with freshmen with only one quarter's experience, some predictions had been made about the quality of the work. Before describing the predictions, the errors, the casual conversation alluded to above, and the research, we first describe the handcarving problem itself.

The Handcarving

The handcarving has been discussed in the literature in design.[2] It is often one of the three dimensional problems in a basic design curriculum. At the University of Cincinnati it is the first. As such, it carries particular weight and many special meanings. The block is a mass in space. The student's task is to create spaces within the mass which relate to one another and the mass(es).

The scale is general; although a six-inch wood cube is used, the design can be envisaged as smaller, life size, or larger. The carver is also faced with other new problems. The natural material gives limits--certain forms and shapes cannot be imposed on or obtained from it. The finished product is visual and tactile. The usual orientations of objects--right and left, up and down--are avoided.

The major theme of the handcarving is the integration of mass and space. Carvings are usually assessed on this dimension. However, this problem has in it all the complexity of any design problem. For many carvings the basic principles of mass-space relations seem to be irrelevant or are relevant because they are violated. Some carvings also focus attention on the texture and grain of the wood in relation to the form; others excel because of the grace and integration of the object as a geometric entity. Yet another dimention is the maintenance of a relationship to the original form, the cube. This issue is relevant for the future architect because it focuses his attention on the relationship of the spaces to the masses within a specific structure.

The First Study

These ideas lead to the notion of limiting the boundary conditions of the cube itself. At one extreme, the design staff reasoned, with no edges or boundaries there would be many possible forms to carve, with no natural limits (edges). The task should thus be very difficult. At the other extreme, if 12 edges were to be left intact, the boundary problem was reversed. The extreme boundaries would leave little room for creating spaces. Thus, carving is difficult.

From these two arguments, it was generalized that easing either of these restrictions would make the problem progressively more amenable to solution. By adding or subtracting three edges to the 0 or 12 limitation, the task should change. At the six intact, six removed condition, the ideal combination of freedom and limits, the most straightforward problem would be presented and the best carvings obtained.

This was the reasoning about the handcarving which led to the first study. The data did not support the hypothesis. The best carvings, on the average, were 9's (nine edges intact) with the 0's close behind. The worst were 12's, next worse 6's, and 3's were in the middle of the five experimental conditions. The two best carvings, on the average, were 6's. Furthermore, as the students were working, the staff saw several 6's which were excellent at the outset but which were quite poor when completed.

Three other observations are relevant here. 1) The students assigned to the 0 edge intact condition began work almost immediately. 2) Those assigned to 12 edges did not begin for some time, often as long as one week; they seemed to be paralyzed. 3) Upon close study, several 0's were graded for their beauty as forms although the student had omitted an essential relationship to the original cube.

Theoretical Issues: Design and Motivation

So much for the data. The design staff did not have an explanation of these results. The staff was more or less fixed on the notion of motivation which was suggested above. Extreme limits or lack of limits create tension which blocks or inhibits attempts to relieve it.

This line of thinking fits a well-known and well-accepted psychological view of motivation called the "tension-reduction model". It is held by orthodox Freudian psychoanalysis on the one hand and more modern behavioral learning theorists on the other. Essentially, it says that man seeks quiescence; action occurs to relieve tensions, whatever their source. However, when tensions are extreme, actions do not flow as smoothly or effectively as they do at some optimal tension level. Too much tension gets in the way. The "golden mean" of tension yields the most effective behavior.

Soon after the carvings were completed, the study, the predictions, and the results were casually mentioned at the authors' first meeting. The conversation that ensued led to extended discussion about the first study[3] and the research efforts described below.

The first steps were having the handcarvings judged by independent raters and a re-examination of the data. At first the psychologist author understood and agreed with the predictions which had been made. The data, however, were incontrovertible; thus, we began to look more carefully at the reasoning behind the predictions.

In brief, we finally realized that the predictions had ignored the key issue in the handcarving, the integration of mass and space from the point of view of the complexity of the object itself. The "tension reduction" theory refers only to the frustrations in the student, the carver. Furthermore, the psychologist had agreed with the tension-reduction model of motivation until the two, architect and psychologist, talked at length about the design curriculum and the problem itself.

The "break" in understanding occurred when both, architect and psychologist, became aware that they knew of another set of motivation theories called the "new look" views[4]; these theories make different assumptions about behavior: the person's behavior "is not merely the result of inner tension systems (too few or too many restrictions and their attendant frustrations) but is responsive to external stimuli and special properties of the stimuli. In order to predict behavior, the researcher must assess the degree of stimulus complexity and the person's needs and abilities in relation to that complexity".[3]

This theory helped us to make sense out of the data. The freshmen had been unsuccessful at the six-intact condition because this task was too complex for them; on the average, they lacked the skill and critical judgment to solve the problem. The nine-intact condition was handled best because a relatively straightforward solution was available: make the mass dominant and the space secondary. The opposite condition, three-intact, was very difficult because of the strangeness of having the space as primary.

The theory not only clarified the data; it also led us to several new questions. To date, we have data on some, speculations on others, and only fantasies for some. In the remainder of this paper, we will sample each group.

Can other variables besides overall carving grade and the integration of mass and space be rated reliably? Yes! We have demonstrated that integration of material and form[5] and universality of form can be rated reliably. In a series of related studies, Peterson and his students[6] have shown that these variables can be rated reliably from slides and that such ratings correlate significantly with independent ratings of the carvings themselves.

What edges would students leave intact if given a free choice? The theoretical issue is the "naturalness" of the nine-edge intact condition as a solution to the handcarving problem. The "new look" in motivation focuses on complexity in the environment and desires for complexity in the individual. According to the theory, the person will not perform well in tasks which are either too boring or too difficult. In general, he will tend to work at tasks he can master with some challenge and, at times, go beyond this level to more difficult problems. One often studied source of variation in persons is the amount of challenge they will set for themselves.

We reasoned that few students would decide to complete carvings with all 12 edges, but that they would do so if they had successfully solved the problem at that level. We also expected quite a few students to attempt the six edge condition, again, if they were successful. Furthermore, to sample two other predictions of this particular study: (1) we expected most students to end up with zero edges intact either because of original decisions or errors in their attempts at more complex carvings; and (2) we were open to the possibility that no students would do nine edges intact because the condition is not all that challenging.

In general, we were correct. Most carvings were 0's; there were no 9's. The 12's were, on the average, better than the others except for the 6's which were also excellent. However, the 12's outnumbered the 6's four to one.[5]

What are the relationships between handcarvings and other design and academic work? The first exploration of this question was part of the study just mentioned. We looked at the relationships between the carvings and the students' grades in design and in other subjects during their first quarter. Students who successfully selected difficult edge combinations were also those who had been doing well in design during the first quarter of their architecture program. Furthermore, these students were also doing well in their other subjects.

Although it appears that the first quarter's work can be useful in predicting later effort, the relationships are complex. In data not yet completely analyzed, we have three tentative findings. First, the first quarter's design grades predict design grades during the third and fourth year; second, students who worked with difficult edge combinations (3, 4, 6, or 12) obtained better design grades in later years than students who worked with easier combinations (0's); third, more of the difficult combination group remained in the program.

How do students make their design decisions? The third study took us back to three particular limitations or edges to be left intact and to a look at the students' accounts of their design decisions via diaries. Here we randomly assigned different groups to either the 3, 6, or 9 intact condition. The critical finding to date is that diaries do not help us to distinguish excellent from poor work. Two positive results from the diaries are: (1) students vary tremendously in their articulateness, none of them actually describing their design decisions; and, (2) decisions and plans most often follow actions. First, students carve or make some commitment; then they realize implications and work out the solution. These findings should not have surprised us. The beahvior is commonplace in the daily work of both architect, psychologist, and most other professions. As is often the case in research and other learning experiences, we had to "discover" the phenomena for ourselves.[7]

What learning experiences and design strategies might affect how well students perform on the carvings and other problems? The fourth study focused on the six-intact condition because the first three studies pointed to it as the most exciting, complex, difficult, and challenging for the student. Indeed, contrary to the alternate hypotheses of the original study, we now realize that there are more edge configurations available in the six-intact, six removed condition. Although the six-six is the most difficult among the others we had assigned, the variability within the six-six offered both more research control and the possibility of discovering new dimensions of complexity and difficulty within this single problem.

The students were given the task under three different conditions. One group was given the usual instruction, removing and leaving six edges. Each student in the second group was also asked to do a transparent grain structure diagram of his block before he began carving. Each student in the third group was asked to do several charcoal mass-space sketches of a possible handcarving.

The blocks were assessed on overall excellence, integration of mass and space, and integration of form and material. In addition, the configuration of edges was carefully noted.

As expected, there was no significant difference in overall excellence among the groups; unexpectedly, there were no significant differences on the integration of mass and space and integration of form and material. Perhaps, we reasoned, the intervention of the drawing had been too minimal; however, this interpretation had to be ruled out when it was found that there was a rather consistent trend relating the edge configurations to the handcarving scores. Using a preliminary classification of the configurations, we found that one set of configurations was successful for the transparent grain structure group on the integration of mass and space variable, while the no drawings (control) group performed less well when they used this same set of configurations. Further, the students in the no drawing group performed well on integration of mass and space when they chose a distinctly different set of configurations. These, and other findings were post hoc and not all of them reached "respectable" levels of statistical significance.[8]

We also looked at the relationships between the handcarvings and previous work in design[9] Scores on handcarvings by the charcoal sketches group were significantly correlated with previous design grades whereas scores for the other groups were not. It seems as if the charcoal sketches most resembled the two-dimensional tasks of the first quarter and/or that the design commitments made in the sketches affected the designing done during the handcarving. For the other groups the previous design experience did not seem to relate to critical judgment on the handcarving.

Previously in this paper, we noted other relationships between the handcarving and previous design grades. These latest findings again show the complexity of the six-six condition. The two experiments, one giving free choice on edges and the other requiring different design strategies before carving, produced different results.

Are there some "popular" and/or more and less difficult-complex configuration in the six-six condition? There seem to be some popular patterns. This past year (January, 1969), we again focused on the six-six condition. The pattern of various configurations almost exactly matches that for the previous year. However, in all candor, neither architect nor psychologist has discovered a meaningful scheme for classifying the configurations.

We will re-examine all our findings after we analyze the data collected at the first EDRA conference. We asked each conferee to select the most and least complex configurations of six edges intact by marking the edges on sketches of the cube. If there are common choices, we may then have a clearer view of the students' choices as they carve.

What other problems in the curriculum might be amenable to simple and fruitful manipulations? Here we have only talked about the motion cube and the lighted space. In time we intend to explore these and other problems.

What can we conclude and where do we go next? These questions are always with us. At this writing, we think we have two handles on the issues of critical judgment. One is to look for manipulations with other problems; the other is to pursue the meaning of the configurations within the six-six handcarving. We also want to look at all the design and other data on the students we have studied in the past five years. These searches will, we hope, lead us to more criteria against which we can judge the effectiveness of educational innovations. For example, a good test for critical judgment and skill in integrating mass and space is the result on the six-intact handcarving.

Another implication is that our experimental approach itself is teaching us about critical judgment and creativity. We have been tape recording our discussions in the hope that they will also provide clues about the vicissitudes of critical judgment.

References

1   This research has been supported in part by a grant from the Bettman Foundation, Cincinnati, Ohio.

2   Moholy-Nagy, L., *Vision In Motion,* (Chicago: Paul Theobald, 1947).

3   Lansky, L. M., and Peterson, J. M., "Effect of Instructions on a Creative Task: Discipline and Form in Hand Sculptures," *Perceptual and Motor Skills,* (1966), pp. 22, 943-950.

4   Dember, W. N., "The New Look in Motivation," *American Scientists,* (1965), pp. 53, 409-427.

5   Lansky, L. M., and Peterson, J. M., "Stimulus Complexity and the Cube: Mass and Space in a Handcarving," *Perceptual and Motor Skills,* (1968a), pp. 27, 967-974.

6   Peterson, J. M., Eaton, R., and Woodman, D., "Critical Judgment Based on Direct Versus Indirect Experience: Photographs Versus Reality," (1969), University of Cincinnati, mimeo, (submitted *Art Journal*).

7   Peterson, J. M., and Lansky, L. M., "Design Decisions During a Handcarving Task: Wordless Images?", *The Barat Review: A Journal of the Liberal Arts,* (1968), pp. 3, 33-40.

8   Lansky, L. M., and Peterson, J. M., "The Effect of Prior Specific Experience on the Quality and Pattern of Hand Sculptures." Paper presented to the annual meeting of the American Psychological Association, (San Francisco, August, 1968b).

9   Lansky, L. M., Leonard, W. A. and Peterson, J. M., "The Effects of Various Design Approaches on Critical Judgment in Handcarving," University of Cincinnati, *American Educational Research Journal,* (1969), in press.

George Bireline
School of Deisgn
North Carolina State University

TOWARD AN EXPERIMENT IN DESIGN EDUCATION [1]

Today's youth feel alienated from their environment because their potential contributions are not considered relevant by the established community; the community does not value the skills they possess, and before they are allowed to become involved in community problems, young people are expected to first fit themselves to pre-determined roles that the community has designated as relevant for them.

Added to this, it is difficult for a young person to state problems based upon *his* experience in the world that has any meaning for the institutionalized community. He has been brought up in a technology and environment different from that of his elders and his perceptions of the problems of our society and the meaningful direction for societal goals diverge considerably. It appears, however, that his view will only be acknowledged when he fits the established mold or when the problems he has identified are consistent with what is perceived of as socially admissable.

As a tool of the established community, our educational institutions attempt to maneuver students into already defined patterns which youth see as created by political and economic forces having questionable social consequences. These external forces operate in contradiction to the value structure of young people and the resultant feeling of frustration encourages negative attitudinal development toward both education and society.

These frustrations are not peculiar to our youth, however; many educators have been asking similar questions about their responsibilities to themselves, their peers and society. However, where the past generation was ready to make concessions which they felt were necessary in order to adapt, today's youth question the necessity of conforming when it is difficult to participate in the formation of societal goals that will ultimately influence the shape and quality of their lives. It seems the fundamental question both students and educators are asking is: "How can the social structure accomodate the participation of all its members in the process of self-determination?"

One answer is to allow young people to participate fully in the patterning of the one realm of experience most accessible to them—education. By allowing the student to become the initiator and director of his own education, we are preparing him to act upon and take responsibility for the implementation of his ideas in the changing world beyond the university. If we really believe that youth has a unique and valid world view which can possibly contribute to the changing needs of our society, then as educators we must help him develop the means to make his ideas operational. This implies a program for education which, by virtue of its differing goals will diverge considerably from the traditional programs for learning.

What kind of learning program is it which encourages the student to explore the meaning of his objectives in relation to the world? Foremost, any such program must begin and grow from the previous experiences and development of the learner. Learning occurs only in relation to what is already known. The student will be able to assimilate what is new only when he is able to make an association with what already exists. Any program must help the student make conscious and build upon what he already knows, so that it may be subjected to verification, reordering and, ultimately, extension.

A productive learning program must provide experiences in areas where the learner may be deficient. Generally, learning occurs by giving structure to what we have done or what we have experienced. Knowledge pursued in the abstract, through a book or a lecture, is experience which has already been structured through the perceptions of some other learner. Learning occurs most furitfully when the student himself becomes involved in the experience, and then through his own unique perceptions and awareness, he structures and patterns these experiences into meaningful statements about the world.

A good learning program must be responsive to the interests of the students. Learning only occurs when it is appropriate to some issue which the learner has named as relevant to his own purposes. Interests develop out of a learner's positive experiences of the world. And for as many students as there are, each with his unique set of experiences, there will be as many varying and divergent interests. When the student has become able to clearly express and identify his interests, the program structure should be flexible to accommodate his explorations.

Clearly, an effective program must provide the emotional climate conducive to productive learning experiences. The atmosphere must foster the attitude of creativity and initiative, self-confidence and self-reliance. This means a change in the traditional role of the teacher from an authoritarian exponent of an existing body of knowledge, to a person whose task it is to encourage, to pose questions, to act as a resource for tools of awareness and communication, and ultimately, to act as a learner himself.

Learning in this way, the student begins to understand that all knowledge is the result of a process coming out of himself. He perceives that, in order to grow, he must extend beyond himself; he must put his ideas into the world to be implemented and tested for their effectiveness. As he gains confidence in his abilities, he becomes increasingly willing to take responsibility for his thoughts and actions. In short, he is in the process of becoming a unique, self-determined individual.

Such a learning program has special relevance for design education today. Design fields traditionally have taken the limited role of resolving problems in some physical form, problems whose solutions are largely already determined by non-design oriented decision makers before the task is introduced to the designer. As a result, today we find designers having little impact on the quality of the physical environment. Other disciplines and interests are identifying and defining the problems to be attacked, their decisions influencing the design often to a far greater degree than the designer himself. For designers to begin to take a more determinant role in environmental quality, they must develop the ability to define issues independently rather than being constrained by the stereotypical images of a designer inherited from the past. A design education program should attempt to extend the concern of future designers to the environment as a whole. It should help to develop the capacity to perceive new patterns and relationships within the environment which may contribute to emerging problems and it should develop techniques for identifying and structuring those problems.

In seeking meaningful new roles as a change agent in the environment, designers are becoming aware of the impact they have on the behavior of users by means of the manipulation of physical relationships. Understanding the behavioral consequences of physical form can become a major contribution of the future designer toward the quality of the environment. A design education program should provide the opportunity for observation and discovery of the various interfaces that occur between people's activity and the physical environment. In the same way, the student should learn to discriminate between existing patterns of behavior within the environment; he should acquire the ability to forecast physical form relationships which would produce specified behavior; and he should learn not to impose personal values and judgments in his designs, but rather, be able to predict the behavioral consequences of what he proposes.

A design education program has been developed at the introductory year in the School of Design for the purposes of exploring and developing these new roles for designers. The educational goals established for this program are: 1) that the student learn to identify forces exerted in the community so that he can begin to perceive the problems to which they contribute; 2) that the student learn to identify the problems in the community, to state them and to begin to propose alternative resolutions; 3) that the student have real and transferable experiences to understand user behavior and needs, both social and physical; 4) that the student develop an attitude of anticipator of the consequences of his action; and 5) that the student become consciously aware of the design process with special attention to predictive thinking, testing and evaluation and goal structuring.

The method for achieving these goals was the simulation of a community. With relatively few problem constraints, each student designed and built for his own comfort and use, a work-study station within one large studio area. This community provided a common base of experience to which every member of the community could refer; it established the overall context for discovery and learning in which both student and faculty could explore aspecto of the community and learn from one another's experience. The community in its variety of transformations over time provided the experiential framework to involve the students directly with the issues of community and environment.

As the students settled down into their newly constructed community, they began to make certain unsettling discoveries. One student had inadvertantly built his station over the most logical pathway leading in from the front entrance. Not only was he harrassed by people walking by and through, his structure was also obstructing the major route by which to get materials in and out of the studio. Many students discovered that, in trying to build a cozy hide-away, the physical barriers they erected succeeded instead in severing their contact with the rest of the community. This fragmentation of individuals eventually caused a breakdown in group feeling and communication. As the first class met after the construction of the work-study stations, the students discovered that they had not considered leaving a space for the class to gather as a group. Blackboards, water fountains and sinks had all been incorporated into someone's private station and were inaccessible. In short, the students were experiencing the consequences of living in a community built without plan and without consideration for the needs of the community as a whole. The students began to understand the need for regulation in a community and for techniques of group organization and cooperation.

The congestion and clutter which accumulated in the course of the design and construction of the work-study stations finally became overwhelming and demanded action. The students and faculty began a series of tasks that began to order the community in terms of accessibility, circulation, orientation, designation of public and private spaces and visual quality. Various structuring tools were employed to help the student to see qualities of the environment which he may have been unaware of previously. For example, the semantic differential[2] was used as a tool to evaluate the image quality of the environment; sociograms were used to evaluate the structure of personal and group relationships, and so forth. Later, as the more obvious problems of the community were resolved, the effort was continued to help the student see problems which are less apparent but which may have powerful effect on the quality of the community environment. Some of these problems were identified as:

The isolation of an individual within a community because of differences in demographic and attitudinal characteristics.

The problems of congestion and the resulting physical and psychological stresses.

The loss of identity and the depression caused by frequent movement and relocation.

The effect of a totally homogeneous neighborhood.

The solutions to these problems were less accessible to the students; but the very awareness of them may be considered the first step in a search for solutions.

It can be said that order and meaning came to the experience as the student began to identify and define the problems of their new community. Alternative courses of action were proposed, changes were implemented and their effects experienced directly so that the student derived conclusions about the results of actions taken, both in terms of his own behavior changes and in terms of the effects on others in the community.

Further structuring of the learning experience was developed by having the students and faculty develop analogies between the problems encountered in the learning community and the problems already existant in actual communities. This assumed that the student would have enough perception of the actual community to make such a transfer possible.

Another means of giveing the experience of the community, structure and meaning, was through faculty who served to guide the student by suggesting the observation of various phenomena of human and physical organization. The difference between going through a situation without learning to see it and dealing with the situation effectively lies in the ability of a perceptive guide to ask at an appropriate time, "What is happening?" In this way, the student is not programmed according to predetermined viewpoints, but is placed in a situation in which his own way of seeing, interpreting, and acting is nurtured.

Most opportunities to pattern and give meaning to the student's experience arose out of day-to-day occurrances in the community. Sometimes specific tasks were developed as the chance appeared to introduce any of the organizational phenomena. One such task allowed the student to experience the strong cultural impact the phenomena of territoriality has on man's behavior in the community. The territory the student was to use for his work-study station was acquired in a "land rush." This was an event in which all of the students at a set time ran for and grabbed claims for the best spaces and physical amenities. Once the claims for the best spaces were made, defense of the spaces began immediately. Many heated encounters occurred and students experienced on a very emotional level, the power of territoriality as an environmental influence.

Another opportunity for giving meaning to the community experience occurred at mid-semester when a new group of first semester design students moved into the studio. The problems of immigrant populations attempting to assimilate into an already existing culture became apparent as the efforts of new students to find their place in the community were met alternately with indifference and hostility. The concurrent feelings of confusion and lack of acceptance gave new understanding to some of the emotional forces which impel the newly arrived to band together, sometimes forming enclaves.

In another instance, an opportunity to develop analogies between the learning community and problems in the real community occurred when the university maintenance department and the fire marshal declared the community an unsafe fire hazard and ordered it torn down. The faculty were able to introduce tasks that dealt with the general problem of communicating with the established authorities whose aims were quite different from those of the community; tasks that dealt with planning toward specific objectives. In addition, students developed insights into the human problems that arise when an area is to be destroyed and the inhabitants relocated.

Some examples of other phenomena of human and physical organization suggested to the students included:

Personal space—that distance which an animal customarily places between itself and other animals of the same species.

Autonomy—the degree to which group members function independently of other groups. The degree to which it generates its own activity.

Control—the degree to which a group regulated the behavior of individuals while they are functioning as group members.

Stratification—the degree to which the group orders its members into status hierarchy, that is, power, privileges, and obligation distribution.

Sense of Place—all of those elements in the environment, physical, psychological and social, that give the user a sense of belonging, being part of, being comfortable in, familiar with.

Proxemic—the orientation to changing relationships by users of the environment.

Sense of Control—the ability to form and control the quality of the environment.

Fundamental questions may be asked of this program: 1) To what extent does the students' behavior concur with the five stated objectives for the program? and 2) What are the transfer effects to later performance as a design student and ultimately as a professional? A thorough study will be necessary to test the real effects of this design education program.

This design program for first year environmental design students is in its initial stages of development, having been in operation only two years in the School of Design. At this time, there exists no conclusive evidence that the program attained its stated objectives. By observation, however, it appeared that the students were beginning in a consistent manner to question assumptions and to pursue the issues behind the emerging problems. Students appeared to be highly aware of both their physical and social environment and involvement was at once, highly emotional and intellectual.

References:

1   The author wishes to thank Lynne Meyer Gay, a colleague in the Environmental Design Program, for her assistance in editing this manuscript.

2   Charles Osgood, George Suci, and Percy Tannenbaum, *The Measurement of Meaning,* University of Illinois Press, Urbana, 1957.

Marvin Sevely
College of Architecture
Virginia Polytechnic Institute

WORKSHOP ON DESIGN EDUCATION

For a conference addressing itself to the subject of environmental design research, it was disappointing that both papers delivered in the field of Design Education failed to reflect the increasing awareness of both the environmental professions and education of the rapidly changing character of the built environments and the professions serving them.

Architectural design and urban planning have a role to play in the world that we are approaching, but the role of the traditional view of these studies is ever-diminishing. As an analogue of the real world, environmental education must define a wider spectrum of concern, a more meaningful interface of parts, a more dynamic, more mutable view of our own world. The studies and methodologies that even a few years ago seemed not to fit into a curriculum in architecture, that were defined as nothing, have become the most significant nothing of the present. Even the words "architecture" and "city planning" are obsolete, increasingly having less relevance to the reality of the frontiers of environmental concern.

In one paper an attempt was made to set up a scientific method for predicting the performance of architectural students therr, four or five years hence. This work was carried out with solid experimental techniques and over the years produced a prolific display of design from the students. This is meaningful if everything stands still; but the students involved are not standing still, are not static, and their world is not static. The opinion has been that if you can't count something, it doesn't count. Here we have seen an attempt to count the uncountable.

Whatever we think is the most advanced, the most enlighyened insight of yesterday or even today is with certainty going to be in the professional backwater of tomorrow. Often students broken by the established academies of a few years ago are, or could be, at the dynamic center of the environmental professions todya. Many who were at the center of "design" schools have been moved to the periphery.

The second paper concerned itself with the effort to awaken in the first year design students a personal self-discovery. Its main objective was to shock the student into an awareness of the inner self. It was reported that the fortuitous collision between this discovered inner world and the reality of the outer world developed the most creative situations. The methods reported must certainly be regarded as successful in awakening the qualities thought essential for design students. But what happens to those students who will fill research, analysis and policy-making roles in forming the built-environments and who do not fit into this mold?

What we look for and should have had in this conference is a report on environmental design education as the innovative, dynamic, expanded set of methodologies and techniques for exploring the revolutionizing environments of the last third of the twentieth century. There are beginnings, but only beginnings in the schools of "architecture" pointing toward this goal. This is the area of development in environmental education in our time.